Ecological Studies, Vol. 125

Analysis and Synthesis

Edited by

M.M. Caldwell, Logan, USA
G. Heldmaier, Marburg, Germany
O.L. Lange, Würzburg, Germany
H.A. Mooney, Stanford, USA
E.-D. Schulze, Bayreuth, Germany
U. Sommer, Kiel, Germany

Ecological Studies

Volumes published since 1989 are listed at the end of this book.

Springer

New York
Berlin
Heidelberg
Barcelona
Budapest
Hong Kong
London
Milan
Paris
Santa Clara
Singapore
Tokyo

Fritz L. Knopf   Fred B. Samson

Editors

Ecology and Conservation of Great Plains Vertebrates

With 59 illustrations

Springer

Fritz L. Knopf
National Biological Service
4512 McMurry Avenue
Ft. Collins, CO 80525-3400
USA

Fred B. Samson
U.S. Forest Service
Region One
Missoula, MT 59807
USA

QL
606
.52
.G74
E36
1997

Cover illustration: The cover illustration was prepared by Dale E. Crawford.

Library of Congress Cataloging-in-Publication Data
Ecology and conservation of Great Plains vertebrates/[edited by] Fritz L. Knopf
    and Fred B. Samson.
        p.   cm.—(Ecological studies; 125)
    Includes bibliographical references and index.
    ISBN 0-387-94802-3 (hc: alk. paper)
    1. Vertebrates—Ecology—Great Plains.   2. Wildlife conservation—Great
Plains.   I. Knopf, Fritz L.   II. Samson, Fred B.   III. Series: Ecological
studies: v. 125.
QL606.52.G74E36   1996
596′.045′0978—dc20   96-18352

Printed on acid-free paper.

Production coordinated by Princeton Editorial Associates and managed by Francine McNeill; manufacturing supervised by Jeffrey Taub.
Typeset by Princeton Editorial Associates, Princeton, NJ.
Printed and bound by Braun-Brumfield, Inc., Ann Arbor, MI.
Printed in the United States of America.

9 8 7 6 5 4 3 2 1

ISBN 0-387-94802-3 Springer-Verlag New York Berlin Heidelberg   SPIN 10523644

Preface

The frontier images of America embrace endless horizons, majestic herds of native ungulates, and romanticized life-styles of nomadic peoples. The images were mere reflections of vertebrates living in harmony in an ecosystem driven by the unpredictable local and regional effects of drought, fire, and grazing. Those effects, often referred to as ecological "disturbances," are rather the driving forces on which species depended to create the spatial and temporal heterogeneity that favored ecological prerequisites for survival. A landscape viewed by European descendants as monotony interrupted only by extremes in weather and commonly referred to as the "Great American Desert," this country was to be rushed through and cursed, a barrier that hindered access to the deep soils of the Oregon country, the rich minerals of California and Colorado, and the religious freedom sought in Utah. Those who stayed (for lack of resources or stamina) spent a century trying to moderate the ecological dynamics of Great Plains prairies by suppressing fires, planting trees and exotic grasses, poisoning rodents, diverting waters, and homogenizing the dynamics of grazing with endless fences—all creating boundaries in an otherwise boundless vista.

Historically, travelers and settlers referred to the area of tallgrasses along the western edge of the deciduous forest and extending midway across Kansas as the "True Prairie." The grasses thinned and became shorter to the west, an area known then as the Great Plains. Today, the entire region labeled the Great Plains encompasses tallgrass prairies on the east and shortgrass prairies on the west and a band of interspersed, mixed grasses through the central states.

In 1994, we coauthored a paper in *BioScience* arguing that the North American prairie, the largest biological province on the continent, remains its most significantly threatened ecosystem. Despite numerous volumes on the ecology and physiological ecology of grasses, however, there has been little attempt to evaluate major changes in Great Plains ecosystems as they relate to the ecology of endemic vertebrates. Nowhere is there a volume that specifically focuses on the distribution and basic ecology of the native vertebrates. This volume includes chapters prepared by currently active, leading authorities with long-term research programs on the Great Plains.

The volume is divided into three sections. The first presents three overview chapters addressing the vegetative template of vertebrate habitats: the grasslands, wetlands, and streamside woodlands. The grassland chapter evaluates the roles of fire, grazing, and drought in the patterning of grasslands on different spatial and temporal scales; the wetland chapter synthesizes information on the dynamics of wetland ecology as vertebrate habitats; and the streamside forest chapter addresses the ecological succession that has occurred along prairie rivers with water management and fire suppression. A fourth chapter in this section discusses the comparative ecology of the historical native and contemporary introduced ungulates.

The second section begins with a summary of a recent field effort to document 100-year changes in landscapes and the vertebrate fauna along the Niobrara River of northern Nebraska. Subsequent chapters present original data and topical syntheses of the ecology of fishes indigenous to the western Great Plains, avian assemblage responses to grassland management activities in the tallgrass prairie of Kansas and mixed prairie of North Dakota, and the ecology of small mammal assemblages in prairie landscapes. A final chapter in this section reminds us that highly mobile species require conservation actions on the Great Plains to incorporate regional or larger perspectives to accommodate unique ecological requirements.

The volume concludes with a reminder that the profession of natural resource conservation in North America evolved from the specific interests of two key individuals, George Bird Grinnell and Theodore "Teddy" Roosevelt, whose friendship developed around their love for, and frontier adventures on, the Great Plains. After a discussion of their early initiatives, we close with a synthesis of presented and additional information to define a contemporary, holistic approach to conservation of prairie vertebrates. It is our hope that this volume will stimulate more inquiry into the historical ecology and future conservation of the Great Plains and its endemic biota.

Preparation of a volume such as this one requires much technical assistance. The editors thank Rob Garber and his staff at Springer-Verlag for the invitation to assemble this volume and for their oversight of its production. Barbara A. Knopf edited all original manuscripts for style, diction, and usage and provided assistance in reading proofs. We also thank Peter Strupp and his staff at Princeton Editorial Associates for their exemplary production assistance and Dale E. Crawford for preparing the cover illustration.

Fritz L. Knopf
Fred B. Samson

Contents

Contributors

Gregor T. Auble

Midcontinent Ecological Science Center, National Biological Service, Fort Collins, CO 80525-3400, USA

Kevin R. Bestgen

Department of Fisheries and Wildlife Biology, Colorado State University, Fort Collins, CO 80523, USA

Michael A. Bogan

Museum of Southwest Biology, National Biological Service, University of New Mexico, Albuquerque, NM 87131, USA

Scott L. Collins

Division of Environmental Biology, National Science Foundation, Arlington, VA 22230, USA

Kurt D. Fausch

Department of Fisheries and Wildlife Biology, Colorado State University, Fort Collins, CO 80523, USA

Leigh H. Fredrickson

Gaylord Memorial Lab, University of Missouri, Puxico, MO 63960, USA

Jonathan M. Friedman

Midcontinent Ecological Science Center,
National Biological Service, Fort Collins,
CO 80525-3400, USA

D.C. Hartnett

Division of Biology, Kansas State
University, Manhattan, KS 66506-4901,
USA

K.R. Hickman

Division of Biology, Kansas State
University, Manhattan, KS 66506-4901,
USA

Douglas H. Johnson

Northern Prairie Science Center,
National Biological Service, Jamestown,
ND 58401-9736, USA

Donald W. Kaufman

Division of Biology, Kansas State
University, Manhattan, KS 66506-4901,
USA

Glennis A. Kaufman

Division of Biology, Kansas State
University, Manhattan, KS 66506-4901,
USA

Fritz L. Knopf

Midcontinent Ecological Science Center,
National Biological Service, Fort Collins,
CO 80525-3400, USA

Murray K. Laubhan

Midcontinent Ecological Science Center,
National Biological Service, Fort Collins,
CO 80525-3400, USA

Fred B. Samson

Northern Region, U.S.D.A. Forest
Service, Missoula, MT 59807, USA

Michael L. Scott

Midcontinent Ecological Science Center,
National Biological Service, Fort Collins,
CO 80525-3400, USA

Susan K. Skagen

Midcontinent Ecological Science Center,
National Biological Service, Fort Collins,
CO 80525-3400, USA

A.A. Steuter   Niobrara Valley Preserve, The Nature
                                   Conservancy, Johnstown, NE 69214,
                                   USA

Mary Ann Vinton   Department of Biology, Creighton
                                   University, Omaha, NE 68178, USA

John L. Zimmerman   Division of Biology, Kansas State
                                   University, Manhattan, KS 66506-4901,
                                   USA

1. The Great Plains Landscape as Vertebrate Habitats

1. Landscape Gradients and Habitat Structure in Native Grasslands of the Central Great Plains

Mary Ann Vinton and Scott L. Collins

Introduction

Habitat variables have a powerful influence on the distribution and abundance of organisms, and many organisms directly affect the physical structure of their local environment. These interactions produce a complex feedback system that drives community dynamics. This scenario is particularly true in grasslands where the effects of "keystone engineers," such as the North American bison (*Bison bison*), have a tremendous influence on ecosystem structure and function (Collins and Benning 1996). This interplay between organisms and habitat structure in grasslands occurs within a climatic regime that varies dramatically from one year to the next. In addition, the relative influence of different habitat variables changes at different spatial and temporal scales.

This chapter describes the general large-scale gradients in habitat structure in the central Great Plains of North America. We illustrate the broad-scale east-to-west habitat gradients in the Great Plains and then discuss how these gradients affect, and are affected by, fire and grazing by ungulates. We then analyze some of the factors affecting small-scale gradients in habitat structure in grasslands. Our overall approach is quite general; other chapters in this volume address the relationship between specific animal populations and communities, and how they relate to habitat structure in grasslands.

As grassland habitat decreases and habitat quality declines in the United States (Samson and Knopf 1994) it is imperative that the mechanisms generating habitat structure and associated biodiversity within the remaining patches of the once-extensive North American prairie are better understood. Indeed, the use of combinations of natural disturbances, including grazing by either cattle or bison (Plumb and Dodd 1993) as management practices (Steuter et al. 1990), provides key mechanisms for maintaining local and regional biodiversity in grassland systems. Like other ecosystems, prairies are threatened by the encroachment of development, deposition of airborne pollutants, and the invasion of non-native species (Wilson and Belcher 1989). These environmental threats often have negative impacts on native biodiversity. Thus, our focus is to describe the large- to small-scale factors that generate and maintain the diversity of grassland ecosystems. We hope that an understanding of these factors will lead to management guidelines that can produce cost-effective approaches to sustainable regional land use practices in the Great Plains.

Broad-Scale Patterns in Habitat Structure

Habitat structure influences the distribution and abundance of animals at different spatial scales (Cody 1981). In grassland communities, however, more focus has been placed on the impacts of animals on vegetation structure as opposed to how vegetation structure affects animal communities. At least two factors contribute to this disparity. First, grassland communities appear to be structurally simple compared with forest environments. Thus, gradients in habitat structure are less obvious. Such gradients do exist, however, and their broad-scale effects on species' distribution and abundance have received some attention (Wiens and Rotenberry 1981, Kemp et al. 1990).

How does vegetation and associated vertebrate habitat vary in the seemingly uniform plains? The vertical and horizontal vegetation structure have received much attention as factors affecting vertebrate habitat (e.g., MacArthur and Mac-Arthur 1961, Rotenberry and Wiens 1980, Cody 1985). In addition to these physical dimensions of vegetation, the biological dimensions of vegetation productivity and nutrient quality for vertebrate consumption are also important habitat features. The physical and biological dimensions vary along spatial gradients in the Great Plains region and are manifest in the climatically determined vegetation types, shortgrass prairie in the western, driest portion of the Plains, and tallgrass prairie in the eastern, most mesic portion of the Plains. Prairies in the northern Plains are dominated by cool-season grasses and forbs (Teeri and Stowe 1976), and the proportional contribution of warm-season species, especially grasses, to overall production increases from north to south (Epstein et al. in press).

Topography and other important physical features, such as wetlands, generate mesoscale heterogeneity in grassland landscapes. Despite the term "Great Plains," most grasslands throughout this region occur in gently to steeply rolling topography. These topographic gradients can be strong enough to induce measurable changes in vegetation composition and ecosystem function (Barnes et al. 1983, Gibson and Hulbert 1987, Schimel et al. 1991). Numerous wetland habitats also occur throughout the prairie region as a consequence of topographic variation. In the northern Plains, prairie potholes (Dix and Smeins 1967) support wetland ecosystems that serve as important resources for wildlife, and they increase regional biodiversity by increasing habitat heterogeneity. Naturally occurring wetland habitats are less common in the southern portion of the Great Plains, where wetland plant communities are often limited to intermittent drainages through lowland prairie. Nevertheless, these primarily linear landscape features contain plants and animals that are not found in upland prairies, thus increasing regional biodiversity. In the southwestern High Plains, playa lakes contain adequate relief to create a moisture gradient that supports vegetation different from surrounding uplands (Bolen et al. 1989, Hoagland and Collins in press). Thus, in the southwestern Plains, playas act as a comparable habitat to that found in the northern prairie potholes.

The second factor that explains why emphasis has been placed on animal impact on vegetation structure is that the dynamics of grassland ecosystems are disturbance driven, and many of these disturbances are animal mediated (Collins and Glenn 1988). Given the recent emphasis on disturbance theory in ecology, much current research has focused on understanding the effects of animals on grassland vegetation, rather than the converse. These factors are not independent. For example, fire affects plant community structure and composition, which, in turn, affect the abundance, habitat structure, and foraging behavior of animals in grassland environments. Disturbance regimes interact in complex ways with animal habitat in Great Plains grasslands, and these interactions take place against a backdrop of climatically imposed gradients in vegetation types and habitat structure.

Vegetation Structure

The general features of vegetation physiognomy vary across a gradient from short- to tallgrass prairie. In this section, we describe these patterns of vegetation structure and discuss how these patterns are altered by fire and grazing. Finally, we illustrate some current observations and general theory development about animal responses to habitat structure and end with an example from the Konza Prairie Research Natural Area in northeastern Kansas that illustrates the effects of fire and topography on mesoscale variability in vegetation physiognomy, habitat structure, and bird response.

Within a given site, vertical vegetation structure is determined mainly by the height, productivity, and life-form of the plants. This feature of vegetation obviously changes from short- to tallgrass prairie, a pattern attributed to the effects of precipitation on both production and species composition (Fig. 1.1). More precipitation in mesic tallgrass prairie not only results in taller individual plants and growth forms, but also results in more variation in canopy structure, with additional canopy layers and substantial litter layers beneath the canopy. Lane (1995) documented this increasing canopy complexity from shortgrass steppe to tallgrass

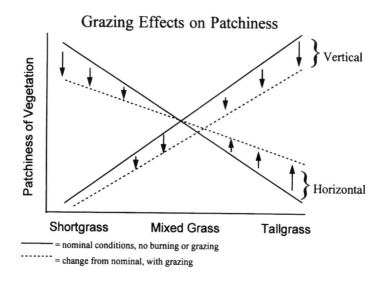

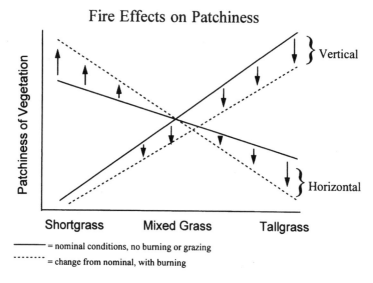

Figure 1.1. Characteristic vertical and horizontal patchiness of vegetation in central Great Plains grasslands, and the proposed effects of fire and grazing on patchiness.

prairie and its effects on light distribution in the canopy. Shortgrass steppe had fewer layers and less light attenuation at the soil surface than did tallgrass prairie. The accumulation of litter is minimal in shortgrass prairie but dramatic in tallgrass prairie (Knapp and Seastedt 1986), with far-reaching effects on both abiotic conditions (Hulbert 1988) and plant growth and nutrient cycling processes (Ojima et al. 1990).

Horizontal vegetation structure is influenced mainly by the growth form and productivity of the plants. In the absence of grazing, mesic tallgrass prairie vegetation is relatively productive and develops a "closed canopy," and several of the dominant grass species, such as big bluestem (*Andropogon gerardii*) and indiangrass (*Sorghastrum nutans*), are rhizomatous. All these factors contribute to the apparent horizontal uniformity of the tallgrass prairie vegetation (Fig. 1.1). Productivity is a function of time since burning, and productivity declines over time as litter and standing dead material accumulate (Knapp and Seastedt 1986). Clearly, litter and standing dead vegetation are important components of vertical structure, which may enhance habitat quality. However, there is a concomitant decrease in resource availability (e.g., leaf and seed production) in the absence of fire, which may reduce habitat quality.

By contrast, the shortgrass prairie is dominated by the bunchgrass, blue grama (*Bouteloua gracilis*), and is relatively low in productivity. In this case, production appears to be coupled to precipitation events (Sala and Lauenroth 1982), and there may be little or no reduction in productivity as litter accumulates. Thus, the horizontal structure of the vegetation is patchy as a function of the distribution of the dominant bunchgrasses, with more bare ground between plants than occurs in mesic tallgrass prairie (Fig. 1.1). The presence of shrubs also enhances horizontal heterogeneity in shortgrass systems. Shrubs are common on coarse-textured soils in the shortgrass steppe (Lauenroth and Milchunas 1991), and their presence appears to be important in determining the abundance patterns of many small mammal species (Stapp 1996).

The dynamic boundary or ecotone between the tallgrass and shortgrass prairies represents an important large-scale component of regional habitat structure. Given that grassland production is correlated with rainfall across the west-to-east gradient (Webb et al. 1983, Sala et al. 1988), changes in climate and associated vegetation response will affect broad-scale patterns of productivity in the Great Plains. In an analysis of studies along the shortgrass/tallgrass ecotone, Brown (1993) concluded that a continental westward movement of this boundary by as much as 240 km has been occurring since the Little Ice Age. This movement would suggest that the shortgrass-to-tallgrass ecotone is not in equilibrium with climate and that large-scale structural changes in vegetation composition and physiognomy will continue to occur. These changes, as well as directional, human-induced climate change, will have important implications for the distribution and abundance of plant and animal species across the Great Plains.

The obvious patchiness of vegetation structure in tall- and shortgrass regions influences soil nutrients as well. Vinton (1994) showed a clear decrease in horizontal vegetation and soil patchiness along a precipitation gradient from shortgrass to tallgrass prairie. In the shortgrass prairie, resources such as total soil carbon and nitrogen differed significantly between bare areas and areas occupied by blue grama (Vinton and Burke 1995). The relatively high horizontal patchiness of vegetation in shortgrass steppe is also influenced by the presence of shrubs on relatively coarse-textured soils (Lauenroth and Milchunas 1991). In the arid southwestern United States, establishment of woody vegetation led to the development

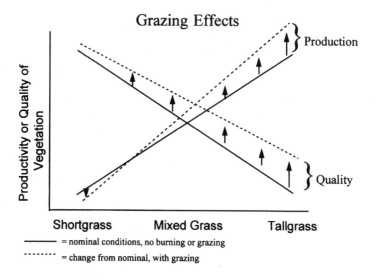

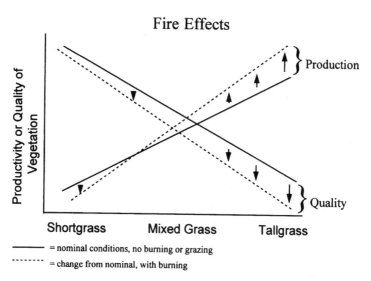

Figure 1.2. Proposed pattern in vegetation productivity and tissue quality for herbivores in central Great Plains grasslands.

of resource islands, creating positive feedback loops that further accelerated the change of habitat structure from grassland to woodland (Schlesinger et al. 1990). Within the tallgrass region, Wedin and Tilman (1990) showed that grass species differentially affected the quality and quantity of nutrients in soils. These species effects may tend to promote local dominance by strong competitors, with the result that small-scale patchiness decreases but large-scale patch structure may increase in the absence of disturbance.

The last two important habitat characteristics of Great Plains vegetation are the quality and quantity of food for vertebrates. Productivity and species composition are important in determining quantity and quality of food resources (Coley et al. 1985). In the shortgrass prairie, productivity is relatively low, but vegetative tissue quality may be fairly high, whereas the reverse is true in tallgrass prairie (Fig. 1.2). The reason for the inverse relationship between plant tissue quantity and quality likely involves trade-offs between water and nitrogen as limiting resources for plants (Field et al. 1983). Where production is limited by moisture, nitrogen is relatively concentrated in plant tissues, whereas in areas where water is less limiting and production is high, tissues are relatively coarse and low in nitrogen. This trend was apparent in grass tissue collected at the conclusion of the growing season from various locations in the central grasslands, with shortgrass tissue having relatively low C:N ratios and tallgrass tissue having much higher C:N ratios (Vinton 1994).

Life-forms growing within a site also may vary in nutritional quality. Grasses, shrubs, and forbs may differ in nitrogen content as well as in levels of defensive compounds. Within grasses, defensive compounds, other than silica, are rare, and so the inverse relationship between precipitation and quality may hold (McNaughton et al. 1985, Brizuela et al. 1986). The nutritional quality of forbs and shrubs varies widely, depending on nutrient concentrations, presence of lignin, and the presence of secondary compounds such as alkaloids and tannins. Within a particular plant group, nutritional quality for vertebrates can vary widely with plant part. Highest concentrations of nitrogen generally occur in plant parts with high turnover rates, such as flowers, fruits, seeds, leaves, and cambial tissues, whereas lower concentrations occur in more quiescent tissue such as stems and roots (Mattson 1980).

Effects of Fire and Grazing on Vegetation Structure

Undisturbed grassland types in the Great Plains have characteristic physiognomies as previously described, determined mainly by the height, life-form composition, and productivity of the vegetation. However, these physical and biological features of habitat may be altered drastically by the frequency of, and the response to, fire and grazing. The dotted lines and arrows in Figures 1.1 and 1.2 summarize our ideas about how fire and grazing alter habitat characteristics and, furthermore, how these fire- and grazing-induced alterations vary regionally.

The major effect of fire is to change the canopy radically, so that the litter layer is reduced, standing dead stems are eliminated, and more radiation reaches the surface of the ground (Mushinsky and Gibson 1991). In mesic prairies, burning may stimulate rapid regrowth and canopy development of grasses, whereas some forbs and shrubs will be eliminated, depending on the frequency of burning (Gibson and Hulbert 1987). Thus, burning, especially spring fires that favor regrowth of the dominant warm-season grasses, would decrease the patchiness of tallgrass prairie vegetation (Fig. 1.1). Although fire is not as common in shortgrass prairie due to fuel load constraints, burning would likely increase horizontal patchiness because vegetation regrowth would likely be limited by available moisture (Fig. 1.1).

The effect of fire on production and species composition is associated with its effect on food quality and quantity for herbivores. Enhanced production with spring fires in tallgrass prairie is associated with decreases in C:N ratios of dominant grasses (Seastedt et al. 1991) (Fig. 1.2). Thus, in grasslands that are burned frequently, soil nitrogen availability decreases, but productivity remains high relative to unburned prairie (Briggs and Knapp 1995). Nutrient quality of grasses declines because C:N ratios increase as a consequence of plants fixing more carbon on less available nitrogen. As time since burning increases, however, a thick litter layer develops, and resource limitation on primary production shifts from nitrogen to light (Seastedt and Knapp 1993). In the absence of burning, leaf nitrogen content increases, but tiller density and palatability decrease, leading to a greater overall decrease in resource availability in unburned prairie.

The effects of grazing on habitat structure for vertebrates depends strongly on the response of the dominant plants to tissue removal. Milchunas et al. (1990) showed that both above- and belowground distribution of plant tissue was more continuous on grazed than on ungrazed shortgrass prairie. Apparently, the dominant species in shortgrass prairie, blue grama, responds to grazing by tillering on the periphery of the clone, which results in dramatic smoothing of the horizontal patchiness inherent in this system (Fig 1.1). However, in tallgrass prairie, some of the dominant grass species do not flourish under grazing pressure so that grazing tends to open the canopy, create more patchiness (Weaver and Tomanek 1951, Conant and Risser 1974, Vinton et al. 1993), and increase species diversity (Collins 1987) (Fig. 1.1). Furthermore, large herbivores tend to graze areas preferentially; this patch-grazing in tallgrass prairie likely creates more obvious patches than patch-grazing in shortgrass prairie.

Grazing-induced changes in tallgrass prairie are further complicated by the feedbacks between burning and grazing. Recent and frequently burned grasslands tend to be fairly homogeneous, yet these areas are attractive to large herbivores because of improved productivity and less standing dead vegetation relative to unburned areas. The increased grazing intensity on burned areas may introduce more patchiness early in the growing season. Subsequent grazing often occurs on previously grazed patches, so the patterns of heterogeneity that start to develop early in the growing season may persist throughout the year and even between growing seasons until the area is burned again. Within the tallgrass prairie, we predict that the effects of grazing on patch structure would be most apparent on frequently burned lowland prairie, where heterogeneity is lowest in the absence of grazing. The interannual variability would also be quite high in these areas because the system would be "reset" by frequent burning, and new patch structure would be established each year due to random processes in selection of local grazing areas.

Wallowing and urine deposition by bison also alter habitat heterogeneity at small spatial scales. Wallows are depressions in the soil formed by the dust-bathing behavior of bison. These depressions collect water in the spring and support vegetation that differs from adjacent undisturbed areas (Collins and Uno 1983). Urine deposition by bison has a cascading effect on heterogeneity and plant community structure (Day and Detling 1990). Primary production and tiller nitrogen content of plants growing on urine deposition areas are higher than on areas without urine (Day and Detling 1990, Jamarillo and Detling 1992a, Steinauer and Collins 1995). These areas serve as the focal point of concentrated grazing activities by both vertebrates and invertebrates later in the growing season (Jamarillo and Detling 1992b, Steinauer 1994). Presumably, these wallowing and urine deposition activities of ungulates cause patchiness in shortgrass prairie as well. The capacity of these activities to increase baseline heterogeneity is likely less in short- than in tall- and mixed-grass prairie, although once created, small soil disturbances tend to persist and become revegetated rather slowly in shortgrass prairie (Coffin and Lauenroth 1990).

There appears to be evidence of positive interactions among different species of grazers in North American grasslands. Based on extensive observational analyses, Coppock et al. (1983b) and Krueger (1986) have shown that bison spend more time than expected, based on random movement patterns, feeding and resting on prairie dog (*Cynomys ludovicianus*) colonies. The apparent mechanism leading to enhanced grazing rates on colonies by bison is the increased nitrogen available in the aboveground vegetation compared with forage quality in off-colony areas (Coppock et al. 1983a). The enhanced grazing activity by bison maintains a low vegetation physiognomy, which is a preferred habitat of prairie dogs. Vegetation in areas with prairie dog colonies has higher plant species diversity than comparable ungrazed areas (Collins and Barber 1985, Archer et al. 1987).

Habitat Structure and Animal Responses

Habitat structure not only affects the types of animals that occur in an area but also how those animals respond to and perceive their environment. For example, foraging movements and grazing behavior of ungulates appear to be determined hierarchically (Senft et al. 1987). Whereas fine-scale (30 × 30 m² resolution) foraging patterns of elk (*Cervus elaphusnelsoni*) and bison in Yellowstone National Park appeared to be random, forage site selection at larger spatial scales was related to resource abundance (Wallace et al. 1995). Similar observations were noted for large-scale movement of bison herds in the Niobrara Valley Preserve in the Sandhills of central Nebraska (Steuter et al. 1995). There, bison essentially tracked spatial variation in resource abundance of C_4 (warm-season) grasses. Pocket gophers (*Geomys bursarius*), however, had greatest densities in areas where forb abundance was high. Given that grazing by ungulates can shift the abundances of grasses and forbs (Vinton et al. 1993), the foraging preferences of bison and pocket gophers leads to a dynamic interaction over time within the prairie landscape (Steuter et al. 1995).

It is often difficult to conduct experimental analyses of the effects of habitat heterogeneity on movement patterns of large animals, such as birds or ungulates,

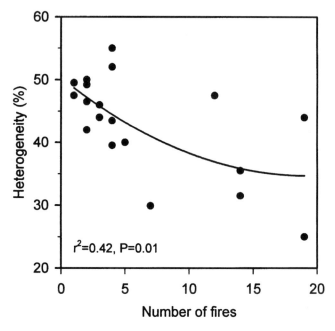

Figure 1.3. Relationship between number of times a site was burned between 1972 and 1990 and site heterogeneity (= 1 - % similarity) in species composition, as measured in 1990 on Konza Prairie Research Natural Area (KPRNA) in northeastern Kansas. (From Collins 1992.)

but detailed analyses of movement patterns of invertebrates in relationship to habitat heterogeneity have been quantified (Crist et al. 1992, Johnson et al. 1992, With 1994, Wiens et al. 1995). These studies may serve as surrogates for understanding movements of larger animals in complex landscapes (Johnson et al. 1992). Results of comparative studies demonstrated that invertebrate species with larger body sizes perceived a given environment to be relatively more homogeneous than did species with smaller body sizes (Crist et al. 1992, With 1994, Wiens et al. 1995). Thus, degree of habitat heterogeneity and its consequences are a function of both spatial scale of sampling and the perceptual scaling of organisms. Disparities in measurement and perception scales can lead to biased conclusions regarding the impact of habitat heterogeneity on animal community structure. This area of research deserves more attention, and we believe the complex local and regional gradients in grasslands provide an ideal opportunity for understanding habitat heterogeneity effects on vertebrate population and community ecology.

As an example of how topography and disturbance can interact to affect habitat heterogeneity and community structure, we analyzed fire effects on vegetation and bird species diversity at Konza Prairie Research Natural Area in northeastern Kansas. Analyses are based on permanent vegetation plots located within watersheds subjected to different frequencies of burning. For a description of the Konza Prairie experimental design, see Hulbert (1985). Using these data, we demon-

Figure 1.4. Effect of fire and topography on average cover of grasses, forbs, and woody plants in annually burned and long-term unburned tallgrass prairie at KPRNA. Data are based on 16 transects of five 10-m^2 permanent vegetation quadrats for all except annually burned slope sites, which had 8 transects. Error bars are 95% confidence intervals.

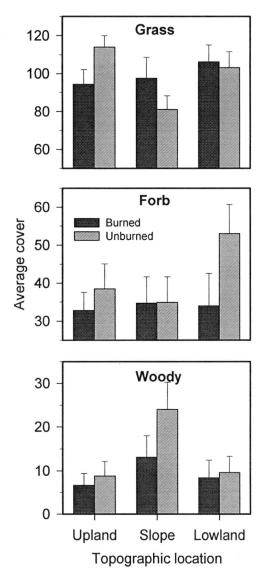

strated that frequent burning reduces plant species diversity (Collins et al. 1995) and plant community heterogeneity (Collins and Gibson 1990, Collins 1992) (Fig. 1.3). Plant community composition, however, may not be the best indicator of animal habitat structure at the local scale. Instead, gross differences in plant functional groups may be more relevant to animal habitat selection within a given region.

To test this idea, we compared abundances of plant functional groups at different topographic positions in annually burned and unburned sites at Konza. Fire effects on abundance of plant functional groups differed with topographic position (Fig. 1.4). Cover of grasses was significantly greater in lowland compared with

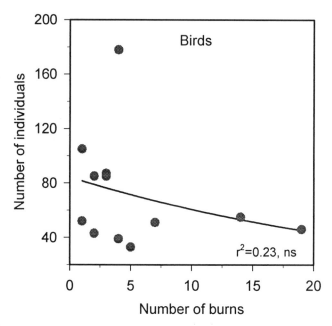

Figure 1.5. Relationship between the number of times a site was burned between 1972 and 1990 at KPRNA and the number of individuals of all birds species combined, exclusive of raptors and vultures.

upland prairie, with intermediate values on slope sites. Grass cover differed significantly among all sites in unburned prairie and was lowest on slopes. Topography had little effect on forb cover except on unburned lowlands, where forb cover was nearly double that of all other sites in both burned and unburned treatments. Woody vegetation was greatest on unburned slope sites. Thus, within a management unit at Konza Prairie, habitat structure varies as a function of topography and fire frequency.

The variable abundances of plant functional groups within a management unit produce heterogeneity in large-scale habitat structure. Is this heterogeneity related to bird abundances and species richness? Unlike the plant community, fire frequency does not seem to have a direct effect on abundance of breeding birds (Fig. 1.5). Although there appears to be generally more individuals in less frequently burned grasslands, the overall relationship between fire frequency and number of individuals of birds is not significant. Thus, direct fire effects on community structure appear to be dissipated at higher trophic levels. Instead, fire effects are perhaps best expressed through their effect on habitat structure, including litter accumulation and cover of woody vegetation.

An interesting trend suggests that bird species richness increases with cover of woody vegetation in these grasslands (Fig. 1.6). As woody cover increases, habitat heterogeneity would likely increase. The relationship is not significant, and the analysis suffers from limited sample sizes and perhaps a disparity in measurement

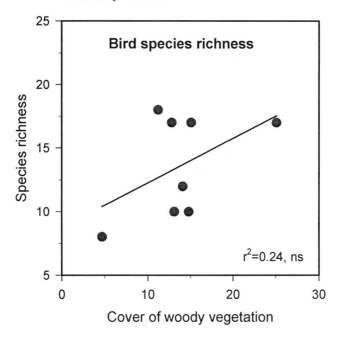

Figure 1.6. Relationship between average cover of woody vegetation (12 vegetation transects per site) and bird species richness at KPRNA.

scales of vegetation structure and bird abundances. Nevertheless, the trend is logical up to a point where sufficient woody cover allows a mixture of ground-nesting grassland species to coexist with species requiring woody vegetation for foraging and/or nest sites.

Our analysis of disturbance effects on community structure exemplifies the complex interactions that occur between disturbance frequency and topography at the mesoscale. Common prairie disturbances, such as fire and grazing, indirectly affect abundances of animals by altering habitat structure and resource abundance. Grazing in tallgrass prairie appears to reduce cover of grasses and litter while increasing the abundances of forbs (Steinauer and Collins in prep). Seed production may be lower in grazed prairie where abundances of flowering tillers are reduced after grazing. These impacts may vary considerably along the gradients discussed earlier in this chapter. More detailed studies are needed to understand fully how these complex interactions change over large spatial gradients.

Summary

Both physical and biological aspects of vertebrate habitat vary in Great Plains grasslands as a result of large-scale, climatically determined gradients in vegetation productivity and species composition, as well as factors varying at the mesoscale, such as topography. Physical habitat structure can be described by the vertical and horizontal patchiness of the vegetation; vertical structure tends to be greatest in more mesic grasslands, whereas horizontal patchiness tends to be greatest in dry grasslands. Biological habitat structure, as indicated by the quantity and quality of food available, also varies with vegetation type and plant life-form. This inherent variability in grassland habitats is influenced by the varying occurrence and effects of two common grassland disturbances, grazing by large ungulates and fire, across the Great Plains region. The effects of these disturbances on grassland habitat appears to be scale and ecosystem-type dependent. Understanding the separate and interacting mechanisms generating variability in habitat structure can shed light on how habitats are likely to change under different environmental conditions or disturbance regimes as well as how human management practices can modify vertebrate habitat in different portions of the Great Plains.

References

Archer, S.R., M.G. Garrett, and J.K. Detling. 1987. Rates of vegetation change associated with prairie dog (*Cynomys ludovicianus*) grazing in North American mixed-grass prairie. Vegetatio 72:159–166.

Barnes, P.W., L.L. Tieszen, and D.J. Ode. 1983. Distribution, production and diversity of C_3 and C_4 dominated communities in a mixed prairie. Can. J. Bot. 61:741–751.

Bolen, E., L.M. Smith, and H.L. Schramm, Jr. 1989. Playa lakes: prairie wetlands of the southern High Plains. BioScience 39:615–623.

Briggs, J.M., and A.K. Knapp. 1995. Interannual variability in primary production in tallgrass prairie: climate, soil moisture, topographic position, and fire as determinants of aboveground biomass. Am. J. Bot. 82:1024–1030.

Brizuela, M. A., J. K. Detling, and M. S. Cid. 1986. Silicon concentration of grasses growing in sites with different grazing histories. Ecology 67:1098–1101.

Brown, D.A. 1993. Early nineteenth-century grasslands of the midcontinent plains. Ann. Assoc. Am. Geogr. 83:589–612.

Cody, M.L. 1981. Habitat selection in birds: the roles of vegetation structure, competitors, and productivity. BioScience 31:107–111.

Cody, M.L., ed. 1985. Habitat selection in birds. Academic Press, New York.

Coffin, D.P., and W.K. Lauenroth. 1990. A gap dynamics simulation model of succession in a semiarid grassland. Ecol. Modelling 4:229–266.

Coley, P.D., J.P. Bryant, and F.S. Chapin III. 1985. Resource availability and plant antiherbivory defense. Science 230:895–899.

Collins, S.L. 1987. Interactions of disturbances in tallgrass prairie: a field experiment. Ecology 68:1243–1250.

Collins, S.L. 1992. Fire frequency and community heterogeneity in tallgrass prairie vegetation. Ecology 73:2001–2006.

Collins, S.L., and S.C. Barber. 1985. Effects of disturbance on diversity in mixed-grass prairie. Vegetatio 64:87–94.

Collins, S.L., and T.L. Benning. 1996. Spatial and temporal patterns in functional diversity. Pp. 253–280 *in* K. Gaston, ed. Biodiversity: a biology of numbers and difference. Blackwell Science, London.

Collins, S.L., and D.J. Gibson. 1990. Effects of fire on community structure in tallgrass and mixed grass prairie. Pp. 81–98 *in* S.L. Collins and L.L. Wallace, eds. Fire in North American tallgrass prairies. Univ. Oklahoma Press, Norman.

Collins, S.L., and S.M. Glenn. 1988. Disturbance and community structure in North American prairies. Pp. 131-143 *in* H.J. During, M.J.A. Werger, and J.H. Willems, eds. Diversity and pattern in plant communities. SPB Academic Publishers, The Hague.

Collins, S.L., S.M. Glenn, and D.J. Gibson. 1995. Experimental analysis of intermediate disturbance and initial floristic composition: decoupling cause and effect. Ecology 76:486-492.

Collins, S.L., and G.E. Uno. 1983. The effect of early spring burning on vegetation in buffalo wallows. Bull. Torrey Bot. Club 110:474-481.

Conant, S., and P.G. Risser. 1974. Canopy structure of a tallgrass prairie. J. Range Manage. 27:313-318.

Coppock, D.L., J.K. Detling, J.E. Ellis, and M.I. Dyer. 1983a. Plant–herbivore interactions in a North American mixed-grass prairie. I. Effects of black-tailed prairie dogs on intraseasonal aboveground plant biomass and nutrient dynamics and plance species diversity. Oecologia 56:1-9.

Coppock, D.L., J.K. Detling, J.E. Ellis, and M.I. Dyer. 1983b. Plant–herbivore interactions in a North American mixed-grass prairie. II. Responses of bison to modification of vegetation by prairie dogs. Oecologia 56:10-15.

Crist, T.O., D.S. Guertin, J.A. Wiens, and B.T. Milne. 1992. Animal movement in heterogeneous landscapes: an experiment with *Eleodes* beetles in shortgrass prairie. Functional Ecol. 6:536-544.

Day, T.A., and J.K. Detling. 1990. Grassland patch dynamics and herbivore grazing preference following urine deposition. Ecology 71:180-188.

Dix, R.L., and F.E. Smeins. 1967. The prairie, meadow and marsh vegetation of Nelson County, North Dakota. Can. J. Bot. 45:21-58.

Epstein, H.A., W.K. Lauenroth, I.C. Burke, and D.P. Coffin. In press. Regional productivity patterns of C_3 and C_4 functional types in the Great Plains of the U.S. Ecology.

Field, C., J. Merino, and H.A. Mooney. 1983. Compromises between water-use efficiency and nitrogen-use efficiency in five species of California evergreens. Oecologia 60:384-389.

Gibson, D.J., and L.C. Hulbert. 1987. Effects of fire, topography and year-to-year climatic variation on species composition in tallgrass prairie. Vegetatio 72:175-185.

Hoagland, B.W., and S.L. Collins. In press. Heterogeneity in shortgrass prairie vegetation: the role of playa lakes. J. Vegetation Sci.

Hulbert, L.C. 1985. History and use of Konza Prairie Research Natural Area. Prairie Scout 75:63-95.

Hulbert, L.C. 1988. Causes of fire effects in tallgrass prairie. Ecology 69:46-58.

Jamarillo, V.J., and J.K. Detling. 1992a. Small-scale grazing in a semi-arid North American grassland: I. Tillering, N uptake, and retranslocation in simulated urine patches. J. Appl. Ecol. 29:1-8.

Jamarillo, V.J., and J.K. Detling. 1992b. Small-scale grazing in a semi-arid North American grassland: II. Cattle grazing of simulated urine patches. J. Appl. Ecol. 29:9-13.

Johnson, A.R., J.A. Wiens, B.T. Milne, and T.O. Crist. 1992. Animal movements and population dynamics in heterogeneous landscapes. Landscape Ecol. 7:63-75.

Kemp, W.P., S.J. Harvery, and K.M. O'Neill. 1990. Patterns of vegetation and grasshopper community composition. Oecologia 83:299-308.

Knapp, A.K., and T.R. Seastedt. 1986. Detritus accumulation limits productivity of tallgrass prairie. Bioscience 36:662-668.

Krueger, K. 1986. Feeding relationships among bison, pronghorn, and prairie dogs: an experimental analysis. Ecology 67:760-770.

Lane, D.L. 1995. Above-ground net primary production across a precipitation gradient in the central grassland region. M.S. thesis. Colorado State Univ., Fort Collins.

Lauenroth, W.K., and D.G. Milchunas. 1991. Short-grass steppe. Pp. 183-226 *in* R.T. Coupland, ed. Natural grasslands: introduction and western hemisphere. Ecosystems of the world 8A. Elsevier, New York.

MacArthur, R.H., and J.W. MacArthur. 1961. On bird species diversity. Ecology 42:594-598.

Mattson, W.J. 1980. Herbivory in relation to plant nitrogen content. Ann. Rev. Ecol. Systematics 11:119-161.

McNaughton, S.J., J.L. Tarrants, M.M. McNaughton, and R.H. Davis. 1985. Silica as a defense against herbivory and a growth promotor in African grasses. Ecology 66:528-535.

Milchunas, D.G., W.K. Lauenroth, P.L. Chapman, and M.K. Kazempour. 1990. Effects of grazing, topography, and precipitation on the structure of a semiarid grassland. Vegetatio 80:11-23.

Mushinsky, H.R., and D.J. Gibson. 1991. The influence of fire periodicity on habitat structure. Pp. 237-259 in S.S. Bell, E.D. McCoy, and H.R. Mushinsky, eds. Habitat structure: the physical arrangements of objects in space. Chapman & Hall, London.

Ojima, D.S., W.J. Parton, D.S. Schimel, and C.E. Owensby. 1990. Simulated impacts of annual burning on prairie ecosystems. Pp. 118-132 in S.L. Collins and L.L. Wallace, eds. Fire in North American tallgrass prairies. Univ. Oklahoma Press, Norman.

Plumb, G.E., and J.L. Dodd. 1993. Foraging ecology of bison and cattle on a northern mixed prairie: implications for natural area management. Ecol. Appl. 3:631-643.

Rotenberry, J.T., and J.A. Wiens. 1980. Habitat structure, patchiness, and avian communities in North American steppe vegetation: a multivariate analysis. Ecology 61:1228-1250.

Sala, O.E., and W.K. Lauenroth. 1982. Small rainfall events: an ecological role in semiarid regions. Oecologia 53:301-304.

Sala, O.E., W.J. Parton, L.A. Joyce, and W.K. Lauenroth. 1988. Primary production of the central grassland region of the United States. Ecology 69:40-45.

Samson, F.B., and F.L. Knopf. 1994. Prairie conservation in North America. BioScience 44:418-421.

Schimel, D.S., T.G. F. Kittel, A.K. Knapp, T.R. Seastedt, W.J. Parton, and V.B. Brown. 1991. Physiological interactions along resource gradients in a tallgrass prairie. Ecology 72:665-671.

Schlesinger, W.H., J.F. Reynolds, G.L. Cunningham, L.F. Huenneke, W.M. Jarrell, R.A.Virginia, and W.G. Whitford. 1990. Biological feedbacks in global desertification. Science 247:1043-1048.

Seastedt, T.R., J.M. Briggs, and D.J. Gibson. 1991. Controls of nitrogen limitation in tallgrass prairie. Oecologia 87:72-79.

Seastedt, T.R., and A.K. Knapp. 1993. Consequences of non-equilibrium resource availability across multiple time scales: the transient maxima hypothesis. Am. Nat. 141:621-633.

Senft, R.L., M.B. Coughenour, D.W. Bailey, L.R. Rittenhouse, O.E. Sala, and D.M. Swift. 1987. Large herbivore foraging and ecological hierarchies. BioScience 37:789-799.

Stapp, P. 1996. Determinants of habitat use and community structure of rodents in northern shortgrass steppe. PhD dissertation. Colorado State Univ., Fort Collins.

Steinauer, E.M. 1994. Effects of urine deposition on small-scale patch structure and vegetative patterns in tallgrass and sandhills prairies. PhD dissertation Univ. Oklahoma, Norman.

Steinauer, E.M., and S.L. Collins. 1995. Effects of urine deposition on small-scale patch structure in prairie vegetation. Ecology 76:1195-1205.

Steuter, A.A., C.E. Grygiel, and M.E. Biondini. 1990. A synthesis approach to research and management planning: the conceptual development and implementation. Nat. Areas J. 10:61-68.

Steuter, A.A., E.M. Steinauer, G.L. Hill, P.A. Bowers, and L.L. Tieszen. 1995. Distribution and diet of bison and pocket gophers in a sandhills prairie. Ecol. Appl. 5:756-766.

Teeri, J.A., and L.J. Stowe. 1976. Climatic patterns and distribution of C_4 grasses in North America. Oecologia 23:1-12.

Vinton, M.A. 1994. The influence of individual plants on soil nutrient dynamics in the central grassland region of the United States. PhD dissertation. Colorado State Univ., Fort Collins.

Vinton, M.A., and I.C. Burke. 1995. Interactions between individual plant species and soil nutrient status in shortgrass steppe. Ecology 76:1116–1130.

Vinton, M.A., D.C. Hartnett, E.J. Finck, and J.M. Briggs. 1993. Interactive effects of fire, bison (*Bison bison*) grazing and plant community composition in tallgrass prairie. Am. Midl. Nat. 129:10–18.

Wallace, L.L., M.G. Turner, W.H. Romme, R.V. O'Neill, and Y. Wu. 1995. Scale of heterogeneity of forage production and winter foraging by elk and bison. Landscape Ecol. 10:75–83.

Weaver, J.E., and G.W. Tomanek. 1951. Ecological studies in a midwestern range: the vegetation and effects of cattle on its composition and distribution. Univ. Nebraska Conserv. Surv. Div. Bull. 31.

Webb, W.L., W.K. Lauenroth, S.R. Szarek, and R.S. Kinerson. 1983. Primary production and abiotic controls in forests, grasslands, and desert ecosystems of the United States. Ecology 64:134–151.

Wedin, D.A., and D. Tilman. 1990. Species effects on nitrogen cycling: a test with perennial grasses. Oecologia 84:433–441.

Wiens, J.A., T.O. Crist, K.A. With, and B.T. Milne. 1995. Fractal patterns of insect movement in microlandscape mosaics. Ecology 76:663–666.

Wiens, J.A., and J.T. Rotenberry. 1981. Habitat associations and community structure of birds in shrubsteppe environments. Ecol. Monogr. 51:21–41.

Wilson, S.D., and J.W. Belcher. 1989. Plant and bird communities of native prairie and introduced Eurasian vegetation in Manitoba, Canada. Conserv. Biol. 3:39–44.

With, K.A. 1994. Using fractal analysis to assess how species perceive landscape structure. Landscape Ecol. 9:25–36.

2. Wetlands of the Great Plains: Habitat Characteristics and Vertebrate Aggregations

Murray K. Laubhan and Leigh H. Fredrickson

Introduction

Grassland, the largest biome in North America, is characterized by the uniform presence of perennial grasses that originally supported extensive populations of native ungulates (e.g., buffalo [*Bison bison*], pronghorn [*Antilocapra americana*]) and burrowing mammals (Shelford 1949). The Great Plains, a component of the northern temperate grassland, encompasses about 20% (200 million ha) of the land mass in the 48 conterminous United States (Willson 1995). The region receives scant rainfall and exhibits extremes in climatic conditions within and among seasons and years. Although climatic conditions in the Great Plains do not favor the perpetuation of numerous, naturally occurring permanent wetlands, historic records indicate that ephemeral, temporary, and seasonal wetlands once were a prominent feature throughout the region.

Less than 1% of the original grassland remains undisturbed by human activities (Kopatek et al. 1979). Human intrusions such as transportation systems, agricultural developments, and water projects have devastated wetland habitats within the Great Plains. Many wetlands have been destroyed (Dahl 1990), and remaining wetlands, associated species, and ecological processes have been affected (Darnell 1978). As a result, remnant wetlands in the Great Plains are widely scattered and comprise a small area relative to surrounding terrestrial habitats. Unfortunately, destruction of wetland habitats continues. The Garrison Diversion will eliminate 89,000 ha of wetlands (Tiner 1984), and privately supported drainage improvement projects continue to radically change the area of functional wetlands and reduce habitat values for wildlife in many portions of the Great Plains (G. Krapu, personal communication).

Despite a relatively small surface area, wetlands are an important landscape component within the Great Plains that provide resources to myriad wildlife species that reside in or migrate through the area. Although much information exists on Great Plains wetlands, few studies have considered their status and value as habitat for vertebrates. Foremost among these deficiencies is an understanding of the importance of wetlands for resident and migrant vertebrates. Identifying the value of wetlands for many species is complicated because terrestrial habitats of varying quality (size, structure) in close juxtaposition to wetlands are required for completion of specific life cycle events, such as egg-laying (turtles) and nesting (geese, shorebirds, and many dabbling ducks). Thus, this chapter synthesizes existing information regarding climate, geomorphology, type and distribution of wetlands, and plant associations in relation to the value of wetlands as habitat for native vertebrates.

Area of Consideration

The Great Plains region is delineated by several distinct geographic features (Fig. 2.1). The Balcones and Caprock escarpments form the southern boundary, the Missouri Escarpment the northern limit, and the foothills of the Rocky Mountains the western boundary (Shimer 1972). The eastern edge of the region is less well defined geographically, particularly in Kansas and Oklahoma. Thus, many authorities have designated the eastern boundary of the central North American prairie as the transitional area between tallgrass prairie and eastern deciduous forest in Minnesota, Missouri, and Oklahoma (Willson 1995). However, the types and distribution of wetlands and associated vertebrate assemblages along this transitional gradient are not representative of the Great Plains biome. We define the eastern limit of the Great Plains as the boundary between the tallgrass and mixed-grass prairie (e.g., historic Great Plains), which is about 95°W longitude or 500–800 km east from the Rocky Mountain Front Range (Fig. 2.1).

Wetland Characteristics

Formative Processes and Abiotic Components

The Great Plains are primarily composed of the Tallgrass Prairie and Great Plains Shortgrass Prairie provinces (Bailey 1978). The 100th meridian approximates the boundary between the two provinces (Fig. 2.1). Tallgrasses (*Andropogon* spp.) occur east and shortgrasses (*Bouteloua* spp., *Buchloe* spp.) west of this line (Ringelman et al. 1989). However, the demarcation between short- and tallgrass prairie shifts periodically due to changes in rainfall and other factors (Shelford 1949).

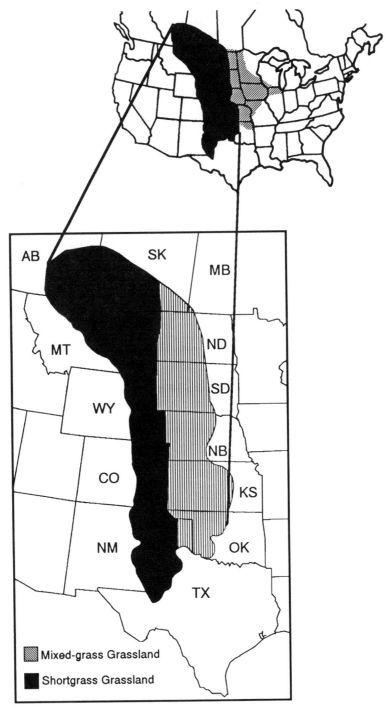

Figure 2.1. Location of the Great Plains within the central plains of North America and distribution of mixed and shortgrass prairie.

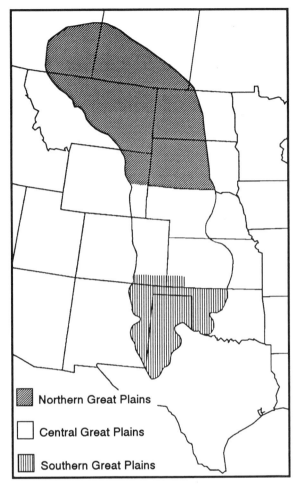

Figure 2.2. Location of northern, central, and southern regions of the Great Plains based on differences in geomorphology and climate.

Wetlands of various types are interspersed throughout both provinces; there is no clear correlation between wetland type and location based on the 100th meridian. Rather, the best separation of wetlands is based on formative processes leading to their development. We divide the Great Plains into three regions (northern, central, and southern) largely based on formative processes (Fig. 2.2). Although boundaries are not distinct, wetlands in each region tend to be similar with regard to geomorphology and physiochemical characteristics, factors that exert tremendous influence on short- and long-term wetland structure and function.

Northern Great Plains

The northern Great Plains are composed of portions of four states and three Canadian provinces (Fig. 2.2). The dominant terrestrial vegetation is shortgrass prairie, topography is irregular, and the climate is dry and subhumid with precipitation substantially lower than potential evapotranspiration (Borchert 1950). Summers are hot (average daily July temperature is 21°C) and winters cold (average daily January temperature is -12°C). Annual precipitation occurs mostly (approximately 66%) during the growing season and ranges from 34 cm in the northwest to 44 cm in the southeast (Winter 1989). The average annual growing season varies from fewer than 120 days in the north to 188 days in the south (Chapman and Sherman 1975). The dominant pedogenic process is calcification, but salinization is dominant in poorly drained sites. Mollisols are typical, but soils along river floodplains are mosaics of sand, silts, and clays in various combinations and thicknesses (Johnson et al. 1976, Bailey 1978). North of the Missouri River (northern Montana and northwestern North Dakota), the region is glaciated and wetlands include lakes, swamps, moraines, and various meltwater deposits. From the Missouri River south to roughly the South Dakota–Nebraska line, there is greater evidence of erosion. Several rivers and associated tributary systems have eroded soil material to form moderately deep valleys in otherwise flat terrain (Shimer 1972).

Wetlands in the glaciated portion of the northern Great Plains primarily are prairie potholes. Formation of the region began about 12,000 years ago as retreating ice from the Wisconsin glacier created millions of depressions in glacial drift that retained rainfall and snowmelt. The entire pothole region comprises about 600,000 km^2, of which 155,000 km^2 are within the United States (U.S. Department of the Interior 1988). Before settlement, there were an estimated 8 million ha of potholes (Frayer et al. 1983), but more than 50% have been converted to other uses, primarily agriculture (Tiner 1984). Further, drainage in some areas continues at an alarming rate (8,100 ha/yr in North Dakota [Louma 1985]). By 1988, only about 2.8 million ha (35%) of the original potholes remained, and only 20% were protected by government ownership, easements, or private conservation group ownership (U.S. Department of the Interior 1988).

The hydrology of wetlands in the pothole region is largely determined by local precipitation. In general, aquatic habitats in the northern Great Plains are small (<8 ha) and tend to develop in snowmelt and rainwater catch basins or shallow depressions where groundwater reaches the surface (Stewart and Kantrud 1972). Most potholes are considered fresh, but saline and alkaline wetlands commonly occur in the western portion of the northern Great Plains. Of the potholes that still remain, tilled basins are most numerous, followed by untilled ephemeral and temporary wetlands (Pederson et al. 1989). Natural lakes of glacial origin occur in this region but are much less abundant compared with glaciated areas of Minnesota and Wisconsin. However, seasonal and semipermanent wetlands encompass a larger area than temporary and ephemeral basins because numerous lacustrine wetlands (reservoirs) have been created by damming rivers and streams. Fort

Peck, Lake Sakakawea, and Oahe are mainstem reservoirs on the Missouri River and represent the largest waterbodies in the region (Pederson et al. 1989).

Central Great Plains

The central Great Plains includes portions of Wyoming, Nebraska, Kansas, and Colorado (Fig. 2.2). Unlike the northern Great Plains, the central and southern Great Plains exhibit a depositional rather than an erosional surface (Shimer 1972). Mixed-grass prairie dominates the eastern portion of the central Plains, whereas shortgrass prairie is the predominant terrestrial vegetation in the west (Fig. 2.2). Topography within the tallgrass prairie is characterized as flat to rolling plains with relief less than 90 m (Bailey 1978). The climate is subhumid, with evapotranspiration usually exceeding precipitation in most years (Bailey 1978). Average annual precipitation ranges from 63 to 81 cm, and average annual temperatures range from 12°C in Nebraska to 18°C in Kansas (Bailey 1978, Jones et al. 1983). Loess and sand deposits cover most of the area. Soils primarily are Mollisols, with isolated areas of Entisols and Vertisols. By contrast, the shortgrass prairie is semiarid continental with evapotranspiration exceeding precipitation (Bailey 1978). Topographic relief is moderate, with rolling plains and tablelands as characteristic landscape features. Winters are cold and dry; summers are warm to hot. Average annual precipitation is lower compared with the mixed-grass prairie, ranging from 46 to 62 cm. Average annual temperatures range from 10 to 15°C.

Wetland types, and the processes leading to wetland formation, are varied in the central Great Plains. In some cases, multiple hypotheses have been proposed regarding the origin of specific wetland types or a particular basin. For example, millions of depressional wetlands once occurred throughout this region. The origin of such depressions is not understood completely, but formative processes may have included deflation, differential compaction of unconsolidated materials, and solution of soluble layers (Shimer 1972). Cheyenne Bottoms, a 1,600-ha basin in west central Kansas, is another excellent example. Some early investigators suggested this basin was formed by stream erosion, whereas others have stated that removal of salts by solution and subsequent subsidence formed Cheyenne Bottoms (Vogler et al. 1987). Another theory is that Cheyenne Bottoms resulted from the combined actions of subsidence and subsequent stream erosion (Vogler et al. 1987). Most recently, structural movement in the subsurface between the early-late Cretaceous and early Tertiary (i.e., late Pliocene Epoch) periods has been suggested as the most plausible mechanism (Vogler et al. 1987). Riverine wetlands occur along major rivers, such as the Platte, Arkansas, and Republican and their large tributaries. In some cases, these water courses have eroded surface material to form broad flat-floored valleys (Shimer 1972), some of which retain water.

Regardless of formative processes, many palustrine wetlands in the central Plains have been drained and converted to other uses. Perhaps the most notable area of agricultural conversion is the Rainwater Basin. Located in south central Nebraska, the Rainwater Basin encompasses an area of about 10,878 km^2 south of the Platte River (Ducks Unlimited 1994). Topographically, the area is characterized by flat to gently rolling loess plains formed by deep wind deposits of silt-loam soils. Wetlands typically develop in depressions where leaching concentrates clay particles in the subsoils (Erickson and Leslie 1987). The hydrology of these wetlands largely is determined by surface runoff and precipitation. Approximately 4,000 individual wetland basins totaling 38,000 ha originally occurred in the basin. Most were 0.4–16 ha, but some exceeded 400 ha, and flood duration ranged from temporary to semipermanent. Destruction of wetlands to facilitate farming began in the early 1900s and accelerated after large earth-moving equipment became available after World War II (Raines et al. 1990). Conversion of wetlands to other uses greatly altered wetland distribution and size. By 1984, approximately 375 basins and 8,100 ha of wetlands remained (Nebraska Game and Parks Commission 1984). In addition, upland farming practices have accelerated siltation of remaining basins (Raines et al. 1990). McPherson Wetlands in central Kansas is another highly modified area of seasonal wetlands similar to the Rainwater Basin. Currently, attempts are being made to restore the hydrology of these areas by reclaiming farmed wetlands and installing pumps and networks of levees.

Another important wetland feature in the central Plains is the Sandhills region of Nebraska. Located north of the Platte River, the Sandhills encompass about 51,000 km² and were formed mostly during the past 8,000 years as sand, originating from the weathering of the Ogallala formation, accumulated against low hills and ridges, and was subsequently stabilized by grasses (Tiner 1984). Topography varies greatly; dunes may be up to 122 m in height and 32 km in length, with slopes of 25% (Bleed and Flowerday 1990). The Loup and Elkhorn rivers, tributaries of the Platte, originate in the Sandhills and flow southeastward across central Nebraska. The general climate is semiarid, exhibiting low precipitation and high evaporation. Average annual precipitation ranges from 59 cm in the east to greater than 43.5 cm in the west (Bleed and Flowerday 1990). The growing season is about 120 days (Nebraska Game and Parks Commission 1995). Sand is the predominant soil type, and the primary land use is cattle ranching. Habitats in the Sandhills vary greatly, ranging from communities typical of arid areas to wetlands. Wetlands include permanent lakes and marshes (some of which are alkaline), semipermanent and ephemeral basins, and wet meadows that may cover 121,000 ha of the landscape (Sanderson 1976). Wetland type and distribution are largely determined by the dynamics of a large underground reservoir (estimated 863–987 × 10⁹ m³) that is part of the Ogallala Aquifer (Bleed and Flowerday 1990). Sandhill wetland hydrology depends on groundwater recharge in valleys, but perched marshes develop in sites having poorly drained soils on upper contours (Tiner 1984).

Man-made reservoirs also are an important aquatic habitat in the central Great Plains. Large (>200 ha) reservoirs are one of the few aquatic habitats that have increased in area (>500,000 ha from the mid-1950s to mid-1970s) since 1945 (Tiner 1984). For terrestrial vertebrates, the most valuable wetlands associated with reservoirs usually are adjacent to and along small streams at the point where they enter the reservoir. Surveys conducted in the early 1980s indicate 171 reservoirs greater than 200 ha have been constructed in the High Plains, including 59 reservoirs comprising 112,000 ha of wetland habitats in Colorado, Kansas, and Nebraska (Ploskey and Jenkins 1980, Jenkins et al. 1985). Much of this habitat is deepwater that varies in value for wetland-dependent wildlife (Ringelman et al. 1989). Although many of these reservoirs are located outside the Great Plains, they still affect wetland habitats. For example, water projects constructed in western Colorado divert an estimated 617×10^6 m^3 of water annually from the western slope into eastern Colorado (Ringelman et al. 1989). The Arkansas River is the prime example of changes in riverine systems related to diversions of surface water and extraction of groundwater for irrigation (Vogler et al. 1987). Annual fluctuations in flow from snowmelt no longer occur, and use of groundwater within the floodplain has been so extensive that during extended dry periods little or no surface flow occurs in the Arkansas River in central Kansas.

Most reservoirs in this region are associated with river systems that provide wetlands in the form of riverine marshes, riparian habitats, and oxbow lakes. Therefore, reservoirs often constitute a unique wetland type that is incorporated into an existing complex of wetlands exhibiting various hydrologic regimes and associated plant assemblages. For example, Kirwin Reservoir in northern Kansas also is operated as a national wildlife refuge for the purpose of improving habitat for wildlife. However, construction of many reservoirs has resulted in the destruction of extensive areas of native floodplain habitats, and deepwater habitats often become the predominant or only wetland type remaining in areas that once had diverse wetland resources. In other cases, reservoirs have not resulted in direct destruction of wetlands but have significantly altered environmental conditions that have a negative effect on adjacent wetland types (see Moore and Mills 1977).

Southern Great Plains

Southeast Colorado, eastern New Mexico, the Texas and Oklahoma panhandles, and southwest Kansas compose the southern Great Plains (Fig. 2.2). Topographic relief is minimal, and the region is one of the flattest, most wind-swept, and featureless areas of North America (Wind Erosion Research Unit 1995). Geologically, the region was created during the Pleistocene by outwash from several river systems originating in the Cordillera that were subsequently overlain with an eolian mantle (Frye and Byron 1957). Thus, the area is characterized by a thin mantle of sediment overlying limestone caprock. After formation, geologic events altered the course of early rivers and left the flat tablelands dry; the portion of the Ogallala Aquifer underlying the southern Plains thus receives little recharge (Bolen et al. 1989). Sandy-loam soils dominate but are interspersed with soils of higher clay content. The climate is semiarid and historically supported shortgrass prairie vegetation (Bailey 1978). In the Texas Panhandle, average evaporation may range from 4.5 to 10 cm (Guthery et al. 1981), whereas precipitation averages 45 cm, with peaks in April–May and October (Bolen et al. 1989). Although the average temperature throughout the entire Shortgrass Prairie Province is 8˚C, average temperature in the southern Plains may reach 15˚C (Bailey 1978). The frost-free season is more than 200 days.

Playas are the predominant palustrine wetland in the southern Great Plains. Thought to be wind-deflated depressions (Reeves 1966), playas typically are shallow, circular or oval wetlands. Precipitation is the only hydrologic input. Unlike the loam soils that dominate terrestrial areas, soils underlying playas are high in clay content and were formed in the reworked sediment of the surrounding eolian mantle (Bolen et al. 1989). Historically, precipitation was the only water source. Estimates (10%) of percolation loss in Texas (Dvoracek 1981) indicate that soils are only slowly permeable and that most water is lost via evapotranspiration (Aronovici et al. 1970). Thus, the hydrology of unmodified playas typically is determined by precipitation and evapotranspiration rates. However, the southern Great Plains is one of the most intensively farmed areas of the United States, and many playas have been destroyed. Cotton, wheat, and grain sorghum are the dominant crops and often are planted in dry playa basins. Recent information suggests that playas number between 20,000 and 30,000 and that potential aquatic habitat ranges from 152,000 to 168,000 ha (Ward and Huddleston 1972). The greatest number (19,300) and density $(33-139/km^2)$ of playas occur in Texas (Guthery and Bryant 1982), where a survey of 52 counties indicated that 64% of playas were at least partially tilled and 36% were completely tilled (Guthery and Bryant 1982). Further, the hydrology of many remaining playas has been modified. Some playas receive water via irrigation runoff and diversion of surface water by levees or trenches. In addition, deepwater pits often are constructed to concentrate water within a smaller surface area. Estimates of modification vary, ranging from 33% (Guthery and Bryant 1982) to more than 85% (Bolen et al. 1979). Although such modifications reduce the evaporation rate as much as 90% and provide a source of irrigation water as well as deepwater habitat for some vertebrates during summer, steep contours diminish the biotically productive littoral zone (Bolen et al. 1979).

Numerous small rivers and streams traverse the southern Great Plains. The primary river system is composed of the Canadian River and associated tributaries, including the Rita Blanca, North Canadian, and Punta de Agua. Dams have been constructed across most of the major rivers and tributaries to form lacustrine habitats (lakes, reservoirs) for the primary purposes of water supply, irrigation, and recreation.

Wetland Dynamics

Historic Hydrology

The hydrology of many palustrine wetlands in the Great Plains originally was determined primarily by precipitation cycles and evapotranspiration rates. In general, precipitation decreased from east to west, and evapotranspiration rates increased from north to south (Table 2.1). Thus, the hydrology of potholes in the northern Great Plains, depressional wetlands in the central Great Plains, and playas in the southern Great Plains often differed. Because the watersheds of most wetlands in the northern Great Plains were small, surface runoff was minimal and

Table 2.1. Factors Influencing Wetland Dynamics in the Great Plains

	Great Plains Region		
	Northern	Central	Southern
Precipitation			
Amount (cm)	34–44	46–81	22–45
Time	66% Growing season	Growing season	April–May, October
Growing season (days)	120–160	160–200	200–280
Formative processes			
Primary	Glaciation	Wind deflation, differential solution of soluble soil material	Wind deflation
Secondary	Riverine	Riverine,	
Dominant wetland type	Semipermanent	Seasonal	Seasonal
Annual hydroperiod			
High water	Spring	Spring/summer	Winter/spring
Low water	Fall	Fall/winter	Mid- to late summer
Dominant vegetation			
Semipermanent wetlands	Cattail/bulrush	Cattail/bulrush	Cattail/bulrush
Seasonal wetlands	Grasses/sedges	Grasses/spikerush	Grasses/spikerush

hydrologic input was dominated by local events. By contrast, wetland hydroperiods in the Rainwater Basin and Sandhills of Nebraska were influenced primarily by groundwater hydrology (recharge, discharge) of the Ogallala Aquifer and to a lesser extent by local events. In many years, hydrologic outputs exceeded inputs in many shallow palustrine wetlands because of high evapotranspiration during summer. This pattern of high evapotranspiration was particularly true of playas in the semiarid southern Great Plains. Consequently, playas dried seasonally more frequently than potholes, and the frequency of natural drawdowns in palustrine wetlands of the central Great Plains was intermediate. By contrast, the larger, deeper glaciated prairie potholes historically dried less frequently and often exhibited 5- to 20-year marsh cycles stimulated by periodic droughts (Weller and Spatcher 1965, van der Valk 1989).

The hydrology of riverine systems originally was determined by climatic events within a watershed. Snowmelt from the Rocky Mountains, precipitation events, and groundwater were the primary hydrologic inputs, whereas primary outputs were surface discharge, percolation, and evapotranspiration. Although highly variable, volume and flow of water in riverine systems tended to decrease with increasing distance from the source. The distribution and types of wetlands also varied but included permanent, semipermanent, and intermittent streams as well as oxbows and sloughs.

Current Hydrology

Today, the hydrology of remnant palustrine wetlands in the Great Plains is affected either directly or indirectly by human activities. Although clines in precipitation and evaporation rates remain similar to historic patterns, man-made ditches and trenches, roads, sedimentation caused by agricultural tillage practices, intensive wetland management on federal and state lands, and myriad other factors have compromised the hydrology of potholes, depressional wetlands, and playas. As a result, wetlands with less permanent flooding largely are disrupted because they have been drained into larger, deeper wetlands or tillage practices have either filled the basin or shortened the duration of flooding. In some cases, the hydrology of sizable basins has changed so extensively that little evidence of wetland features remains. One such basin is Hackberry Flats near Frederick in southwestern Oklahoma. Historically, this 1,600-ha wetland, including 240 ha of permanent water, was at the lowest elevation in a 3,600-ha closed drainage system. The site was once so productive of wild game that it attracted the attention of President Theodore Roosevelt, who spent several weeks there in 1905. In the early 1900s, all surface water was diverted into an adjacent drain system via a deep ditch. The drainage was so effective that wetland vegetation was nonexistent throughout most of this century (A. Stacey, personal communication), and wetland values were not apparent for decades. The value of the site for wetland restoration was realized in the early 1990s when historical information indicated the potential to restore the basin for wetland wildlife (A. Stacey, personal communication). Other wetland depressions that could provide resources for wetland wildlife likely were widely distributed in the Great Plains, but as time passes, the abundance of wildlife associated with these natural features is forgotten.

Another widespread change in Great Plains hydrology is the declining ground-water table. This is the case for Nebraska's Rainwater Basin and Sandhills (Mitsch and Gosselink 1993) and much of southwestern Kansas (Sadeghipour and McClain 1987). In the late 1980s, flow in the Arkansas River almost stopped at Dodge City, and at Kinsley it was reduced to 5% of 1940s flow. Extraction of groundwater for irrigation is a primary cause of these changes, but land-use practices, including watershed dams, terracing, and stubble-mulching, also are contributing factors (Welker 1987a). Likewise, agricultural developments, such as construction of dugouts, further compromise the historic hydrology (Flake 1979).

Hydroperiods of riverine wetlands in the Great Plains have a fate analogous to nonriverine basins even though riverine systems are regulated by different mechanisms than isolated wetlands. Construction of dams and levees on mainstem rivers and large tributaries throughout the Great Plains not only has destroyed many valuable riverine wetlands, but also effectively altered the hydroperiod, causing deleterious effects in remaining wetlands. Many mainstem reservoirs have a negative effect on riparian habitats and associated floodplain wetlands by altering retention time, depth, frequency, flow rate, and quality of water discharged through the system (Belt 1975, Moore and Mills 1977). For example, approximately 70% of the annual flow in the Platte River is withdrawn before reaching south central Nebraska (Kroonemeyer 1978). Such modifications cause drastic changes in channel width and available wetland habitat (Williams 1978), plant communities (Crouch 1978, Graul 1982) and concomitantly, vertebrate assemblages (Knopf 1994).

The hydroperiod of reservoirs is determined largely by the purpose for which they were built. Constructed for flood control, irrigation, municipal water supply, and power generation among other purposes, water-level fluctuations in most reservoirs and large lakes tend to be minimal except during extreme environmental events, such as droughts or floods when large volumes of water are released to maintain dam integrity. Minimal fluctuations that expose the shoreline sometimes occur in summer when outputs (human consumption, irrigation, evapotranspiration) exceed inputs (precipitation, stream inflow).

In summary, hydrology has been modified in all wetland types throughout the Great Plains, which has had a significant effect on vegetation composition on a continental scale. Historically, although the distribution and number of wetland types varied among regions, wetlands exhibiting different hydroperiods were located in close juxtaposition on a regional or local scale. Ephemeral, temporary, and seasonal wetlands supported rich communities of annuals (and varying amounts of some biennials) with some perennials that provided a rich food base in the form of seeds and substrates for invertebrates. By contrast, semipermanent and permanent basins largely were dominated by perennials, except in drought years when mudflats are exposed and annuals germinate. As a result, all life requisites required by vertebrate species inhabiting a region were available during most years. Today, however, the Great Plains landscape is dominated by permanent and semipermanent basins and man-made lacustrine habitats, and the distribution and area of ephemeral, temporary, and seasonal basins is much reduced (Ringelman et al. 1989). The hydroperiod of shallow wetlands no longer exhibits regional differences nor emulates historic patterns because human decisions regarding water (e.g., need for irrigation water) determine hydroperiods rather than natural events. Local wetland mosaics with varying hydrology and vegetation types have been disrupted, and the potential to support diverse assemblages of vertebrates has been jeopardized. In addition to development of monotypic vegetation on a local scale, many hydrologic modifications have caused shifts in the timing, duration, and frequency of flooding. These modifications reduce habitat values and decrease the quality and availability of wetlands for vertebrate activities.

Plant Communities

Native Vegetation

In general, interior freshwater wetlands are dominated by grasses, sedges, and other emergent hydrophytes (Kantrud et al. 1989, Mitsch and Gosselink 1993). The specific vegetation that becomes established primarily is determined by the composition of the seedbank (seeds and other reproductive propagules) and abiotic factors. Hydrology is foremost among abiotic factors because plant composition and structure are directly and indirectly influenced by the annual water budget

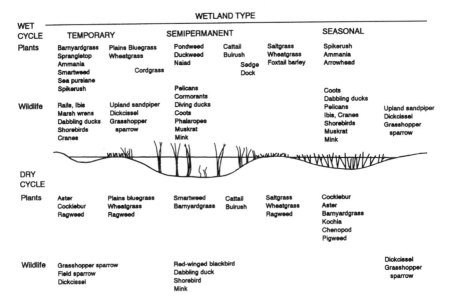

Figure 2.3. Distribution of plants and wildlife in temporary, seasonal, and semipermanent wetlands in the central Great Plains and changes in composition between wet and dry hydroperiods.

(van der Valk 1981, Mitsch and Gosselink 1993). Direct hydrologic effects, including frequency, time, depth, and duration of flooding, influence nutrient cycling, decomposition rates, and transport of plant reproductive structures. Several theoretical models, such as the environmental sieve model (van der Valk 1981), have been developed to predict wetland vegetation based largely on water dynamics. However, other abiotic (e.g., soil type, water chemistry, salinity, climate) and biotic (e.g., herbivory) factors often play a significant role in mediating hydrologic effects (Fredrickson and Laubhan 1994).

Historically, the composition and structure of wetland vegetation communities varied among the Great Plains regions, even though there were similarities among seedbanks. In palustrine wetlands, plant species often are established in zones determined by flooding gradients (Mitsch and Gosselink 1993) (Fig. 2.3). Typical species include relatively flood-intolerant grasses (*Poa* spp., *Graminae* spp.) at high elevations that are flooded only periodically, whereas annual and perennial moist-soil vegetation (e.g., sedges [*Carex* spp.], smartweeds [*Polygonum* spp.], barnyard grasses [*Echinochloa* spp.], and arrowhead [*Sagittaria* spp.]) dominate midelevation sites where soils remain saturated for longer periods (Fig. 2.3). Emergent macrophytes, such as bulrushes (*Scirpus* spp.) and cattails (*Typha* spp.), occurred at lower elevations that dried only periodically and often remained flooded throughout the growing season. Therefore, differences in wetland plant communities among regions largely resulted from differences in geomorphology in combination with climatic events that influenced regional and local hydroperiods.

The type of vegetation that becomes established in wetlands also varies depending on permanency of water. Ephemeral and temporary palustrine wetlands (e.g., shallow playas, depressional wetlands, and potholes), intermittent streams, and rivers that fluctuate seasonally are typically dominated by annual vegetation, whereas semipermanent and permanent palustrine (e.g., deep playas, depressional wetlands, and potholes), lacustrine (e.g., reservoirs, lakes), and riverine (e.g., oxbows, sloughs) wetlands are dominated by perennials. However, vegetation composition within a wetland type varies depending on long-term precipitation cycles. Thus, the plant community can change among years, depending on the timing of drying, soil salinity levels, and other environmental factors. During wet years, plant communities in temporarily flooded wetlands often include obligate wetland plants, particularly those characteristic of seasonal flooding such as dock, smartweeds, and barnyard grass. However, these species are often replaced by aster (*Aster* spp.), kochia (*Kochia scoparia*), or other more xeric species during dry periods. Seasonally flooded basins tend to be dominated by spikerush (*Eleocharis* spp.) communities during wet years. If the wet period continues for more than 1 year, the most water-tolerant perennial vegetation (cattails and bulrushes) may colonize. When the dry cycle returns, little or no evidence of wetland vegetation may be present in seasonal basins, and cocklebur (*Xanthium strumarium*), kochia, and aster become dominant. By contrast, wetlands that retain water longer are dominated by robust emergent perennials, including bulrushes and cattails, and less frequently contain annual vegetation.

Exotic Vegetation

Another ubiquitous change in wetland plant communities of the Great Plains is the invasion of non-native plants. Unfortunately, the range of wetland-associated species includes both annuals and perennials, including the difficult-to-control saltcedar (*Tamarisk* spp.), Russian-olive (*Elaeagnus angustifolia*), and purple loosestrife (*Lythrum salicaria*). With the disruption of native systems, the potential for additional exotics to encroach on wetlands and surrounding upland habitat in the Great Plains is high. Species of concern in uplands include crested wheatgrass (*Agropyron cristatum*), leafy spurge (*Euphorbia esula*), Canada thistle (*Cirsium arvense*), musk thistle (*Cirsium flodmanii*), and tall white-top (*Lepidium latifolium*). Unfortunately, initiatives to integrate exotic species into intensive grazing programs continue and often are supported by universities and agricultural experiment stations, even though they are in conflict with current natural resource strategies to maintain biodiversity. The value of exotic plants to Great Plains vertebrates is minimal compared with values associated with native vegetation. Further, control of exotics is often costly and ineffective, and continued modification of wetland hydrology likely will continue to promote local invasions. The increasing spread of exotics places an additional burden on federal, state, and private wildlife organizations that are trying to provide a diversity of dynamic wetland habitats with food and cover for vertebrates in a greatly diminished wetland landscape.

Value of Wetlands as Vertebrate Habitat

The importance of wetlands for vertebrates is variable and depends on the life history of individual species. Although wetlands are important in providing food and cover for a wide range of species, a single wetland type rarely provides all the resources required by all species (Laubhan and Fredrickson 1993) (Fig. 2.3). Ephemeral, temporary, and seasonal wetlands are important sources of food for waterbirds because seed-producing annual plants and a wide variety of invertebrates occur in these basins when they are flooded. Such basins also provide free water and foraging sites for some mammals, as well as egg-laying sites for amphibians. By contrast, robust emergents in semipermanent and permanent wetlands provide the structural components required for nesting habitat by diving ducks, Coots (*Fulica americana*), grebes, rails, waders, and marsh-nesting passerines. Basins with longer hydroperiods also provide benefits to many wetland-dependent mammals, fishes, and herptiles.

Other chapters in this book address habitat requirements of vertebrate species representative of specific taxa or assemblages. Herein, we present a conceptual framework to facilitate meaningful discussion of wetland values at different landscape scales. In fragmented, greatly modified landscapes such as the Great Plains, the distribution of wetland types is an important factor determining wetland value for vertebrates. The widespread loss of wetland habitat has reduced habitat availability for many species (Tiner 1984, Bolen et al. 1989, Ringelman et al. 1989). Reduced wetland habitat area and quality influence not only regional population size but also the potential for less mobile forms to find suitable habitat to assure survival. Therefore, we classify vertebrate species into the following broad groups based on mobility: continental (move between states and countries [e.g., waterfowl, shorebirds]), regional (movements range from >3 km^2 to <100 km^2 [e.g., river otter [*Lutra canadensis*], coyote [*Canis latrans*])], and local (movements restricted to <3 km^2 [e.g., amphibians, reptiles]). This classification permits a more cohesive discussion of the importance of wetlands as vertebrate habitats in the fragmented landscape of the Great Plains.

Continental Mobility

Members of the Class Aves are the predominant vertebrates exhibiting continental mobility. Most wetlands in the northern and central Great Plains have ice-cover for at least 3–5 weeks annually, and wetland-dependent birds require open water for foraging and other activities. Although many species of birds occur on the Great Plains, only 12 are endemic to the Great Plains and an additional 25 have centers of evolution in grasslands (Knopf 1994). Of these, the Marbled Godwit (*Limosa fedoa*) and Wilson's Phalarope (*Phalaropus tricolor*) are the only wetland-associated species. Thus, the reproductive success to sustain populations of these species is related directly to wetland conditions in the Great Plains. However, 330 of the 435 bird species breeding in the United States have bred in the Great Plains (Knopf 1994). Many of these species, including shorebirds and waterfowl, are

Table 2.2. Relative Potential for Present-Day Wetland Resources to Meet Avifauna Needs in Wetlands of the Great Plains[a]

	Great Plains Region		
	Northern	Central	Southern
Breeding habitat			
Overwater	+++	+	+
Upland/wetland interface	+++	++	+
Upland			
Vegetated	+++	+	+
Nonvegetated	+	++	++
Food resources			
Breeding			
Plant			
Seeds	+	++	+
Tubers/roots	++	++	+
Browse	++	++	+
Invertebrates	+++	+	+
Migration			
Plant			
Seeds	+	+++	+++
Tubers/roots	++	++	++
Browse	++	++	++
Invertebrates	+++	+++	+++
Wintering			
Plant			
Seeds	NA	+++	+++
Tubers/roots	NA	++	++
Browse	NA	+	++
Invertebrates	NA	+++	+++

[a] +, Some resources available, area of habitat or amount of food limited; ++, moderate levels of resources, good distribution of habitat or moderate amount of food; +++, high levels of resources, widespread distribution of habitat or large amount of food; NA = not applicable.

dependent on wetlands in the southern, central, and northern Plains for some event in the annual cycle (Table 2.2). Others, such as the Red-winged Blackbird (*Agelaius phoeniceus*), although not completely dependent on wetlands, use vegetation types often associated with riverine or palustrine wetlands.

A high degree of mobility is critical for species that exploit unreliable and widely spaced seasonal environments (e.g., flooded agricultural fields, farmed wetlands, ephemeral basins). Highly mobile species generally have been more successful in maintaining populations than less mobile vertebrates with more restricted habitat requirements. For example, wetland-dependent bird species endemic to the Great Plains (Wilson's Phalarope, Marbled Godwit) have not exhibited population declines (Knopf 1994) characteristic of less mobile species, such as the federally threatened Concho water snake (*Nerodia paucimaculata*), which is endemic to a single river drainage in central Texas (Corn and Peterson 1996).

Foremost among avian groups exhibiting high mobility are the waterbirds, which have representatives in the following ten orders: Gaviiformes (loons), Podicipediformes (grebes), Pelecaniformes (pelicans and cormorants), Ciconiiformes (herons, bitterns, and ibis), Anseriformes (ducks, geese, and swans), Falconiformes (hawks, ospreys, and eagles), Gruiformes (cranes, coots, and rails), Charadriiformes (shorebirds and gulls), Coraciiformes (kingfishers), and Passeriformes (blackbirds, marshwrens). Species belonging to these orders require wetland habitats or the foods provided in wetlands for survival and are capable of acquiring resources at many different locations during the annual cycle (Table 2.2). Body size often is important in determining the ability of a species to acquire necessary resources. Birds with large bodies have the ability to store more energy for use in migration, reproduction, and other critical life history events. Thus, large birds often can exploit wetland resources that are more widely spaced (Alisaukas and Ankney 1992). Body morphology (bill type and length, tarsus length) dictates the types of foods and wetlands that a species can exploit successfully. Small sandpipers that forage by probing can capture small prey only in shallow wetlands, whereas wading species can consume a greater variety of foods across a wider range of water depths.

Waterfowl and Shorebirds

Some of the most successful long-distance migrants are waterfowl. Of the 45 species of waterfowl native to North America (Bellrose 1980, Baldassare and Bolen 1994), 22 regularly use wetlands in the Great Plains (Bellrose 1980, Heitmeyer and Vohs 1984, Bolen et al. 1989, Ringelman et al. 1989). Dabbling ducks and geese are the most common members of the Anseriformes that use wetlands across the Plains. Mallard (*Anas platyrhynchos*), Northern Pintail (*A. acuta*), Blue-winged Teal (*A. discors*), Green-winged Teal (*A. crecca*), American Wigeon (*A. americana*), Northern Shoveler (*A. clypeata*), and Gadwall (*A. strepera*) are the most abundant dabbling ducks (Bolen et al. 1989, Ringelman et al. 1989). Cinnamon Teal (*A. cyanoptera*) are less abundant, and ducks that use deeper, open water or are associated with forests tend to be less abundant or rare. Thus, whistling ducks, Wood Ducks (*Aix sponsa*), sea ducks, mergansers, and Ruddy Ducks (*Oxyura jamaicensis*) are not dominant groups in the Plains. Because shallow, temporary wetlands also are attractive to migrating geese, Canada (*Branta canadensis*), Greater White-fronted (*Anser albifrons*), and Lesser Snow (*Anser caerulescens*) Geese are temporally abundant on some wetland areas in the Great Plains. Tundra Swans (*Cygnus columbianus*) stop during migration but neither breed nor winter in the region.

Wetlands across the Great Plains are used extensively by migrating waterfowl, with breeding being most important in the northern and central Plains. Canada Geese are the only residents numerous locally in the northern Plains (Ringelman et al. 1989), but 12 species of ducks regularly nest in potholes (Batt el al. 1989). Some waterfowl winter throughout the Great Plains, but the southern regions are far more important for wintering than the central or northern regions (Heitmeyer and Vohs 1984, Pederson et al. 1989, Ringelman et al. 1989). By contrast, large numbers of Lesser Snow Geese use the Rainwater Basin in spring, but only small numbers winter in the southern Plains. A few Greater White-fronted Geese winter in the southern Plains, but large aggregations of migrating geese occur in the central and northern Plains. Use by dabbling ducks and geese is high because of their affinity for grassland habitats and adaptations that allow exploitation of shallow wetlands where seasonal flooding promotes plant communities with high seed production or an abundance of macroinvertebrates (Fredrickson and Reid 1988). However, use of wetlands by these species varies from north to south across the Great Plains and is dependent on differences in habitat characteristics as well as the timing and requirements of annual life cycle events for each species (Fredrickson and Reid 1988, Tables 2.2 and 2.3). Although disruption of wetland functions has compromised seed production in the playas of the Southern Plains, dabbling ducks that use resources in playas also regularly feed on waste grain in agricultural fields to acquire the high-energy foods required during winter and migration (Bolen et al. 1989, Ringelman et al. 1989).

Nearly 40 species of shorebirds use wetlands in the Great Plains (Hoffman 1987, Skagen and Knopf 1993); of these, 7 are rare. Fifteen species (White-rumped Sandpiper [*Calidris fuscicollis*], Pectoral Sandpiper [*C. melanotus*], Dunlin [*C. alpina*], Semipalmated Sandpiper [*C. pusilla*], Stilt Sandpiper [*C. himantopus*], Baird's Sandpiper [*C. bairdii*], Sanderling [*C. alba*], Long-billed Dowitcher [*Limnodromus scolopaceus*], Wilson's Phalarope, Red-necked Phalarope [*Phalaropus lobatus*], Lesser Golden Plover [*Pluvialis dominica*], Snowy Plover [*Charadrius alexandrinus*], Lesser Yellowlegs [*Tringa flavipes*], American Avocet [*Recurvirostra americana*], and Hudsonian Godwit [*Limosa haemastica*]) are the most common shorebirds in the Great Plains during spring migration. Of these species, several are long-distance migrants that move between continents, including the Stilt Sandpiper, White-rumped Sandpiper, Pectoral Sandpiper, and Lesser Golden Plover (Skagen and Knopf 1993).

Most shorebirds select wetlands where abundant food is readily available on mudflats or in shallow water and little or no vegetation is present (Fredrickson and Reid 1986, Hoffman 1987, Helmers 1991). However, there are exceptions, such as the Wilson's Phalarope that can use deep water and Common Snipe (*Gallinago gallinago*) that often use habitats with dense cover. Chironomids are among the most important food source for shorebirds. At Cheyenne Bottoms, about 6,800 ha of wetlands produced approximately 90 metric tons dry weight of chironomids in a single year (Welker 1987b). This biomass decreased by 80% during spring shorebird migration. Foraging by shorebirds accounted for much of this decrease (Helmers 1991). Chironomids and other invertebrates are readily available in shallow seasonal wetlands in the Great Plains because they are within the foraging range of shorebirds. For most shorebirds, ideal foraging depths are less than 5 cm (Fredrickson and Reid 1986), because invertebrate availability is related to area of

Table 2.3. General Values of Great Plains Wetlands for Selected Avifauna[a]

	Wintering	Spring Migration	Breeding	Fall Migration
Eared Grebe[b]	0	++	+	++
American White Pelican[b]	0	++	+	++
American Bittern[b]	0	++	+	++
Greater White-fronted Goose	+	++	0	++
Lesser Snow Goose	+	++	0	++
Green-winged Teal	++	++++	0	++++
Mallard	++	++++	+	++++
Northern Pintail	+	+++	+	+++
Blue-winged Teal	0	++++	++	++++
Cinnamon Teal	+	+	+	+
American Wigeon	+	+	0	+
Sora[b]	+	++	+	++
American Coot	0	+++	+	+++
Sandhill Crane	0	++++	0	++++
Whooping Crane	0	++++	0	++++
Black-necked Stilt[b]	0	+	+	+
American Avocet	0	++	++	++
Lesser Yellowlegs	0	++	0	++
Western Sandpiper[b]	0	++	0	++
White-rumped Sandpiper	0	++++	0	0
Wilson's Phalarope	0	++++	++++	++++

[a]0, no use; +, limited use, low numbers; ++, moderate use, low numbers; +++, consistent use, moderate numbers; ++++, consistent use, high numbers.
[b]Eared Grebe (*Podiceps nigricollis*); American White Pelican (*Pelecanus erythrorhynchos*); American Bittern (*Botarus lentiginosus*); Sora (*Porzana carolina*); Black-necked Stilt (*Himantopus mexicana*); Western Sandpiper (*Calidris mauri*).

shallow water. As water levels change as a result of evaporation, slow drainage, or wind-driven seiches, new foraging areas become available. Wind fetch is particularly important on large, shallow wetlands. When water is moved by wind, new foraging areas are created daily and sometimes hourly.

Both shorebirds and waterfowl are well adapted to exploit dynamic wetland habitats in the Great Plains where food is seasonally abundant (Table 2.3). Foods stored as endogenous reserves, primarily visceral fat, are available to meet the costs of migration or early phases of the breeding period when food is limited in northern environments (Raveling 1979). Further, cross-seasonal use of habitat types allows these highly mobile species to exploit northern wetlands effectively during the short arctic summer because resources can be acquired from distant habitats, as is the case with the Mallard (Heitmeyer and Fredrickson 1981, Kaminski and Gluesing 1987). When ideal wetland conditions exist in winter (extensive flooding from fall through spring), Mallard age ratios the following fall reflect more successful reproduction. Therefore, wetland habitats in the Great Plains have the potential to supply critical resources to many waterbirds that may only use these dynamic prairie habitats for a short period in the annual cycle.

Other Waterbirds

Sandhill Cranes (*Grus canadensis*) that nest in Alaska and arctic Canada (National Audubon Society 1985) and winter in the Great Plains and along the Gulf Coast are part of the midcontinent population estimated to be 540,000 individuals (National Audubon Society 1985). Sandhill Cranes are attracted to grassland habitats where they feed on large seeds, underground tubers, macroinvertebrates, and earthworms (National Audubon Society 1985, Reinecke and Krapu 1986). These foods commonly are associated with shallow palustrine wetlands characteristic of the Great Plains. Wetland use in the northern and central Plains occurs during migration, but sizable numbers (360,000) winter in the southern Plains (Bolen et al. 1989). The endangered Whooping Crane (*G. americana*) is another long-distance migrant dependent on Great Plain wetlands for survival. As these cranes move from the breeding grounds in Wood Buffalo Park in northern Alberta to the wintering grounds on the Texas Gulf Coast, they use wetlands along their primary migration corridor through the eastern portion of the Great Plains (National Audubon Society 1985).

One of the most important migration sites for Sandhill and Whooping cranes is the Platte River in Nebraska, especially the "Big Bend" segment between Overton and Chapman (National Audubon Society 1988). The open channel and associated riverine wetlands are identified as critical habitat for these species, and efforts are underway to protect the habitat and instream flows critical to maintaining these dynamic wetland areas.

Wading birds often can obtain food resources from a great diversity of wetland types, ranging from permanently flooded to ephemeral basins. Although few waders are abundant in the Great Plains compared with the Southeast or coastal areas, Cattle Egrets (*Bubulcus ibis*) and White-faced Ibis (*Plegadis chihi*) have sizable local populations. Cattle Egrets are similar to geese in that they are adapted to exploit more terrestrial habitats. Large populations use wetlands and associated upland habitats surrounding Cheyenne Bottoms in Kansas (Hoffman 1987). White-faced Ibis respond well to shallowly flooded basins with scattered spikerushes or sedges where crayfish are abundant. These examples of avifauna exhibiting continental mobility identify the value of different wetlands for different groups of birds and emphasize the importance of protecting and restoring wetland complexes as a strategy to maintain avian biodiversity.

Regional Mobility

Species of the Class Mammalia are the most common vertebrates that exhibit regional mobility (Table 2.4). Although most mammals occurring in the Great Plains exhibit widespread distributions, individual organisms are restricted to much smaller areas. The home ranges of species in this group are variable, ranging

Table 2.4. Distribution and Home Range of Some Mammalian Species Using Wetlands in the Great Plains That Exhibit Regional Mobility[a]

Species	Distribution[b]	Home Range (ha)
Beaver	N,C,S,	
Coyote	N,C,S	1036–4146
Long-tailed weasel	N,C,S	12–122
Mink	N,C,S	8–142
Opossum	N,C,S	4–41
Pronghorn	N,C,S	259
Raccoon	N,C,S	972–1943
Red fox	N,C,S	259–777
River otter	N,C,S	
White-tailed deer	N,C,S	259

[a]Data from Bee et al. 1981, Jones et al. 1983, Jones et al. 1985, Caire et al. 1989.
[b]N, northern Great Plains; C, central Great Plains; S, southern Great Plains.

from 4 to 1,900 ha (Jones et al. 1983, Caire et al. 1989). However, all species are capable of at least short movements to locate suitable habitat. As with continentally mobile species, a variety of wetland types is necessary to support the entire range of regionally mobile vertebrates. However, the reduced mobility of this group dictates that the distribution of wetland types must be closer to one another if recolonization is to occur after major changes in hydrology caused by extensive drought or floods. The river otter, beaver (*Castor canadensis*), mink (*Mustela vison*), and fishes are examples of regionally mobile species dependent on aquatic habitats to meet some or all annual cycle events. Beavers tend to move along water courses, and movements of 7–241 km have been reported (Libby 1957, Knudsen and Hale 1965, Leege 1968). Fish differ from mammals because movements between wetlands must be linked by water. Although wetland mammals can move overland, connectivity among drainage systems enhances their dispersal. Conversion of native habitats to agricultures and development of the extensive road system in the Great Plains have compromised connectivity between wetlands and the success of movements among wetlands.

Other mammals are associated with wetlands because aquatic habitats support important prey bases or vegetation that provides food or cover. Thirty-nine mammal species have been reported as being associated with prairie potholes (Fritzell 1989). Species such as the raccoon (*Procyon lotor*) and opossum (*Didelphis virginiana*) often are associated with riparian systems in the Great Plains because this habitat type provides denning sites. By contrast, the home ranges of species such as the red fox (*Vulpes vulpes*) and long-tailed weasel (*Mustela frenata*) often include palustrine wetlands because preferred prey items (small mammals, birds) tend to be abundant around water. Wetlands also serve as a source of free water necessary for large mammals, such as pronghorn and white-tailed deer (*Odocoileus virginianus*), which often have home ranges of about 2.6 km² if undisturbed (Jones et al. 1983).

Most avian species associated with wetlands are continentally mobile, but a few regionally mobile birds, such as the ring-necked pheasant (*Phasianus colchicus*), regularly exploit wetlands. Conversion of the prairie to agriculture, although providing an important food source for pheasants, has destroyed native upland cover. During dry years, vegetation in palustrine wetlands throughout the Plains provides cover that supports high pheasant densities.

Local Mobility

Species exhibiting limited mobility tend to be small and include amphibians, reptiles, and many mammals (Table 2.5). Of 142 species of herpetofauna occurring in the central plains of North America (historic Great Plains and tallgrass prairie), 38 are primarily aquatic or require permanent water at some stage during their life cycle (Corn and Peterson 1996). Among amphibians occurring in the Great Plains, the bullfrog (*Rana catesbeiana*), plains spadefoot (*Scaphiopus couchi*), tiger salamander (*Ambystoma tigrinum*), woodhouse's toad (*Bufo woodhousii*), and Great Plains toad (*B. cognatus*) have a ubiquitous distribution, whereas other species are more regional (Conant 1975) (Table 2.5). In general, all amphibians depend on wetland habitats for reproduction, and many species also use wetlands as foraging sites (Clark 1979, Hammerson 1986). The bullfrog has a 2-year life cycle and inhabits permanently flooded wetlands during both the breeding and nonbreeding seasons. By contrast, other amphibians inhabiting the Great Plains (e.g., northern [*Rana pipiens*] and plains [*R. blairi*] leopard frog, Great Plains toad, Texas toad [*Bufo speciosus*], and red-spotted toad [*B. punctatus*]) require wetlands for breeding but are more terrestrial in their habits during the nonbreeding season (Conant 1975). These latter species depend on still or slowly moving water or moist ground for egg laying and maturation (Grzimek 1974a). Such species often are capable of exploiting seasonally flooded wetlands because of behavioral adaptations that enable survival during drought. For example, the leopard frog can survive dry periods by migrating short distances or remaining in moist depressions (Grzimek 1974a). The Great Plains toad, Couch's spadefoot (*Scaphiopus couchi*), and western spadefoot (*S. hammondi*) breed in temporarily flooded playas that retain water after summer thunderstorms (Hammerson 1986). Consequently, potholes, depressional wetlands, and playas are important habitats. However, the more terrestrial frogs and toads often stay close to breeding wetlands during the nonbreeding season (Marshall and Buell 1955).

Fifty-four reptilian species are common in the Great Plains (Conant 1975). Of these, 33 species are associated with wetland habitats (Table 2.5). Some inhabit all three regions of the Great Plains, but species richness increases slightly from north to south. The pond slider (*Chrysemys scripta*), painted turtle (*C. picta*), Blanding's turtle (*Emydoidea blandingii*), mud turtle (*Kinosternon flavescens*), smooth (*Trionyx muticus*) and softshell (*T. spiniferus*) turtles, snapping turtle (*Chelydra*

Table 2.5. Distribution of Some Species Using Wetlands in the Great Plains That Exhibit Local Mobility

Species	Distribution[a]	Species	Distribution[a]
Amphibia[b]		Reptilia (*continued*)	
Bullfrog	N,C,S	Skink (great plains)	C,S
Leopard frog (northern)	N,C	Smooth green snake	N,C
Leopard frog (plains)	C,S	Turtle (Blanding's)	C
Spadefoot (Couch's)	S	Turtle (mud)	C,S
Spadefoot (plains)	N,C,S	Turtle (ornate box)	C,S
Spadefoot (western)	S	Turtle (painted)	N,C
Tiger salamander	N,C,S	Turtle (pond slider)	C,S
Toad (woodhouse's)	N,C,S	Turtle (snapping)	N,C,S
Toad (great plains)	N,C,S	Turtle (smooth softshell)	N,C,S
Toad (Texas)	S	Turtle (spiny softshell)	C,S
Toad (red-spotted)	S	Water snake (northern)	C
Reptilia[b]		Water snake (blotched)	C,S
Black-masked racer	N,C,S	Water snake (diamondback)	C,S
Bullsnake	N,C,S	Water snake (Grahams)	C,S
Coachwhip	C,S	Water snake (Concho)	S
Earless lizard (lesser)	N,C,S	Water snake (Brazos)	S
Earless lizard (greater)	S	Mammalia[c]	
Garter snake (checkered)	C,S	Cottontail (eastern)	N,C,S
Garter snake (red-sided)	N,C	Meadow jumping mouse	N,C
Garter snake (western plains)	N,C,S	Meadow vole	N,C
King snake	C,S	Mole (eastern)	N,C,S
Milk snake	N,C,S	Muskrat	N,C,S
Prairie racerunner	N,C,S	Shrew (masked)	N,C
Ratsnake	S	Shrew (Merriam's)	N
Ringneck snake	C,S	Shrew (northern short-tailed)	N,C
Skink (many-lined)	C,N		

[a]N, northern Great Plains; C, central Great Plains; S, southern Great Plains.
[b]Data from Conant 1975.
[c]Data from Bee et al. 1981, Jones et al. 1983, Jones et al. 1985, Caire et al. 1989.

serpentina), northern water snake (*Nerodia sipedon*), Concho water snake (*N. paucimaculata*), Brazos water snake (*N. harteri*), blotched water snake (*N. erythrogaster*), grahams water snake (*N. grahami*), and diamondback water snake (*N. rhombifera*) represent the only truly aquatic and semiaquatic species, relying on wetlands for essential life requisites. These species tend to leave water only to bask, lay eggs (turtles), or give birth (snakes). Slow-moving water is preferred; thus, these species commonly occur in intermittent streams and larger, deeper palustrine and lacustrine wetlands. By contrast, the western box turtle (*Terrapene ornata*) primarily is terrestrial and may enter water only to hibernate (Grzimek 1974b). As with amphibians, the ability of turtles to migrate is limited.

For example, the western box turtle can travel 60–90 m/day and has a home range of about 2 ha (Hammerson 1986). Therefore, the presence of more than one wetland is preferable for long-term population stability. Ideally, a group of permanently flooded basins along with several other wetlands assures the long-term integrity of local populations.

Other reptiles associated with wetlands primarily use aquatic environments as foraging sites but are not strictly dependent on such habitats (Clark 1979). Thus, all wetland types potentially are important. The checkered (*Thamnophis marcianus*), red-sided (*T. sirtalis*), and western plains (*T. radix*) garter snake occur in association with all wetland types, whereas the smooth green snake (*Opheodrys vernalis*), ringneck snake (*Diadophis punctatus*), king snake (*Lampropeltis triangulus*), and ratsnake (*Elaphe guttata*) tend to occur along streams to a greater extent. The bullsnake (*Pituophis catenifer*), coachwhip (*Masticophis flagellum*), black-masked racer (*Coluber constrictor*), and milk snake (*L. triangulum*) often inhabit riparian sites. Lizards are rare in freshwater wetlands, but the lesser (*Holbrookia maculata*) and greater (*H. texana*) earless lizard, prairie racerunner (*Cnemidophorus sexlineatus*), Great Plains skink (*E. obsoletus*), and many-lined skink (*E. multivirgatus*) often occur in the vicinity of streams (Conant 1975). Primary food items include amphibians, fish, and invertebrates. However, most species are capable of foraging for terrestrial prey if wetlands are not available.

Small mammals associated with wetlands include insectivores, lagomorphs, and rodents (Table 2.5). The muskrat (*Ondatra zibethicus*) has specialized morphological adaptations (partially webbed feet, laterally compressed tail) that enable efficient exploitation of wetlands. Lodges and feeding houses that provide access to air when foraging beneath the ice often are constructed in marshes that contain abundant emergent macrophytes and aquatic invertebrates, which are the primary sources of food. The meadow vole (*Microtus pennsylvanicus*), which forages on sedges, rushes (*Juncus* spp.), and grasses, also requires water (Jones et al. 1985). In the northern and western Plains, this species typically is associated with riparian habitats. The meadow jumping mouse (*Zapus hudsonius*), an excellent swimmer and diver, occurs along the borders of marshes and streams.

Although not dependent on wetland environments throughout their range, some mammals are restricted to areas influenced by aquatic habitats in the Great Plains to provide necessary foods or habitats (Jones et al. 1985). The distribution of the eastern mole (*Scalopus aquaticus*) is limited by soil moisture and is largely restricted to stream borders and lake margins in the western Plains. The northern short-tailed shrew (*Blarina brevicauda*) and eastern cottontail (*Sylvilagus floridanus*) are increasingly restricted to riparian habitats that provide the necessary vegetation composition and structure, although habitats used in other geographic areas include grasslands, woodlands, and fencerows. Similarly, Merriam's shrew (*Sorex Merriami*) tends to be associated with wet meadows, but in western states is more common in sagebrush (*Artemisia* spp.) and bunchgrass habitats.

Regardless of taxonomic classification, species capable of only limited mobility are dependent on resources within a small geographic area relative to more mobile species. Fragmentation of the wetland landscape has resulted in isolated populations of some species that require wetland resources. Close juxtaposition of several wetlands with varying hydroperiods is necessary to ensure that some wetlands will be present regardless of long-term fluctuations common in wetland habitats in the Great Plains.

Conclusions

Wetlands represent a critical habitat that directly or indirectly influences vertebrate assemblages in the Great Plains. Although wetlands are not the dominant landscape feature, they provide food and cover for many resident and transient species. In addition to providing critical habitat for waterbirds and amphibians, wetlands also are a source of food, cover, and/or free water for species ranging from small mammals to large ungulates.

Loss and modification of wetlands has had a profound effect on the current distribution and abundance of many vertebrate species in the Great Plains. Undoubtedly, wetlands will play a significant role in determining the status of many animal populations in the future. Of paramount importance in meeting life history needs of many species is the distribution of different wetland types at different geographic scales. Although the existing distribution of wetlands may provide sufficient resources for maintaining waterbird and large mammal populations at current levels, the number, distribution, and quality of remnant wetlands may be inadequate to permit population increases and to ensure the survival and genetic diversity of less mobile species. Efforts should be undertaken not only to protect remnant wetlands but also to promote the restoration of wetland complexes composed of different wetland types at local and regional scales.

References

Alisaukas, R.T., and C.D. Ankney. 1992. The cost of egg laying and its relationship to nutrient reserves in waterfowl. Pp. 30–61 *in* B.D. J. Batt, A.D. Afton, M.G. Anderson, C.D. Ankney, D.H. Johnson, J.A. Kadlec, and G.L. Krapu, eds. Ecology and management of breeding waterfowl. Univ. Minnesota Press, Minneapolis.

Aronovici, V.S., A.D. Schneider, and O.R. Jones. 1970. Basin recharging the Ogallala Aquifer through Pleistocene sediments. Pp. 182–192 *in* Ogallala Aquifer Symp. ACCUSALS Spec. Rep. 39. Texas Tech Univ., Lubbock.

Bailey, R.G. 1978. Description of the ecoregions of the United States. U.S. Dept. Agric., For. Serv., Intermountain Reg., Ogden, UT.

Baldassarre, G.A., and E.G. Bolen. 1994. Waterfowl ecology and management. John Wiley, New York.

Batt, B.D. J., M.G. Anderson, C.D. Anderson, and F.D. Caswell. 1989. The use of prairie potholes by North American ducks. Pp. 204–267 *in* A.G. van der Valk, ed. Northern prairie wetlands. Iowa State Univ. Press, Ames.

Bee, J.W., G. Glass, R.S. Hoffmann, and R.R. Patterson. 1981. Mammals in Kansas. Publ. Educ. Ser. 7. Univ. Kansas Print. Press, Lawrence.

Bellrose, F.C. 1980. Ducks, geese and swans of North America. Third ed. Stackpole Books, Harrisburg, PA.

Belt, C.B., Jr. 1975. The 1973 flood and man's constriction of the Mississippi River. Science 189:681–684.

Bleed, A., and C. Flowerday, eds. 1990. An atlas of the Sandhills. Resour. Atlas 5a. Conserv. and Surv. Div., Inst. of Agric. and Nat. Resour., Univ. Nebraska, Lincoln.

Bolen, E.G., G.A. Baldassare, and F.S. Guthery. 1989. Playa lakes. Pp. 341–365 in L.M. Smith, R.L. Pederson, and R.M. Kaminski, eds. Habitat management for migratory and wintering waterfowl in North America. Texas Tech Univ. Press, Lubbock.

Bolen, E.G., C.D. Simpson, and F.A. Stormer. 1979. Playa lakes: threatened wetlands on the southern Great Plains. Pp. 23–30 in Riparian and wetland habitats of the Great Plains. Publ. 91. Great Plains Agric. Counc., U.S. Dept. Agric., For. Serv., Fort Collins, CO.

Borchert, J.R. 1950. The climate of the central North American grassland. Annu. Assoc. Am. Geogr. 40:1–39.

Caire, W., J.D. Tyler, B.P. Glass, and M.A. Mares. 1989. Mammals of Oklahoma. Univ. Oklahoma Press, Norman.

Chapman, J.D., and J.C. Sherman, eds. 1975. Oxford regional economic atlas: the United States and Canada. Oxford Univ. Press, London.

Clark, J. 1979. Fresh water wetlands: habitats for aquatic invertebrates, amphibians, reptiles, and fish. Pp. 330–343 in P.E. Greeson, J.R. Clark, and J.E. Clark, eds. Wetland functions and values: the state of our understanding. Proc. Natl. Symp. on Wetlands. Tech. Publ. TPS79-2. Am. Water Resour. Assoc., Minneapolis, MN.

Conant, R. 1975. A field guide to reptiles and amphibians: eastern and central North America. Second ed. Houghton Mifflin, Boston.

Corn, P.S., and C.R. Peterson. 1996. Prairie legacies—amphibians and reptiles. Pp. 125–134 in F.B. Samson and F.L. Knopf, eds. Prairie conservation: preserving North America's most endangered ecosystem. Island Press, Covelo, CA.

Crouch, G.L. 1978. Effects of protection from livestock grazing on a bottomland wildlife habitat in northeastern Colorado. Pp. 118–125 in W.D. Graul and S.J. Bissell, coords. Lowland river and stream habitat in Colorado: a symposium. Colorado Chapter, The Wildl. Soc. and Colorado Audubon Counc., Denver.

Dahl, T.E. 1990. Wetland losses in the United States: 1780's to 1980's. U.S. Fish and Wildl. Serv., Washington, DC.

Darnell, R.M. 1978. Impact of human modification on the dynamics of wetland systems. Pp. 200–209 in P.E. Greeson, J.R. Clark, and J.E. Clark, eds. Wetland functions and values: the state of our understanding. Am. Water Resour. Assoc., Minneapolis, MN.

Ducks Unlimited. 1994. Continental conservation plan: an analysis of North American waterfowl populations and a plan to guide the conservation programs of Ducks Unlimited through the year 2000. Prep. for the Board of Directors of Ducks Unlimited Canada, Ducks Unlimited, Inc., and Ducks Unlimited de Mexico.

Dvoracek, J.J. 1981. Modification of playa lakes in the Texas panhandle. Pp. 64–82 in J.S. Barclay and W.V. White, eds. Proc. Playa Lakes Symp. FWS/OBS-81/07. U.S. Dept. Inter., Fish and Wildl. Serv.

Erickson, N.W., and D.M. Leslie. 1987. Soil-vegetation correlations in the Sandhills and Rainwater Basin wetlands of Nebraska. Biol. Rep. 87(11). U.S. Dept. Inter., Fish and Wildl. Serv., Washington, DC.

Flake, L.D. 1979. Perspectives on man-made ponds and waterfowl in the northern prairies. Annu. Meet. For. Comm., Great Plains Agric. Counc. 31:33–36.

Frayer, W.E., T.J. Monahan, D.C. Bowden, and F.A. Grayhill. 1983. Status and trends of wetlands and deepwater habitats in the conterminous United States: 1950's to 1970's. Dept. For. Wood Serv., Colorado State Univ., Fort Collins.

Fredrickson, L.H., and M.K. Laubhan. 1994. Managing wetlands for Wildlife. Pages 623–647 in T.A. Bookhout, ed. Research and management techniques for wildlife and habitats. Fifth ed. Wildl. Soc., Bethesda, MD.

Fredrickson, L.H., and F.A. Reid. 1986. Wetland and riparian habitats: a nongame management overview. Pp. 59–96 in J.B. Hale, L.B. Best, and R.L. Clawson, eds. Management of nongame wildlife in the Midwest: a developing art. N. Central Sect., The Wildl. Soc., West Lafayette, IN

Fredrickson, L. H., and F. A. Reid. 1988. Waterfowl use of wetland complexes. Fish and Wildl. Leaflet 13.2.1. U.S. Dept. Inter., Fish and Wildl. Serv., Washington, DC.

Fritzell, E.K. 1989. Mammals in prairie wetlands. Pp. 268-301 in A.G. van der Valk, ed. Northern prairie wetlands. Iowa State Univ. Press, Ames.

Frye, J.C., and L.A. Byron. 1957. Studies of Cenozoic geology along the eastern margin of the Texas High Plains, Armstrong to Howard counties. Univ. Texas Bur. Econ. Geol. Rep. 32. Austin.

Graul, W.D. 1982. Lowland riparian cottonwood community study. Program narrative. Colorado Div. Wildl. Fed. Aid Proj. W-136-R and W-124-R, Job 4, Denver.

Grzimek, B., ed. 1974a. Grzimek's animal life encyclopedia. Vol. 5, Fishes II and amphibia. Van Nostrand Reinhold, New York.

Grzimek, B., ed. 1974b. Grzimek's animal life encyclopedia. Vol. 6, Reptiles. Van Nostrand Reinhold, New York.

Guthery, F.S., and F.C. Bryant. 1982. Status of playas in the southern Great Plains. Wildl. Soc. Bull. 10:309-317.

Guthery, F.S., F.C. Bryant, B. Kramer, A Stoecker, and M. Dvoracek. 1981. Playa assessment study. Water and Power Resourc. Serv., Amarillo, TX.

Hammerson, G.A. 1986. Amphibians and reptiles of Colorado. Second printing. Colorado Div. of Wildl., Dept. of Nat. Resourc., Denver.

Heitmeyer, M.E., and L.H. Fredrickson. 1981. Do wetland conditions in the Mississippi Delta hardwoods influence mallard recruitment? Trans. N. Am. Wildl. Nat. Resour. Conf. 46:44-57.

Heitmeyer, M.E., and P.A. Vohs. 1984. Distribution and habitat use of waterfowl wintering in Oklahoma. J. Wildl. Manage. 48:51-62.

Helmers, D.L. 1991. Habitat use by migrant shorebirds and invertebrate availability in a managed wetland complex. MS thesis. Univ. Missouri, Columbia.

Hoffman, W. 1987. The birds of Cheyenne Bottoms. Pp. 433-550 in M. Adkins-Heljeson, ed. Cheyenne Bottoms: an environmental assessment. Unpubl. rep. Kansas Biol. Surv. and Kansas Geol. Surv., Univ. Kansas, Lawrence.

Jenkins, R.M., L.R. Aggus, and G.R. Ploskey. 1985. Inventory of United States reservoirs. U.S. Dept. Inter., Fish and Wildl. Fed. Aid to Fish Restor. Act. Admin. Fund, Washington, DC.

Johnson, W.C., R.L. Burgess, and W.R. Keammerer. 1976. Forest overstory vegetation and environment on the Missouri River floodplain in North Dakota. Ecol. Monogr. 46:59-84.

Jones, J.K., Jr., D.M. Armstron, and J.R. Choate. 1985. Guide to mammals of the plains states. Univ. Nebraska Press, Lincoln.

Jones, J.K., Jr., D.M. Armstron, R.S. Hoffmann, and C. Jones. 1983. Mammals of the Northern Great Plains. Univ. Nebraska Press, Lincoln.

Kaminski, R.M., and E.A. Gluesing. 1987. Density and habitat-related recruitment of mallards. J. Wildl. Manage. 51:141-148.

Kantrud, H.A., J.B. Miller, and A.G. van der Valk. 1989. Vegetation of wetlands of the Prairie Pothole Region. Pp. 132-203 in A.G. van der Valk, ed. Northern prairie wetlands. Iowa State Univ. Press, Ames.

Knopf, F.L. 1994. Avian assemblages on altered grasslands. Stud. Avian Biol. 15:247-257.

Knudsen, G.J., and J.B. Hale. 1965. Movements of translocated beavers in Wisconsin. J. Wildl. Manage. 29:685-688.

Kopatek, E.V., R.J. Olson, C.J. Emerson, and J.L. Jones. 1979. Landuse conflicts with natural vegetation. Environ. Conserv. 6:191-200.

Kroonemeyer, K.E. 1978. The U.S. Fish and Wildlife Service's Platte River national wildlife study. Pp. 29-32 in J.C. Lewis, ed. Proc. 1978 Crane Workshop. Natl. Audubon Soc., Tavenier, FL.

Laubhan, M.K., and L.H. Fredrickson. 1993. Integrated wetland management: concepts and opportunities. Trans. N. Am. Wildl. Nat. Resour. Conf. 58:323-334.

Leege, T.A. 1968. Natural movements of beavers of southeastern Idaho. J. Wildl. Manage. 32:973-976.

Libby, W.L. 1957. Observations of beaver movements in Alaska. J. Mammal. 38:269.

Louma, J.R. 1985. Twilight in pothole country. Audubon 87:67-85.

Marshall, W.H., and M.F. Buell. 1955. A study of the occurrence of amphibians in relation to a bog succession, Itasca State Park, Minnesota. Ecology 36:381-387.

Mitsch, W.J., and J.G. Gosselink. 1993. Wetlands. Second ed. Van Nostrand Reinhold, New York.

Moore, R., and T. Mills. 1977. An environmental guide to western surface mining. Part two: impacts, mitigation, and monitoring. FWS/OBS-78/04. U.S. Dept. Inter., Fish and Wildl. Serv., Biol. Serv. Prog., Washington, DC.

National Audubon Society. 1985. Audubon Wildlife Report 1985. R. L. DiSilvestro, ed. National Audubon Society, New York.

National Audubon Society. 1988. Audubon Wildlife Report 1988/89. R. L. DiSilvestro and L. Libate, eds. Academic Press, New York.

Nebraska Game and Parks Commission. 1984. Survey of habitat work plan K-83. Lincoln, NE.

Nebraska Game and Parks Commission. 1995. The state of Nebraska. Internet URL http://ngp.ngpc.state.ne.us/gp.html.

Pederson, R.L., D.C. Jorde, and S.G. Simpson. 1989. Northern Great Plains. Pp. 281-310 in L.M. Smith, R.L. Pederson, and R.M. Kaminski, eds. Habitat management for migratory and wintering waterfowl in North America. Texas Tech Univ. Press, Lubbock.

Ploskey, G.R., and R.M. Jenkins. 1980. Inventory of United States reservoirs. U.S. Dept. Inter., Fish and Wildl. Serv., Natl. Reservoir Res. Prog., Fayetteville, AR.

Raines, R.R., M.C. Gilbert, R.A. Gersib, W.S. Rosier, and K.F. Dinan. 1990. Regulatory planning for Nebraska's Rainwater Basin wetlands: advanced identification of disposal areas. Prep. for the Rainwater Basin Advanced Identification Study. U.S. Environ. Protect. Agency, Region VII, Kansas City, KS, and U.S. Army Engineer Dist., Omaha, NE.

Raveling, D.G. 1979. The annual cycle of body composition in Canada geese with special reference to the control of reproduction. Auk 96:234-252.

Reeves, C.C., Jr. 1966. Pluvial lake basins of west Texas. J. Geol. 74:269-291.

Reinecke, K.J., and G.L. Krapu. 1986. Feeding ecology of sandhill cranes during spring migration in Nebraska. J. Wildl. Manage. 50:71-79.

Ringelman, J.K., W.R. Eddelman, and H.W. Miller. 1989. High plains reservoirs and sloughs. Pp. 311-340 in L.M. Smith, R.L. Pederson, and R.M. Kaminski, eds. Habitat management for migratory and wintering waterfowl in North America. Texas Tech Univ. Press, Lubbock.

Sadeghipour, J., and T. McClain. 1987. Analysis of surface water and climatic data for Cheyenne Bottoms. Pp. 55-90 in M. Adkins-Heljeson, ed. Cheyenne Bottoms: an environmental assessment. Unpublished report. Kansas Biol. Surv. and Kansas Geol. Surv., Univ. Kansas, Lawrence.

Sanderson, G.C. 1976. Conservation of waterfowl. Pp. 43-58 in F.C. Bellrose, ed. Ducks, geese, and swans of North America. Second ed. Stackpole Books, Harrisburg, PA.

Shelford, V.E. 1949. The ecology of North America. Univ. Illinois Press, Urbana.

Shimer, J.A. 1972. Field guide to landforms in the United States. Macmillan, Inc., New York.

Skagen, S.K., and F.L. Knopf. 1993. Towards conservation of mid-continent shorebird migrations. Conserv. Biol. 7:533-541.

Stewart, R.E., and H.A. Kantrud. 1972. Vegetation of prairie potholes, North Dakota in relation to quality of water and other environmental factors. U.S. Geol. Surv. Prof. Pap. 585-D. U.S. Gov. Print. Off., Washington, DC.

Tiner, R.W., Jr. 1984. Wetlands of the United States: current status and recent trends. Natl. Wetlands Invent., Washington, DC.

U.S. Department of the Interior. 1988. The impact of federal programs on wetlands. Vol. I. A report to congress by the Secretary of Interior, Washington, DC.

van der Valk, A.G. 1981. Succession in wetlands: a Gleasonian approach. Ecology 62:688–696.

van der Valk, A.G. 1989. Northern prairie wetlands. Iowa State Univ. Press, Ames.

Vogler, L., G. Fredlund, and W. Johnson. 1987. Cheyenne Bottoms geology. Pp. 15–29 in Cheyenne Bottoms: an environmental assessment. Unpubl. rep. Kansas Biol. Surv. and Kansas Geol. Surv., Univ. Kansas, Lawrence.

Ward, C.R., and E.W. Huddleston. 1972. Multipurpose modification of playa lakes. Pp. 203–286 in Playa Lakes Symp. ICASALS Publ. 4. Texas Tech Univ., Lubbock.

Welker, G. 1987a. The effect of land use on runoff in the Cheyenne Bottoms watershed. Pp. 121–148 in Cheyenne Bottoms: an environmental assessment. Unpubl. rep. Kansas Biol. Surv. and Kansas Geol. Surv., Univ. Kansas, Lawrence.

Welker, G. 1987b. Invertebrates. Pp. 317–362 in M. Adkins-Heljeson, ed. Cheyenne Bottoms: an environmental assessment. Unpubl. rep. Kansas Biol. Surv. and Kansas Geol. Surv., Univ. Kansas, Lawrence.

Weller, M.W., and C.S. Spatcher. 1965. Role of habitat in the distribution and abundance of marsh birds. Iowa State Univ. Agric. and Home Econ. Exp. Sta. Spec. Rep. 43, Ames, IA.

Williams, G.P. 1978. The case of the shrinking channels—the North Platte and Platte Rivers in Nebraska. U.S. Geol. Surv. Circ. 781. Reston, VA.

Willson, G.D. 1995. The Great Plains. Pp. 295–296 in E.T. LaRoe, G.S. Farris, C.E. Puckett, P.D. Doran, and M.J. Mac, eds. Our living resources: a report to the nation on the distribution, abundance, and health of U.S. plants, animals, and ecosystems. U.S. Dept. Inter., Natl. Biol. Serv., Washington, DC.

Wind Erosion Research Unit. 1995. U.S. Dept. Agric., Agric. Res. Sta., Cropping Sys. Res. Lab., Lubbock, TX. Internet URL http://lbk131.ars.usda.gov/weru/weru.html.

Winter, T.C. 1989. Hydrological studies of wetlands in the northern prairie. Pp. 16–54 in A.G. van der Valk, ed. Northern prairie wetlands. Iowa State Univ. Press, Ames.

3. Water Management and Cottonwood Forest Dynamics Along Prairie Streams

Jonathan M. Friedman, Michael L. Scott, and Gregor T. Auble

Introduction

Because riparian ecosystems are the principal natural forest in the prairie, they provide important habitat for many vertebrates (Brinson et al. 1981). Thus changes in the abundance and patterns of riparian forest affect the fauna of the Great Plains. Water management in the Great Plains has had important and variable impacts on riparian vegetation. For example, reductions in peak streamflow have increased the area occupied by bottomland forest in some cases and decreased it in others. This variability can be explained by placing the relation between flow and tree establishment in the appropriate geomorphic context.

Herein, we examine some of the factors that influence establishment and survival of riparian trees, with an emphasis on fluvial geomorphic processes. We discuss how floods, drought, water management, and physiographic setting affect these processes. Our review focuses on cottonwood (*Populus* spp.), a dominant early successional riparian tree throughout the Great Plains and often the only tree in bottomlands of the western Great Plains.

Effects of Water Management on Regeneration

Suitable Sites for Cottonwood Reproduction

The conditions associated with successful establishment of cottonwoods and other members of the willow family (*Salicaceae*) are typical of bottomland pioneer

Figure 3.1. Cottonwood seedlings on bare sand along Coal Creek, Arapahoe County, Colorado, on September 27, 1990. The exposed roots extend to the water table as indicated by standing water in the pit. Seedlings in this photo are 2–5 months old and average about 20 cm tall.

species (White 1979). Abundant wind- and water-dispersed seeds are released early each summer in association with peak flow in snowmelt-dominated watersheds (Scott et al. 1993, Johnson 1994). The seeds lose germinability within a few weeks (Moss 1938, Ware and Penfound 1949). Freshly deposited alluvium typically provides ideal substrate for germination and establishment. Because cottonwoods are intolerant of shade and germinate poorly in plant litter, they rarely become established from seed under an existing stand of trees (Johnson et al. 1976) or herbs (Friedman et al. 1995). Seedlings require a continuously moist surface during the first week or more of growth (Moss 1938) but, by the end of the first growing season, are able to survive declines in the water table of as much as 1 m (Segelquist et al. 1993) (Fig. 3.1).

Riparian sites bare and moist enough for cottonwood establishment are usually close to the channel and subject to disturbance by the stream. Although cottonwoods are tolerant of burial and able to sprout from stems or roots, floods and ice scour typically cause high mortality of seedlings and saplings (Johnson 1994, Scott et al. in press). Therefore, establishment from seed occurs only in locations that are moist, bare, and protected from intense disturbance (Everitt 1968, Bradley and Smith 1986, Sedgwick and Knopf 1989, Johnson 1994). In humid regions, vegetation is dense and grows rapidly, and as a result, the need for bare surface is more restrictive than the moisture requirement (Johnson 1965). The reverse is true in arid regions (Zimmerman 1969).

Most cottonwoods are capable of forming root sprouts. Because root sprouts can receive moisture and energy from the parent plant, they may become established under a wider range of conditions than do seedlings. Within the genus *Populus,* root sprouting is most common in species of higher elevations and latitudes. Thus, balsam poplar (*P. balsamifera),* narrowleaf cottonwood (*P. angustifolia*), and aspen (*P. tremuloides)* all form root sprouts more readily than plains cottonwood (*P. deltoides* subsp. *monilifera*) (Rood et al. 1994).

How Fluvial Processes Create Suitable Sites for Establishment

Sites suitable for cottonwood seedling establishment are produced by many fluvial processes. We discuss three that are important in the Great Plains—channel narrowing, meandering, and flood deposition (Table 3.1). At any given site, these processes may act alone or in combination; their relative importance depends on geologic and climatic factors, including flow variability, sediment load, and gradient.

Channel morphology—typically braided or meandering—is a useful indicator of the locally dominant fluvial processes. Braided channels are characterized by steep gradient, high flow variability, and sediment load dominated by sand and coarser particles (Osterkamp 1978). Width-depth ratios are high because coarse sediment must be carried along the bed and forms banks that are easily eroded. Meandering channels are characterized by shallow gradient, low flow variability, and sediment load dominated by silt and finer particles. Width-depth ratios are low because fine sediment can be carried in the water column and forms cohesive

Table 3.1. Fluvial Processes Producing Cottonwood Regeneration Sites in the Great Plains[a]

Fluvial Process	Flow	Landform	Community Patterns
Narrowing	One to several years of flow below that necessary to rework channel bed	Channel bed	Spatial patterns variable Usually not even-aged stands Establishment surface at relatively low elevation of former channel bed Sites become safe by accretion, channel movement, or change to milder flow regime
Meandering	Frequent moderate flows	Point bars	Moderate number of arcuate, even-aged stands Strong left-bank, right-bank asymmetry in ages Flood training of stems common Establishment surface of mature trees often well below present ground surface and near channel bed elevation Sites become safe by accretion and channel movement
Flood deposition	Infrequent high flows	Flood deposits	Small number of linear, even-aged stands Flood training of stems rare Establishment surface of mature trees near present ground surface and well above channel bed elevation Sites are initially safe because flows required to disturb them are infrequent

[a]Based on Scott et al. (1996).

banks resistant to erosion. The proportion of braided channels is greatest in the southwestern and west central Plains (Wyoming, Colorado, New Mexico, and the western portions of Nebraska, Kansas, Oklahoma, and Texas) because (1) glaciated valleys to the north tend to be rich in fine sediment and low in gradient, (2) many watersheds in the southwestern and west central Great Plains contain large amounts of sandstone or wind-blown sand, and (3) flow variability increases with aridity (Wolman and Gerson 1978).

Narrowing

Channel narrowing refers to abandonment by the stream of a portion of former channel bed. This includes both reduction in width of a single channel and cessation of flow in one or more channels of a multiple-channel stream. Channel narrowing can occur as a response to flood-induced widening (Schumm and Lichty 1963, Osterkamp and Costa 1987, Friedman et al. 1996), climate change (Schumm 1969), construction of upstream dams (Williams and Wolman 1984), and introduction of exotic bottomland plant species (Graf 1978) or as part of a cyclic, autogenic process (Patton and Schumm 1981). The immediate cause of narrowing is usually a period of one to several years of flow lower than that necessary to rework the entire channel bed. This relatively low flow allows establishment of vegetation on the bed (Fig. 3.2, Table 3.1). The newly established vegetation promotes deposition of fine sediment (Osterkamp and Costa 1987) and increases resistance to erosion (Smith 1976), thus stabilizing the channel at a narrower width. Narrowing may be facilitated by flow-related fluctuations in channel bed elevation (Friedman et al. 1996).

Cottonwoods established during an episode of channel narrowing are often not even-aged because individuals may have been established at any time within a period of several years of relatively low flow (Friedman et al. in press). Stands usually have an irregular shape whose longest axis is parallel to the direction of flow. The germination point of trees and shrubs established during channel narrowing is at the elevation of the channel bed at the time the surface was abandoned by the stream.

Meandering

In the process of meandering, cutbanks on the outside of bends gradually erode outward and downstream while the sediment removed from them is deposited downstream on the inside of bends (point bars, Fig. 3.3). Suitable conditions for establishment occur on point bars in an arc-shaped band parallel to the channel (Table 3.1). The upper and lower limits of this band are determined by moisture scarcity and riverine disturbance (Bradley and Smith 1986) but may also be influenced by nonuniform waterborne seed dispersal associated with strand lines (Craig and Malanson 1993). Progressive channel movement protects cottonwood on point bars from flood disturbance and ice scour.

Most movement of meandering channels is accomplished by moderate-high flows with recurrence intervals less than 5 years (Wolman and Miller 1960). Such

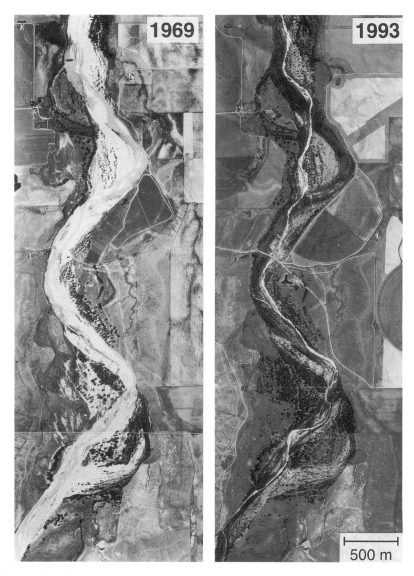

Figure 3.2. Channel narrowing and cottonwood forest establishment along West Bijou Creek, Arapahoe County, Colorado between August 7, 1969, and June 25, 1993. Flow is from bottom to top (northward).

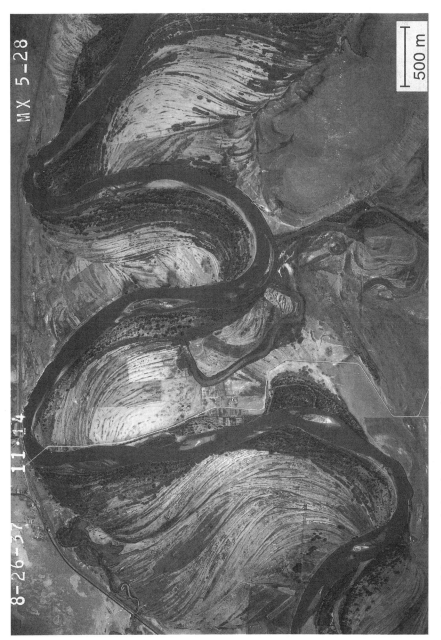

Figure 3.3. A meandering channel: Missouri River, Cascade County, Montana, August 26, 1937. Flow is from left to right.

flows often lead to cottonwood establishment because they form most of the new point bar area and because they moisten high surfaces that are relatively safe from riverine disturbance (Bradley and Smith 1986). Stands produced as a result of meandering generally exhibit a series of parallel, arcuate bands of even-aged trees parallel to the direction of flow at the time of establishment (Everitt 1968, Noble 1979) (Fig. 3.3). On the point bar side of the channel, the age of bands increases with distance from the channel (Everitt 1968, Scott et al. in press). Stands on the cutbank side are older and not necessarily parallel to the present direction of flow (Fig. 3.3). The establishment point of an adult tree is at the elevation of a point bar, typically below the surface of the present floodplain but above the channel bed. Sediment deposition around the tree after establishment raises the elevation of the surface to the level of the floodplain (Everitt 1968, Bradley and Smith 1986).

During channel meandering, bands of trees are formed frequently, and each new band occupies a relatively small portion of the bottomland. Furthermore, the formation of new bands on point bars occurs contemporaneously with removal of old bands at cutbanks. Therefore, meandering is approximately a steady-state process in which channel width, forest area, and forest age structure remain stable. By contrast, channel narrowing involves transient non-steady-state replacement of channel bed by forest (Hughes 1994, Scott et al. 1996).

Flood Deposition

Flood deposition and erosion can produce bare, moist surfaces that are high above the channel bed and therefore relatively safe from future flow-related disturbance. Floods occur along most streams but are especially important for cottonwood establishment where lateral channel movement is constrained by a narrow valley (Scott et al. in press). Because the channel is not moving, the only locations safe from subsequent scouring are high on the bank. Only the greatest flows produce bare, moist substrate at these high positions. Therefore, along a constrained channel, trees occur in a few even-aged groups. The germination point is at a high elevation relative to the channel bed and close to the present ground surface (Scott et al. in press) (Table 3.1).

Floods can cause tree establishment either directly through the process of flood deposition or indirectly by initiating an episode of channel narrowing. For example, along Plum Creek, a sandbed stream in Colorado, deposits left on terraces by a catastrophic flood in 1965 were too high, and therefore too dry, to allow establishment of cottonwood from seed; however, the flood widened the stream, and during subsequent decades of narrowing, extensive establishment occurred at or near the elevation of the channel bed (Friedman et al. in press).

Succession

On the Great Plains, forests on recently formed fluvial surfaces are typically dominated by cottonwood—plains cottonwood at low elevations and latitudes and narrowleaf cottonwood, balsam poplar, and hybrids among the three species at higher elevations and latitudes (Great Plains Flora Association 1986). Willows,

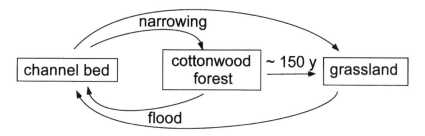

Figure 3.4. Bottomland forest dynamics along ephemeral streams in eastern Colorado. In the absence of flood disturbance, cottonwood forest is replaced by grassland.

especially peachleaf willow (*Salix amygdaloides*), are also locally dominant. The exotic shrub saltcedar (*Tamarix ramosissima*) has become an important pioneer in the southern Great Plains and is naturalized as far north as southern Canada (Great Plains Flora Association 1986, Brock 1994). Because of their high requirements for light and moisture, these genera are unable to reproduce from seed in an existing stand. Reproduction via sprouts from roots or stems may prolong occupation of a site. However, in the absence of physical disturbance, cottonwood and willow are eventually replaced by more shade-tolerant plants.

In the eastern Great Plains, pioneer cottonwood and willow are replaced by a self-sustaining diverse bottomland hardwood forest (Bruner 1931, Weaver 1960, Johnson et al. 1976). For example, in central Kansas, bottomland dominants include green ash (*Fraxinus pennsylvanica*), bur oak (*Quercus macrocarpa*), American elm (*Ulmus americana*), box elder (*Acer negundo*), and hackberry (*Celtis occidentalis*) (Weaver 1960). Compared with cottonwood and willow, these species are typically slower to disperse to new sites and less tolerant of inundation, burial, physical damage, or soils low in organic matter (Johnson et al. 1976, Teskey and Hinckley 1978). Forest diversity decreases westward until only the early successional species, cottonwood and willow, remain. Thus, in much of the western Great Plains, the climax native bottomland community is not forest but shrubland or grassland (Hefley 1937, Lindauer 1983), and maintenance of forest is dependent on physical disturbance (Fig. 3.4). Since the 1800s, fire suppression and decreases in flow variability caused by water development have allowed the exotic Russian-olive (*Elaeagnus angustifolia*) and trees from the eastern Plains, especially green ash and eastern redcedar (*Juniperus virginiana*), to become established in western bottomlands where shade-tolerant trees were formerly absent (Olson and Knopf 1986, Johnson 1992, Shafroth et al. 1995a).

How Water Management Influences Fluvial Processes

Water Management and Channel Narrowing

Dams and diversions typically decrease downstream peak flow and sediment load, thereby reducing both the physical forces exerted by the stream on its bed and the amount of coarse sediment carried along the bed. This situation may lead to channel narrowing as cottonwood and other plants become established on the bed. Channel narrowing after water management is most extensive along formerly wide, shallow, braided channels, which are more common in the southwestern and west central Great Plains than elsewhere in the region. Narrowing related to dams and diversions has been observed along the Arkansas River (Williams and Wolman 1984), the South Platte River (Nadler and Schumm 1981), and the Republican River (Northrop 1965) in Colorado; the North Platte, South Platte, and Platte rivers (Johnson 1994) and the Republican River (Williams and Wolman 1984) in Nebraska; Sandstone Creek (Bergman and Sullivan 1963), Wolf Creek, and the Washita and North Canadian rivers in Oklahoma (Williams and Wolman 1984); the Canadian and Jemez rivers in New Mexico (Williams and Wolman 1984); and the Canadian River and the Rio Grande in Texas (Williams and Wolman 1984). Such channel narrowing is typically associated with establishment of pioneer woody species, especially cottonwood, willow, and saltcedar.

The pulse of cottonwood and willow establishment associated with channel narrowing ends when a new equilibrium width is attained. Further establishment occurs only to the extent that the channel is able to erode existing floodplain. For example, in the mid-1800s the South Platte River, Colorado, was a wide, sandy braided channel bordered by sparse, discontinuous patches of cottonwood (Frémont 1988). Diversion of spring snowmelt flows and occasional droughts allowed narrowing and establishment of cottonwood on the channel bed through the 1930s (Sedgwick and Knopf 1989). About 1940, when narrowing ceased, the channel was less than 15% of its original width, and much of the former channel had been replaced by a continuous cottonwood forest (Johnson 1994). Since that time, cottonwood stands have aged and thinned (Crouch 1979, Sedgwick and Knopf 1989), and some have been replaced by grassland, shrubland, or stands of shade-tolerant trees such as green ash (Snyder and Miller 1991). Because spring flooding still opens new establishment sites, cottonwood will persist indefinitely along the South Platte River but in smaller areas than were occupied in the 1960s.

Water Management and Reduction in Meandering Rate

Reduction in peak flow does not always lead to channel narrowing. In fact, decreases in sediment load downstream of dams often cause erosion and resultant channel widening or deepening (Williams and Wolman 1984). Channel narrowing is least likely to be important along meandering channels, which have a low width–depth ratio. Along a meandering channel, the principal effect of reductions in peak flow and sediment load is reduced rate of meandering, caused by the stream's decreased capacity to erode cutbanks. This effect may be exacerbated by channel downcutting or efforts to stabilize the channel banks. As the rate of formation of establishment sites is decreased, regeneration of pioneer species such as cottonwood is slowed (Johnson et al. 1976, Bradley and Smith 1986, Rood and Mahoney 1990, Scott et al. in press). After many decades, cottonwoods and willows decline and are replaced by more shade-tolerant species (Fig. 3.5). This

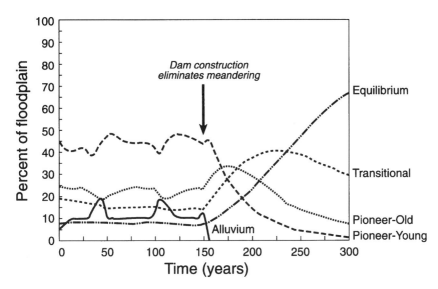

Figure 3.5. Simulated cover type proportions on the Missouri River floodplain in North Dakota. Meandering rate is set to zero at 150 years to simulate effects of dams. Cover types are Alluvium, recently disturbed, nonforested area; Pioneer-Young, cottonwood and willow stands less than 40 years old; Pioneer-Old, cottonwood stands 40–80 years old; Transitional, stands dominated by cottonwood with well-developed midstories of ash, elm, and box elder; and Equilibrium, stands dominated by ash, elm, and box elder. (Redrawn from Johnson 1992.)

syndrome has been most important in the northern and eastern Great Plains, where meandering channels and broad forested bottomlands are common. Occurrences have been reported along the Missouri River in North Dakota (Johnson et al. 1976, Johnson 1992), the Milk (Bradley and Smith 1986) and Marias rivers (Rood and Mahoney 1995) in Montana, the Milk River in Alberta (Bradley and Smith 1986), and the Bighorn River in Wyoming (Akashi 1988).

Water Management and Flood Deposition

Along channels where lateral migration is prevented, flood deposition may be the only process that produces bare, moist sites for cottonwood establishment high enough to survive the effects of future damage from ice and floods. Therefore, if water management along a channel in a narrow valley reduces the variation in the annual peak flow, cottonwood regeneration may decrease. This scenario has been presented as an argument for maintaining the magnitude of peak flows with a recurrence interval of 9 years or greater along a constrained section of the Upper Missouri River flowing through the Missouri Breaks in Choteau, Fergus, and Blaine counties, Montana (Scott et al. in press).

Where meandering is stopped by channel stabilization, flood deposition may be the only remaining fluvial process capable of forming sites suitable for cottonwood establishment. The result will be less frequent establishment and a narrower forested zone.

Water Management and Exotic Species

In the Great Plains and desert Southwest, water regulation has been associated with the spread of exotic species, especially saltcedar and Russian-olive. Like cottonwood and willow, saltcedar produces abundant wind- and water-dispersed seeds capable of becoming established on bare, moist substrates. Compared with cottonwood, adult saltcedar is more tolerant of fire, high salinity, and drought (Lindauer 1983, Busch and Smith 1995) but less tolerant of flood disturbance (Stromberg et al. 1993) and possibly frost (Snyder and Miller 1991). Whereas viable seed of cottonwood is present only for 1-2 months in early summer (Johnson 1994), saltcedar releases seed throughout the summer (Merkel and Hopkins 1957). Therefore, saltcedar often competes successfully against cottonwood, especially in the southern Plains (Hefley 1937) and desert Southwest (Busch and Smith 1995). Water management can foster spread of saltcedar by increasing salinity, decreasing physical disturbance, or delaying peak flows beyond the cottonwood seed viability period (Everitt 1995). By 1994, saltcedar occupied about 500,000 ha of bottomlands in the United States (Brock 1994).

Russian-olive is a large-seeded, small tree tolerant of shade and drought. This exotic has been planted extensively in the Great Plains. Because its seeds are less abundant and less mobile than those of cottonwood, Russian-olive is slower in colonizing recently formed alluvial surfaces. However, shade tolerance enables this species to reproduce from seed in the presence of grass or other trees. As a result, Russian-olive can invade and slowly replace established stands of trees without physical disturbance (Currier 1982, Shafroth et al. 1995a). Water management may favor Russian-olive by reducing the formation of disturbed sites suitable for cottonwood regeneration (Shafroth et al. 1995a). Russian-olive is naturalized throughout the western Great Plains and has been projected to become the dominant woody plant in eastern South Dakota (Olson and Knopf 1986) and a climax dominant species along the Platte River in Nebraska (Currier 1982).

The dependence of cottonwood on flow variability is analogous to similar requirements of some native riverine fish. For example, spring floods stimulate spawning migrations of shovelnose sturgeon (*Scaphirhynchus platorynchus*) and paddlefish (*Polyodon spathula*) (Berg 1981). High flows are also important in creating fluvial features such as side-channels, which are important for spawning and rearing of native fish, including the endangered pallid sturgeon (*Scaphirhynchus albus*) and other native species (Hesse et al. 1993). In intermittent prairie streams, the distribution of fish species is related to their ability to recolonize after floods and droughts (Fausch and Bestgen, this volume). Dam-related reductions in peak flows contribute to replacement of native species by exotic fish (Meffe 1984, Cross and Moss 1987). Therefore, dam-related replacement of cottonwood by Russian-olive can be seen as part of a more general replacement of native riverine species by exotics brought about by decreases in flow variability (Welcomme 1979).

Natural Fluctuations in Establishment of Bottomland Forests

Changes in riparian vegetation caused by water management cannot be assessed properly without an understanding of the changes that are brought about by other causes. Natural variability in the extent and composition of riparian vegetation depends on the fluvial processes that control establishment and survival of vegetation. Along meandering channels, most channel change and formation of cottonwood establishment sites occur during frequent flows, which affect only a small portion of the bottomland (Wolman and Miller 1960). As a result, channel width, forested area, and forest age structure may approach a steady-state reflection of the climate and flow regime.

At the opposite end of the continuum are ephemeral channels in arid locations, where flow variation is more extreme and most channel change occurs as a result of major floods that disturb a large portion of the bottomland (Wolman and Gerson 1978, Hughes 1994). Cottonwood regeneration is infrequent and is typically associated with periods of channel narrowing that follow flood-related widening. Along these streams, channel width, forested area, and forest age structure are all a function of the magnitude of and time since recent floods (Friedman et al. in press). Thus, punctuated age distributions and fluctuations in channel width and forested area are common along ephemeral channels in the western Great Plains (Schumm and Lichty 1963).

Before a major flood in 1914, the Cimarron River in southwestern Kansas was a 16-m-wide stream with few riparian trees. After 28 years of flood-related erosion, width reached 390 m. Narrowing, beginning in 1943, was associated with establishment of an extensive cottonwood forest (Schumm and Lichty 1963). Similarly, early photos of the Arikaree River in eastern Colorado indicate a narrow channel with few riparian trees. A major flood in 1935 widened the channel, and a cottonwood forest developed during subsequent narrowing. Floods resulting from several intense thunderstorms in eastern Colorado in 1935 and 1965 led to synchronous pulses of cottonwood establishment along many ephemeral streams in the area (Friedman et al. in press).

Along small, high-gradient ephemeral channels in the western Great Plains, an intrinsic process of periodic channel incision can lead to cyclic fluctuations in channel width and riparian vegetation in the absence of major floods, climate change, or human disturbance (Patton and Schumm 1981). Even along meandering channels, fluctuations in riparian forest populations can occur as a result of drought-related mortality (Albertson and Weaver 1945) or floods (Everitt 1968). Cottonwood populations in the western Great Plains may have varied during the Holocene in response to changes in the frequency of floods and droughts. In addition, variation in dune activity over the past 1,000 years (Madole 1994) surely influenced the sediment budgets and riparian vegetation of streams.

Effects of flow alteration are often confounded with those of natural processes (Smith 1940). For example, along West Bijou Creek, Colorado, channel narrowing and cottonwood regeneration after a series of major floods in 1935, 1951, and

1965 (Fig. 3.2) have been enhanced by reductions in peak flows resulting from groundwater pumping and small dams on tributary streams. Along the South Platte River, Colorado, channel narrowing associated with irrigation diversions was enhanced by below-average precipitation in the 1930s (Nadler and Schumm 1981). Along the Republican River in Colorado, Nebraska, and Kansas, flood-related widening in 1935 (Smith 1940) led to channel narrowing that has been accentuated by upstream dam construction (Northrop 1965).

Effects of Water Management on Survival

Drought Susceptibility of Cottonwood

Moisture availability strongly influences riparian forests on the Great Plains. Annual growth of trees is correlated with precipitation (Albertson and Weaver 1945). In addition, tree height and growth rate, bottomland forest width, and forest species richness all decline westward with decreasing precipitation (Aikman 1926, Bruner 1931). During the drought of 1933–41, Albertson and Weaver (1945) observed average mortalities of 50–75% for American elm, hackberry, plains cottonwood, and willow along ephemeral streams in western Kansas; mortality was also observed in southeastern Montana (Ellison and Woolfolk 1937) and central Oklahoma (Harper 1940). Mortality was highest for shallow-rooted trees exposed to rapid declines in water table (Albertson and Weaver 1945).

Of all the trees found in the bottomlands of eastern Kansas, only plains cottonwood and, to a lesser extent, peachleaf willow occur in eastern Colorado. In other words, as precipitation decreases, the successional sequence is truncated so that only the pioneer species remain. This trend is surprising considering that (1) cottonwood typically occupies the wettest forested bottomland sites in the eastern Great Plains (Brinson et al. 1981); (2) cottonwood adults planted in shelterbelts are more susceptible to drought mortality than bur oak, hackberry, and eastern redcedar and similar in drought susceptibility to box elder, green ash, American elm, and silver maple (*Acer saccharinum*) (Kaylor et al. 1935, Deters and Schmitz 1936, Bunger and Thomson 1938, Ware and Smith 1939, Albertson and Weaver 1945, Read 1958); (3) cottonwood seedlings require a continuously wet surface for at least 1 week (Friedman et al. 1995); and (4) xylem of adult cottonwood are highly vulnerable to drought-induced cavitation (Tyree et al. 1994). There are two possible explanations for this success of a drought-sensitive tree in a drought-prone region. First, where precipitation is scarce, trees are forced to rely on stream water and the associated alluvial groundwater. The only sites low enough to provide access to this water are highly disturbed by the stream. Thus, those species that require soil organic matter for establishment or are intolerant of flood disturbance do not survive. Second, if occasional droughts, floods, or fires destroy bottomland forests (Albertson and Weaver 1945, Weaver 1960), species that produce abundant, wind-dispersed seeds will return more quickly than species that produce fewer, animal-dispersed seeds.

Because of cottonwood's limited ability to tolerate drought, riparian stands of this species in the western Great Plains are mostly dependent on alluvial ground-water and are, therefore, susceptible to fluctuations in the water table. However, some planted cottonwoods have survived at upland sites where the water table is out of reach (Sprackling and Read 1979). Effects of water table decline on riparian forest are related not only to the magnitude of decline, but also to antecedent conditions, soil texture, and the rate of drawdown.

Water Management and Drought Mortality

Climatic drought affects bottomland trees by reducing soil moisture in the unsatu-rated zone and by decreasing streamflows, which reduces recharge and lowers the alluvial water table. Water management can have similar effects. Decreases in river stage and near elimination of overbank flooding along the Missouri River in North Dakota have reduced growth rates of most floodplain tree species (Reily and Johnson 1982). Damming and diversion of the St. Mary River, Alberta, in 1951 was associated with a 68% downstream mortality in *Populus* by 1985 (Rood et al. 1995). Death of trees was apparently caused by rapid decline in the spring water table and a sustained low summer water table associated with low river water levels. Upstream reaches and adjacent dams without such diversions exhibited no decline. Loss of 67% of area forested by cottonwood on the Arkansas River, Colorado, downstream of John Martin Reservoir between 1949 and 1980 was associated in part with lowered water levels caused by flow diversions and channel downcutting (Snyder and Miller 1991).

Gravel mining in stream beds can kill riparian forest by lowering the water table. Along an ephemeral sandbed stream in eastern Colorado, gravel mining locally lowered the water table by 1.2 m in less than 4 months, resulting in death of 80% of the riparian cottonwoods during the next 3 years (Fig. 3.6). Mortality in unmined reaches during the same period was 2% or less (Shafroth et al. 1995b). The rapid decline in the water table and subsequent tree mortality were similar to effects of climatic drought along ephemeral streams in the 1930s (Albertson and Weaver 1945).

Where aquifers used for groundwater pumping are hydrologically connected to alluvial aquifers, lowering of the water table can reduce streamflow and eliminate riparian forests. Pumping has been associated with decreased streamflows in many Great Plains streams including Frenchman Creek in Nebraska (Leonard and Huntoon 1974) and the Arkansas, Republican, Saline, Smoky Hill, Cimarron, and North Fork Solomon rivers and Rattlesnake Creek in Kansas (Layher 1986, Cross and Moss 1987, Kromm and White 1992). Seven hundred miles of formerly perennial streams in Kansas are now intermittent, in part because of groundwater pumping in the High Plains Aquifer (Layher 1986, Luckey et al. 1988). Flow depletion along the Arkansas River has been associated with death of riparian trees (Kromm and White 1992). Groundwater pumping has eliminated large areas of riparian forest in the desert Southwest (Stromberg 1993).

Figure 3.6. Cottonwood trees killed by groundwater decline associated with gravel mining, Coal Creek, Arapahoe County, Colorado, October 10, 1994. Mining occurred in June 1992.

Other Human Influences on Forest Dynamics

Water management is not the only human alteration that has affected cottonwood forests on the Great Plains. Introduction of the horse (*Equus caballus*) to North America in the 1500s (Winship 1904) may have influenced cottonwood populations because horses changed the hunting of bison and because cottonwood bark was often used to sustain horses in winter (Gregg 1905). Near-elimination of beavers from the watershed of most Plains rivers in the 1800s decreased water storage and may have increased flow variability in those rivers (Knopf and Scott 1990), a change that was later reversed by dam construction. Reductions in fire frequency beginning around the time of settlement in the mid-1800s allowed westward advancement of trees (Bruner 1931), especially fire-sensitive species such as eastern redcedar (Arend 1948). Replacement of free-ranging bison (*Bison bison*) by fenced cattle has affected regeneration of cottonwood (Hartnett et al., this volume). Reservoir construction has probably inundated more than 5% of the length of major streams in the United States (Brinson et al. 1981), eliminating many bottomland forests associated with those streams (Anderson 1971); as reservoirs fill with sediment, new land suitable for tree establishment may be created.

Large areas of bottomland forest in the Great Plains have been cleared for agriculture, mostly along formerly meandering channels in the northern and eastern Plains. Land clearing reduced the forested proportion of the Missouri River floodplain in Missouri from 76% in 1826 to 13% in 1972 (Bragg and Tatschl 1977). Along the Missouri River in North Dakota, between Oahe Reservoir and Garrison Dam, 56% of the bottomland forest was cleared between 1872 and 1979 (Johnson 1992). Along Rush and Wildhorse creeks, Oklahoma, 86% of the bottomland forest was lost between 1871 and 1937 as a result of land clearing and channel straightening (Barclay 1980). In Iowa, the area of forest in 1974 was 23% of the presettlement level (Thomson and Hertel 1981). From before settlement to 1936, Kansas forests declined from 8.5 to 2.4% of the area of the state (Ware and Smith 1939).

Relevance to Vertebrate Habitat

Riparian forests play an important role as vertebrate habitat in landscapes that contain few other trees. In the Great Plains, the proportion of vertebrates found mostly in riparian areas is greater than the proportional representation of these areas in the landscape (Brinson et al. 1981). However, many riparian species are habitat generalists of forest edges that have expanded into the Plains since settlement (Finch 1989). Almost 90% of the 82 breeding bird species predictably present each spring on the steppe of eastern Colorado were not present in 1900 (Knopf 1986). Many grassland specialists are in decline, in part because they are susceptible to predation and competition from animals that use trees (Knopf 1992, Burger et al. 1994).

Three factors have contributed to the increasing abundance of forest vertebrates in parts of the Great Plains. First, channel narrowing as a result of water management has led to expansion of riparian forests. This phenomenon has been limited to parts of the southwestern and west central Plains where wide braided channels were formerly common. In the northern and eastern Plains, losses of bottomland forest due to land clearing and other factors have been greater than gains from channel narrowing, and declines in some forest vertebrates have resulted (Anderson 1971, Barclay 1980, Brinson et al. 1981). Second, shrubs and trees have been planted on uplands to reduce wind damage to agricultural and residential areas. Between 1935 and 1942, the U.S. government facilitated planting of more than 200 million trees and shrubs in the Central Great Plains in shelterbelts totaling 29,800 km and 96,400 ha (Goldsmith 1976). By 1976, the number of trees planted in shelterbelts had more than doubled, although survival was less than 50% (Read 1976). Shelterbelts cover 1.1% of the area of eastern South Dakota (Walker and Suedkamp 1977). Between 1935 and 1958, the Canadian government funded planting of more than 100,000 ha of shelterbelts in Alberta, Saskatchewan, and Manitoba (Pelton 1976). Third, fire suppression has allowed forests to increase, especially in the more humid eastern Plains.

The relative importance of these three factors probably varies among vertebrate species. For example, continuous riparian forests may be important as dispersal corridors, especially for flightless vertebrates. Small patches of forest planted on uplands may be most important on a per area basis for edge species (Martin 1981). A regional examination of these factors could help in distinguishing their effects on vertebrates. For example, in eastern Colorado, forests have increased both in uplands and along major river corridors (Knopf 1992). However, in parts of South Dakota, forests have increased on uplands while declining along river corridors (Anderson 1971). This contrast could be used to explore differential changes in vertebrate populations in the two areas.

Summary

Cottonwood seedling establishment is essentially restricted to bare, moist surfaces relatively safe from disturbance. In a riparian setting, such surfaces are most often produced by channel narrowing, meandering, or flood deposition. The relative importance of these three fluvial processes depends on local geologic and climatic conditions.

Water management may increase or decrease downstream riparian cottonwood forest. Along meandering channels such as much of the Missouri River, a decrease in peak flows slows cottonwood regeneration by reducing the meandering rate. Along braided channels such as the Platte River, a decrease in peak flows can temporarily increase regeneration by bringing about a period of channel narrowing; this phenomenon has been most frequent in the southwestern and west central Plains, where formerly braided channels are most common. In both situations, cottonwood declines over the long term because of the decrease in the physical disturbance that is necessary to produce sites suitable for establishment. Succession in the absence of physical disturbance depends on climate. Where moisture is insufficient for shade-tolerant woody species, cottonwood is replaced by grassland. In more humid regions, cottonwood is replaced by native, and increasingly by exotic, trees and shrubs whose regeneration does not require disturbance. In addition to its effects on seedling establishment, water management can kill existing trees by lowering the alluvial water table or decreasing moisture availability in the unsaturated zone. This phenomenon, common in the desert Southwest, may become more important in the Plains as a result of depletions of the High Plains Aquifer.

Water management is not the only human alteration that has affected riparian forests in the Great Plains. Other important influences include introduction of horses and cattle; elimination of beaver, bison, and fire; and land clearing for agriculture. Along many streams, the effects of human influence have been overwhelmed by floods or droughts. Because such events are often synchronous across a large area, their influence must be taken into account in regional analyses of trends in riparian communities. Increases in riparian forest in parts of the southwestern and west central Plains combined with increases in upland forest caused by fire suppression and tree planting have resulted in increases in populations of generalist vertebrates of forest edges. However, bottomland forest and associated vertebrates are declining in parts of the northern and eastern Great Plains.

Acknowledgments. We thank F. Knopf and P. Anderson for constructive reviews. W. Osterkamp provided insight into Plains physiography.

References

Aikman, J.M. 1926. Distribution and structure of the forests of eastern Nebraska. Univ. Neb. Stud. 26:1-95.

Akashi, Y. 1988. Riparian vegetation dynamics along the Bighorn River, Wyoming. M.S. thesis. Univ. Wyoming, Laramie.

Albertson, F.W., and J.E. Weaver. 1945. Injury and death or recovery of trees in prairie climate. Ecol. Monogr. 15:395-433.

Anderson, B.W. 1971. Man's influence on hybridization in two avian species in South Dakota. Condor 73:342-347.

Arend, J.L. 1948. Influences on redcedar distribution in the Ozarks. Southern Forestry Notes, No. 58, Southern Forest Experiment Station, U.S. For. Serv., New Orleans, LA.

Barclay, J.S. 1980. Impact of stream alterations on riparian communities in southcentral Oklahoma. FWS/OBS-80/17. U.S. Fish and Wildl. Serv. Biol. Serv. Prog., Washington, DC.

Berg, R.K. 1981. Fish populations of the wild and scenic Missouri River, Montana. Restoration Proj. FW-3-R. Job 1-A. Montana Dept. Fish, Wildl. and Parks., Helena, MT.

Bergman, D.L., and C.W. Sullivan. 1963. Channel changes on Sandstone Creek near Cheyenne, Oklahoma. U.S. Geol. Surv. Prof. Pap. 475-C. Washington, DC.

Bradley, C.E., and D.G. Smith. 1986. Plains cottonwood recruitment and survival on a prairie meandering river floodplain, Milk River, southern Alberta and northern Montana. Can. J. Bot. 64:1433-1442.

Bragg, T.B., and A.K. Tatschl. 1977. Changes in flood-plain vegetation and land use along the Missouri River from 1826 to 1972. Environ. Manage. 1:343-348.

Brinson, M.M., B.L. Swift, R.C. Plantico, and J.S. Barclay. 1981. Riparian ecosystems: their ecology and status. FWS/OBS-81/17. U.S. Fish and Wildl. Serv. Biol. Serv. Prog., Washington, DC.

Brock, J.H. 1994. *Tamarix* spp. (Salt Cedar), an invasive exotic woody plant in arid and semi-arid riparian habitats of western USA. Pp. 27-44 *in* L.C. de Waal, L.E. Child, P.M. Wade, and J.H. Brock, eds. Ecology and management of invasive riverside plants. John Wiley & Sons, New York.

Bruner, W.E. 1931. The vegetation of Oklahoma. Ecol. Monogr. 1:99-188.

Bunger, M.T., and H.J. Thomson. 1938. Root development as a factor in the success or failure of windbreak trees in the southern high plains. J. For. 36:790-803.

Burger, L.D., L.W. Burger, Jr., and J. Faaborg. 1994. Effects of prairie fragmentation on predation on artificial nests. J. Wildl. Manage. 58:249-254.

Busch, D.E., and S.D. Smith. 1995. Mechanisms associated with decline of woody species in riparian ecosystems of the southwestern U.S. Ecol. Monogr. 65:347-370.

Craig, M.R., and G.P. Malanson. 1993. River flow events and vegetation colonization of point bars in Iowa. Phys. Geog. 14:436-448.

Cross, F.B., and R.E. Moss. 1987. Historic changes in fish communities and aquatic habitats in plains streams of Kansas. Pp. 155-165 *in* W.J. Matthews and D.C. Heins, eds. Community and evolutionary ecology of North American stream fishes. Univ. Oklahoma Press, Norman.

Crouch, G.L. 1979. Long-term changes in cottonwoods on grazed and ungrazed plains bottomland in northeastern Colorado. Res. Note RM-370. U.S. For. Serv., Fort Collins, CO.

Currier, P.J. 1982. The floodplain vegetation of the Platte River: phytosociology, forest development, and seedling establishment. PhD dissertation. Iowa State Univ., Ames.

Deters, M.E., and H. Schmitz. 1936. Drought damage to prairie shelterbelts in Minnesota. Minnesota Agric. Exp. Stat. Bull. 329.

Ellison, L., and E.J. Woolfolk. 1937. Effects of drought on vegetation near Miles City, Montana. Ecology 18:329-336.

Everitt, B.L. 1968. Use of the cottonwood in an investigation of the recent history of a flood plain. Am. J. Sci. 266:417-439.

Everitt, B.L. 1995. Hydrologic factors in regeneration of Fremont cottonwood along the Fremont River, Utah. Pp. 197-208 *in* J.E. Costa, A.J. Miller, K.W. Potter, and P.R. Wilcock, eds. Natural and anthropogenic influences in fluvial geomorphology. Geophys. Monogr. 89. Am. Geophys. Union, Washington, DC.

Finch, D.M. 1989. Habitat use and habitat overlap of riparian birds in three elevational zones. Ecology 70:866-880.

Frémont, J.C. 1988. The Exploring Expedition to the Rocky Mountains. Smithsonian Inst. Press, Washington, DC.

Friedman, J.M., W.R. Osterkamp, and W.M. Lewis, Jr. 1996. The role of vegetation and bed-level fluctuations in the process of channel narrowing. Geomorphology 14:341-351.

Friedman, J.M., W.R. Osterkamp, and W.M. Lewis, Jr. In press. Channel narrowing and vegetation development following a Great-Plains flood. Ecology.

Friedman, J.M., M.L. Scott, and W.M. Lewis, Jr. 1995. Restoration of riparian forest using irrigation, artificial disturbance, and natural seedfall. Environ. Manage. 19:547-557.

Goldsmith, L. 1976. Action needed to discourage removal of trees that shelter cropland in the Great Plains. Pp. 12-18 *in* R.W. Tinus, ed. Shelterbelts of the Great Plains. Proc. of the Symp. Great Plains Agric. Council Publ. 78. Lincoln, NE.

Graf, W.L. 1978. Fluvial adjustments to the spread of tamarisk in the Colorado Plateau region. Geol. Soc. Am. Bull. 89:1491-1501.

Great Plains Flora Association. 1986. Flora of the Great Plains. Univ. Kansas Press, Lawrence.

Gregg, J. 1905. Reprint of 1845 original in R.G. Thwaites, ed. Early western travels, 1748-1846. Vol. 20. Part 2 of Gregg's Commerce of the prairies, 1831-1839. Arthur H. Clark, Cleveland, OH.

Harper, H.J. 1940. Relation of climatic conditions, soil characteristics, and tree development in the southern Great Plains region. Soil Sci. Soc. Am. Proc. 5:327-335.

Hefley, H.M. 1937. Ecological studies on the Canadian River floodplain in Cleveland County, Oklahoma. Ecol. Monogr. 7:345-402.

Hesse, L.W., G.E. Mestl, and J.W. Robinson. 1993. Status of selected fishes in the Missouri River in Nebraska with recommendations for their recovery. Pp. 327-340 in L.W. Hesse, C.B. Stalnaker, N.G. Benson, and J.R. Zuboy, eds. Restoration planning for the rivers of the Mississippi River ecosystem. Biol. Rep. 19, October 1993. Natl. Biol. Surv., Washington, DC.

Hughes, F.M.R. 1994. Environmental change, disturbance and regeneration in semi-arid floodplain forests. Pp. 321-345 in A.C. Millington and K. Pye, eds. Environmental change in drylands: biogeographical and geomorphological perspectives. John Wiley & Sons, Ltd., New York.

Johnson, R.L. 1965. Regenerating cottonwood from natural seedfall. J. For. 63:33-36.

Johnson, W.C. 1992. Dams and riparian forests: case study from the upper Missouri River. Rivers 3:229-242.

Johnson, W.C. 1994. Woodland expansion in the Platte River, Nebraska: patterns and causes. Ecol. Monogr. 64:45-84.

Johnson, W.C., R.L. Burgess, and W.R. Keammerer. 1976. Forest overstory vegetation and environment on the Missouri River floodplain in North Dakota. Ecol. Monogr. 46:59-84.

Kaylor, J.F., C.C. Starring, and C.P. Ditman. 1935. A survey of past plantings. Pp. 39-47 in Possibilities of shelterbelt planting in the Plains region. U.S. For. Serv., Washington, DC.

Knopf, F.L. 1986. Changing landscapes and the cosmopolitism of the eastern Colorado avifauna. Wildl. Soc. Bull. 14:132-142.

Knopf, F.L. 1992. Faunal mixing, faunal integrity, and the biopolitical template for diversity conservation. Trans. N. Am. Wildl. Nat. Resour. Conf. 57:330-342.

Knopf, F.L., and M.L. Scott. 1990. Altered flows and created landscapes in the Platte River headwaters, 1840-1990. Pp. 47-70 in J.M. Sweeney, ed. Management of dynamic ecosystems. N. Cent. Sec. Wildl. Soc., West Lafayette, IN.

Kromm, D.E., and S.E. White. 1992. Groundwater problems. Pp. 44-63 in D.E. Kromm and S.E. White, eds. Groundwater exploitation in the High Plains. Univ. Kansas Press, Lawrence.

Layher, B. 1986. The four deadly sins. Kansas Wildl. 43:32-35.

Leonard, G.J., and P.W. Huntoon. 1974. Groundwater geology of Southwest Nebraska Ground Water Conservation District. Nebraska Water Surv. Pap. 37. Univ. Nebraska, Lincoln.

Lindauer, I.E. 1983. A comparison of the plant communities of the South Platte and Arkansas River drainages in eastern Colorado. Southwest. Nat. 28:249-259.

Luckey, R.R., E.D. Gutentag, F.J. Heimes, and J.B. Weeks. 1988. Effects of future groundwater pumpage on the High Plains Aquifer in parts of Colorado, Kansas, Nebraska, New Mexico, Oklahoma, South Dakota, Texas, and Wyoming. U.S. Geol. Surv. Prof. Pap. 1400-E. Washington, DC.

Madole, R.F. 1994. Stratigraphic evidence of desertification in the west-central Great Plains within the past 1000 yr. Geology 22:483-486.

Martin, T.E. 1981. Limitation in small habitat islands: chance or competition? Auk 98:715-734.

Meffe, G.K. 1984. Effects of abiotic disturbance on coexistence of predator-prey fish species. Ecology 65:1525-1534.

Merkel, D.L., and H.H. Hopkins. 1957. Life history of salt cedar (*Tamarix gallica* L.) Trans. Kansas Acad. Sci. 60:360-369.

Moss, E.H. 1938. Longevity of seed and establishment of seedlings in species of *Populus*. Bot. Gazette 99:529-542.

Nadler, C.T., and S.A. Schumm. 1981. Metamorphosis of South Platte and Arkansas rivers, eastern Colorado. Phys. Geog. 2:95–115.

Noble, M.G. 1979. The origin of *Populus deltoides* and *Salix interior* zones on point bars along the Minnesota River. Am. Midl. Nat. 102:59–67.

Northrop, W.L. 1965. Republican River channel deterioration. Pp. 409–424 *in* Proc. Federal Interagency Sedimentation Conf. 1963. U.S. Dept. Agric. Misc. Publ. 970. Washington, DC.

Olson, T.E., and F.L. Knopf. 1986. Naturalization of Russian-olive in the western U.S. W. J. Appl. For. 1:65–69.

Osterkamp, W.R. 1978. Gradient, discharge, and particle-size relations of alluvial channels in Kansas, with observations on braiding. Am. J. Sci. 278:1253–1268.

Osterkamp, W.R., and J.E. Costa. 1987. Changes accompanying an extraordinary flood on a sandbed stream. Pp. 201–224 *in* L. Mayer and D. Nash, eds. Catastrophic flooding. Allen & Unwin, Boston.

Patton, P.C., and S.A. Schumm. 1981. Ephemeral-stream processes: implications for studies of quaternary valley fills. Quaternary Res. 15:24–43.

Pelton, W.L. 1976. Windbreak studies on the Canadian prairie. Pp. 64–68 *in* R.W. Tinus, ed. Shelterbelts of the Great Plains. Proc. of the Symp. Great Plains Agric. Council Publ. 78. Lincoln, NE.

Read, R.A. 1958. The Great Plains shelterbelts in 1954. Great Plains Agric. Council Publ. 16. Lincoln, NE.

Read, R.A. 1976. Provenance testing and introductions. Pp. 147–153 *in* R.W. Tinus, ed. Shelterbelts of the Great Plains. Proc. of the Symp. Great Plains Agric. Council Publ. 78. Lincoln, NE.

Reily, P.W., and W.C. Johnson. 1982. The effects of altered hydrologic regime on tree growth along the Missouri River in North Dakota. Can. J. Bot. 60:2410–2423.

Rood, S.B., C. Hillman, T. Sanche, and J.M. Mahoney. 1994. Clonal reproduction of riparian cottonwoods in southern Alberta. Can. J. Bot. 72:1766–1770.

Rood, S.B., and J.M. Mahoney. 1990. Collapse of riparian poplar forests downstream from dams in western prairies: probable causes and prospects for mitigation. Environ. Manage. 14:451–464.

Rood, S.B., and J.M. Mahoney. 1995. River damming and riparian cottonwoods along the Marias River, Montana. Rivers 5: 195–207.

Rood, S.B., J.M. Mahoney, D.E. Reid, and L. Zilm. In press. Instream flows and the decline of riparian cottonwoods along the St. Mary River, Alberta. Can. J. Bot. 73:1250–1260.

Schumm, S.A. 1969. River metamorphosis. J. Hydraulics Div. Am. Soc. Civil Engineers 95:255–273.

Schumm, S.A., and R.W. Lichty. 1963. Channel widening and floodplain construction along Cimarron River in southwestern Kansas. U.S. Geol. Surv. Prof. Pap. 352-D. Washington, DC.

Scott, M.L., G.T. Auble, and J.M. Friedman. In press. Flood dependency of cottonwood establishment along the Missouri River, Montana, USA. Ecol. Appl.

Scott, M.L., J.M. Friedman, and G.T. Auble. 1996. Fluvial process and the establishment of bottomland trees. Geomorphology 14:327–339.

Scott, M.L., M.A. Wondzell, and G.T. Auble. 1993. Hydrograph characteristics relevant to the establishment and growth of western riparian vegetation. Pp. 237–246 *in* H.J. Morel-Seytoux, ed. Proc. 13th Ann. Am. Geophys. Union Hydrology Days. Hydrology Days Publ., Atherton, CA.

Sedgwick, J.A., and F.L. Knopf. 1989. Demography, regeneration, and future projections for a bottomland cottonwood community. Pp. 249–266 *in* R.R. Sharitz and J.W. Gibbons, eds. Freshwater wetlands and wildlife. U.S. Dept. Energy. DE 90005384. NTIS, Springfield, VA.

Segelquist, C.A., M.L. Scott, and G.T. Auble. 1993. Establishment of *Populus deltoides* under simulated alluvial groundwater declines. Am. Midl. Nat. 130:274–285.

Shafroth, P.B., G.T. Auble, and M.L. Scott. 1995a. Germination and establishment of the native plains cottonwood (*Populus deltoides* Marshall subsp. *monilifera*) and the exotic Russian-olive (*Elaeagnus angustifolia* L.). Conserv. Biol. 9:1169–1175.

Shafroth, P.B., M.L. Scott, E.D. Eggleston, G.T. Auble, J.M. Friedman, and L.S. Ischinger. 1995b. Responses of mature plains cottonwood (*Populus deltoides* ssp. *monilifera*) to groundwater level changes. Pp. 55–56 *in* Wetland understanding, wetland education. 16th Ann. Meeting Soc. Wetland Sci. (Abstract), Lawrence, KS.

Smith, D. 1976. Effect of vegetation on lateral migration of anastomosed channels of a glacial meltwater river. Geol. Soc. Am. Bull. 87:857–860.

Smith, H.T.U. 1940. Notes on historic changes in stream courses of western Kansas, with a plea for additional data. Trans. Kansas Acad. Sci. 43:299–300.

Snyder, W.D., and G.C. Miller. 1991. Changes in plains cottonwoods along the Arkansas and South Platte rivers—eastern Colorado. Prairie Nat. 23:165–176.

Sprackling, J.A., and R.A. Read. 1979. Tree root systems in eastern Nebraska. Nebraska Conserv. Bull. 37. Lincoln, NE.

Stromberg, J.C. 1993. Riparian mesquite forests: a review of their ecology, threats, and recovery potential. J. Arizona-Nevada Acad. Sci. 27:111–124.

Stromberg, J.C., B.D. Richter, D.T. Patten, and L.G. Wolden. 1993. Response of a Sonoran riparian forest to a 10-year return flood. Great Basin Nat. 53:118–130.

Teskey, R.O., and T.M. Hinckley. 1978. Impact of water level changes on woody riparian and wetland communities; vol. VI: plains grassland region. FWS/OBS-78/89. U.S. Fish and Wildl. Serv. Biol. Serv. Prog. Columbia, MO.

Thomson, G.W., and H.G. Hertel. 1981. The forest resources of Iowa in 1980. Proc. Iowa Acad. Sci. 88:2–6.

Tyree, M.T., K.J. Kolb, S.B. Rood, and S. Patino. 1994. Vulnerability to drought-induced cavitation of riparian cottonwoods in Alberta: a possible factor in the decline of the ecosystem? Tree Physiol. 14:455–466.

Walker, R.E., and J.F. Suedkamp. 1977. Status of shelterbelts in South Dakota. South Dakota Dept. Game, Fish and Parks Publ. Pierre.

Ware, E.R., and L.F. Smith. 1939. Woodlands of Kansas. Kansas Agric. Exp. Sta. Bull. 285. Manhattan.

Ware, G.H., and W.T. Penfound. 1949. The vegetation of the lower levels of the floodplain of the South Canadian River in central Oklahoma. Ecology 30:478–484.

Weaver, J.E. 1960. Flood plain vegetation of the central Missouri valley and contacts of woodland with prairie. Ecol. Monogr. 30:37–64.

Welcomme, R.L. 1979. Fisheries ecology of floodplain rivers. Longman, New York.

White, P.S. 1979. Pattern, process, and natural disturbance in vegetation. Bot. Rev. 45:229–299.

Williams, G.P., and M.G. Wolman. 1984. Downstream effects of dams on alluvial rivers. U.S. Geol. Surv. Prof. Pap. 1286. Washington, DC.

Winship, G.P. 1904. The journey of Coronado, 1540–1542. A. S. Barnes, New York.

Wolman, M.G., and R. Gerson. 1978. Relative scales of time and effectiveness of climate in watershed geomorphology. Earth Surface Processes 3:189–208.

Wolman, M.G., and J.P. Miller. 1960. Magnitude and frequency of forces in geomorphic processes. J. Geol. 68:54–74.

Zimmerman, R.C. 1969. Plant ecology of an arid basin, Tres Alamos-Redington area southeastern Arizona. U.S. Geol. Surv. Prof. Pap. 485-D. Washington, DC.

4. Comparative Ecology of Native and Introduced Ungulates

D.C. Hartnett, A.A. Steuter, and K.R. Hickman

History of the Great Plains Landscape and Its Ungulates

The defining period of coevolution among Great Plains plant and ungulate species occurred during the past 12,000 years (Mack and Thompson 1982, Axelrod 1985). In the late Pleistocene and early Holocene, a diverse array of large grazers and browsers were reduced to a much smaller group of ungulate species represented by bison (*Bison bison*), pronghorn (*Antilocapra americana*), deer (*Odocoileus hemionus* and *O. virginianus*), and elk (*Cervus canadensis*). These changes occurred in the presence of nomadic humans from the Asian steppe who were immigrating to the Great Plains during the same time. The landscape was characterized by gently rolling interfluvial surfaces covered with perennial herbaceous vegetation. These exposed grasslands were periodically interrupted by more protected wetland, riparian woodland, or scarp woodland habitats. Although wetlands and woodlands occupied less than 7 and less than 3% of the Great Plains, respectively (National Wetlands Inventory, and Nebraska Natural Heritage Program data bases), the heterogeneity that they created at landscape scales played a major role in determining the distribution and abundance of native ungulates. Extreme cold and heat, drought, flood, fire, wind, and countless biotic interactions caused locally short-term fluctuations in ungulate populations and long-term shifts in landscape features. These dynamic temporal changes were overlayed on a multiscale spatial mosaic. Native ungulates were adapted to this landscape.

Although significant changes occurred in the Plains cultures through the Holocene, the human approach to land management was extensive, with few man-made boundaries imposed on the landscape. Yet, along with the native ungulates, humans played a significant role in defining the large-scale landscape structure and diversity through their use of fire, species introductions, organized hunting techniques, and eventually limited cultivation practices. The first widespread introduction of a European ungulate, the horse (*Equus* sp.), was carried out by the Plains Indians during the 1500s through the 1700s. By the first decade of the 1800s, large herds of horses derived from the Spanish stock of the Southwest had transformed the Plains Indian cultures. Providing winter food (e.g., young cottonwood bark, *Populus* sp.) for these animals near riparian encampments may have caused the first significant changes in plant communities by an ungulate of European origin (Lewis and Clark 1893). The introduction of other European ungulates and their managers to the Great Plains took place in the middle to late 1800s. The cattle (*Bos taurus*) and sheep (*Ovis domesticus*) introduced to the Great Plains had undergone thousands of years of coevolution with various human cultures (Morey 1994) in Europe before being introduced on the Great Plains. European cultures had developed intensive management practices (Voisin 1988) to manipulate the spatial and seasonal distribution of forage and protection from predators for previously wild ungulates long since domesticated. These changes occurred within a landscape that was more mesic, wooded, topographically dissected, and humanly subdivided at the hectare-to-kilometer scale than the Great Plains.

In addition to cattle and sheep, the European bison or wisent (*B. bonasas*) and red deer (*C. elaphus*) have generic relatives on the Great Plains. Red deer and American elk have morphology, behavior, and habitat similarities to the point that a debate exists as to whether they are one species (Jones et al. 1973, Schonewald 1994). Contemporary red deer and elk populations are generally associated with wooded habitats and broken or dissected topography. We know little about Great Plains elk populations because they were extirpated early in the period of European exploration (Reiger 1972). By contrast, the European bison survives in a forest landscape more similar to those of the North American wood bison (*B. bison athabascae*) (Krasinska and Krasinski 1995), which may now also be an extinct ecotype (Geist 1991, but also see van Zyll de Jong et al. 1995). Through conservation, plains bison (*B. bison bison*) are once again expanding on private and public grasslands in the Great Plains and now number nearly 150,000 animals.

From an ecological perspective, significant changes in the Great Plains resulted from the replacement of the native ungulates with introduced domestic grazers and the changes in the spatial and temporal patterning of the landscape associated with their management. The replacement of native with European species also resulted in changes in plant habitat selection and other aspects of plant–herbivore interactions at multiple scales ranging from the individual organism to the landscape level across the Great Plains. Other ecological processes and interactions such as fire–grazer interactions, biotic and abiotic disturbance patterns, and relationships among sympatric herbivores have also changed.

In this chapter, we discuss the ecology of plant–ungulate interactions in the Great Plains and compare the ecology of native and introduced species. Because changes in the large grazing animals did not occur independently of other landscape and land management changes, it is often difficult to separate the effects of the two, which poses several challenging questions. Ecologically, are native and introduced ungulates analogous herbivores? How much of the response of the Great Plains grasslands to introduced versus native ungulates can be attributed to fundamental differences in the biology of the animals per se? Differences may be explained by responses to the spatial and temporal patterns of herbivory, yet these space and time relationships are largely dictated by their human managers. What are the important differences in nongrazing activities between native and introduced ungulates, and what are their consequences for disturbance regimes and plant community dynamics? These are important fundamental questions, and their answers also have important implications for the management and conservation of the Great Plains grasslands and other habitats.

Because bison were once the most numerous Great Plains ungulate (numbering 30 million; McHugh 1979) and their replacement by cattle throughout the region was perhaps the most ecologically significant historical change, we emphasize bison versus cattle as a focal comparison. First, we compare grazing and nongrazing activities and their effects on the Great Plains grasslands habitats at various scales. We examine the relationships between these ungulates and vegetation patterns in their context of complex interactions with fire, topography, and other biotic and abiotic forces. Finally, we discuss the ecological implications of intensive ungulate management and the opportunities for managing large ungulate impacts relative to the conservation of the biological diversity and biotic integrity of the Great Plains.

Ungulate Trophic Ecology and Habitat Relationships

Differences among ungulates in body size, mouth morphology, gut morphology and function, thermoregulation, social organization, and environmental tolerance may all result in species differences in diet and grazing patterns at various scales (Schwartz and Ellis 1981, Belovsky 1986, Krueger 1986, Telfer and Cairns 1986). Environmental factors such as forage availability, snow cover, topography, or herbivore density may also influence patterns of ungulate foraging and habitat use (Peden 1976, McNaughton 1979, Tieszen et al. 1980, Van Vuren 1982, 1984, Telfer and Kelsall 1984, Hobbs and Swift 1988), and different herbivore species may respond differently to their influence.

Sympatric native ungulates on the Great Plains that vary in these traits tend to exploit their environment and resources in different ways. The dominant Great Plains ungulates, bison and pronghorn, have had a long association (Fig. 4.1). They have distinct trophic strategies and adaptations to partition use of the grassland forage resource. These are predictable based on their differences in body size, metabolic requirements, and digestive system structure and physiology. Bison

Figure 4.1. (A) Pronghorn (*Antilocapra americana*) on mixed grass prairie, Albany County, Wyoming. (Photograph by F.L. Knopf.) (B) Bison (*Bison bison*) on Sandhills prairie at the Niobrara Valley Preserve, Brown County, Nebraska (Photograph by A.A. Steuter.) (C) Bison grazing tallgrass prairie under light snow cover, Konza Prairie Research Natural Area, Riley County, Kansas. (Photograph by D.C. Hartnett.)

Figure 4.1. (*continued*)

(350–1000 kg) have higher gross intake needs and thus consume large quantities of the abundant and randomly distributed grasses, whereas pronghorn (34–70 kg) forages for a wider variety of species higher in nutrient quality. Bison show low dietary niche breadth, consuming primarily dominant grasses even when they are relatively low in quality. Pronghorn consume a more species-rich diet of forbs and browse (Schwartz and Ellis 1981, Wydeven and Dahlgren 1985). Pronghorn diets on shortgrass steppe are comprised of 63–92% forbs when available and mostly cool-season grasses at other times (Schwartz and Ellis 1981). Smaller ungulates such as pronghorn and deer also show greater seasonal variation in diet quality than bison. As a result, although bison and pronghorn overlap considerably in habitat use, they show significant dietary niche divergence. At an even smaller scale, sympatric herbivores may show divergence in consumption of plant parts. Bison consume leaf blades, sheaths, and stems and rarely discriminate among forage parts, whereas pronghorn and prairie dogs make finer discriminations among various plant parts available and select the higher-quality, more easily digestible components (e.g., leaves and flowers as opposed to stems, and dicots as opposed to monocots) (Krueger 1986).

Similar patterns occur among other vertebrates. Sympatric species diverge in spatial and temporal distributions or in their use of forage classes, vegetation structures, or other habitat components. Guilds of ungulates may develop equilibria with plant communities when left undisturbed for long periods (Caughley 1979). Niche complementarity is evident in that those species with high spatial overlap (e.g., bison and pronghorn) tend to show low dietary overlap, and species with similar diets show little spatial overlap. Mule deer and elk, for example, have similar diets but diverge in habitat use in mixed-grass prairie and forested habitat mosaics, the elk using primarily woodlands and mule deer shrubby draws (Wydeven and Dahlgren 1985). Food habits of co-occurring ungulates in these habitats also vary seasonally. Elk consume primarily cool-season graminoids in the spring, warm-season graminoids in summer, a mix of both in autumn, and graminoids and forbs in winter (Wydeven and Dahlgren 1985). Mule deer select mostly forbs in summer and browse in autumn and winter.

If cattle differ from bison in diet or habitat use, we would expect changes to occur in the overall plant-herbivore dynamics of the Great Plains. Differences may result from basic species traits and management associated with bison and cattle.

Foraging Patterns of Bison and Cattle

Bison and cattle are large generalist ungulate herbivores, are predominantly graminoid feeders, and show generally high dietary overlap. Thus, they can be considered ecological analogs and functionally equivalent. Mack and Thompson (1982) viewed the shift in prominence from bison to cattle in the Great Plains as essentially one bovid replacing another, with cattle grazing providing a continuation of the selective pressures characteristic of the long history of coevolution of vegetation with native ungulates. Ecological similarities may depend on the scale of the interaction. For example, as cattle gradually replaced bison on the Plains, the several dozen species of native dung beetles persisted, shifting to cattle dung as a resource base (Hayes 1927). Also, pasturing of cattle and their frequent congregation can create local site disturbance analogous to severe grazing and wallowing observed in bison herds (England and DeVos 1969; Roe 1970). Sites disturbed by either bison or cattle tend to be recolonized mostly by native annual dicots or by ruderal plant species that are soon replaced by native perennials (Mack and Thompson 1982). By contrast, the intensity of Cowbird (*Molothrus ater*) nest parasitism is probably consistently higher throughout the Great Plains as a result of the shift from nomadic bison herds to sedentary cattle operations. The change from nomadic bison to resident cattle herds also changed the temporal dynamics of otherwise similar local disturbances.

Despite similarities, differences in foraging ecology between bison and cattle are predicted based on differences in their morphology, physiology, social behavior, and environmental tolerance. However, only a few comparative studies of the foraging ecology of bison and cattle in the Great Plains have been conducted, primarily in shortgrass steppe and mixed-grass prairie (Fig. 4.2). A qualitative summary of these comparisons is shown in Table 4.1.

Although bison and cattle both exhibit forage selectivity (i.e., they use plant species and growth forms out of proportion to their availability), cattle are more selective foragers but show greater dietary niche breadth than bison (Peden et al. 1974, Schwartz and Ellis 1981). Among Great Plains ungulates, dietary breadth (forage class diversity) appears to be generally higher for introduced species

Figure 4.2. Cattle (*Bos taurus*) on Great Plains grasslands. (A) Mixed-grass prairie, Albany County, Wyoming (Photograph by F.L. Knopf.) (B) Sandhills prairie, Brown County, Nebraska (Photograph by A.A. Steuter.) (C) Cattle congregating near a watering tank on shortgrass steppe, Weld County, Colorado. (Photograph by F.L. Knopf.)

Figure 4.2. (*continued*)

Table 4.1. Qualitative Comparison of Foraging Ecology Traits of Bison and Cattle

Trait	Bison	Cattle	References[a]
Forage plant selectivity	Lower	Higher	1,9
Use of forbs and browse (% of diet)	Lower (<10%)	Higher (10-20%)	1-9
Use of graminoids (% of diet)	Higher	Lower	1-9
Diet niche breadth (number of available species/growth forms consumed)	Lower	Higher	4
Time allocated to grazing (during the growing season)	Lower	Higher	9
Time allocated to nonfeeding activities	Higher	Lower	9
General diet quality (crude protein, digestibility, cell wall constituents)	Lower	Higher	1,4,9
Digestibility of C_3 and C_4 graminoids	Higher	Lower	9
Mean digesta retention time	Higher	Lower	3

[a] 1, Peden et al. 1974; 2, Kautz and Van Dyne 1978; 3, Schaefer et al. 1978; 4, Schwartz and Ellis 1981; 5, Bennett and Dahlgren 1982; 6, Coppock et al. 1983; 7, Wydeven and Dahlgren 1985; 8, Krueger 1986; 9, Plumb and Dodd 1993.

(cattle and sheep) than for native species such as bison or pronghorn (Schwartz and Ellis 1981). Where they co-occur, native and introduced herbivores also tend to diverge in both their spatial and temporal use of habitats (e.g., pronghorn use ridges whereas cattle use drainages, and pronghorn tend to avoid pastures actively being grazed by cattle).

Bison consume graminoids in higher proportion than their availability, and their diets are characterized by a larger percentage contribution of graminoids and a smaller browse/forb component relative to cattle (Peden et al. 1974, Krueger 1986, Plumb and Dodd 1993). They consume a lower-quality diet but show greater digestion efficiencies for poorer-quality graminoids relative to forbs (Plumb and Dodd 1993). Bison diets on Great Plains grasslands are also temporally dynamic. Foraging is concentrated on warm-season C_4 grasses during the summer, but relative use of C_3 graminoids increases during the cooler seasons (Schwartz and Ellis 1981, Vinton et al. 1993, Steuter et al. 1995). Unlike cattle, bison seldom use the tongue in a horizontal plane for forage prehension (Hudson and Frank 1987), allowing them to graze nearer to the ground than cattle and producing more closely cropped grazing patches. Both herbivores, however, are capable of discriminating among co-occurring bunchgrass and rhizomatous grass species and among plant size classes within species (Pfeiffer and Hartnett 1995). Compared with cattle, bison spend more time on nonfeeding activities such as aggression, play, defense, grooming, and intrasexual competition.

In their analyses of the foraging ecology of bison and cattle on mixed-grass prairie, Plumb and Dodd (1993) interpreted two distinct strategies. Due to the scale of patchiness of mixed prairie vegetation, neither bison nor cattle can consume all plant classes (species/growth forms) at a feeding station. Cattle invest the additional search time necessary to locate and consume the nonrandomly distributed but high-quality forb components and woody species, resulting in a greater dietary niche breadth than bison. Bison invest a significant amount of time in nonfeeding activities, particularly during the rut when social time may significantly limit foraging time (Fuller 1960, Lott 1979, 1981). Thus, bison do not allocate time to search for patchily distributed individual forage classes. Rather, they balance their nutrient requirements with social and other nongrazing time investment by feeding primarily on the abundant matrix grasses.

On tallgrass prairie, bison similarly select patches heavily dominated by abundant C_4 grasses and avoid patches with significant cover of forbs or woody species (Vinton et al. 1993). Thus, cattle display greater small-scale dietary selectivity, whereas bison show greater selectivity at the larger patch scale, and the relationship between social time investment and forage patchiness may be important in explaining differences in their trophic ecology. These differences between bison and cattle in patterns of foraging are also consistent with differences in their diet quality, digestive efficiencies for different forage components, and digesta retention time (Table 4.1). These differences in the spatial scale of selectivity and patterns of herbivory in relation to vegetation patchiness suggest that the historical transition from free-roaming ungulates to semifree-roaming herds confined to fenced areas within a subdivided landscape may have had as great or greater influence as the transition from native to introduced species. Furthermore, relationships between body size, other traits, and animal foraging suggest that bison and cattle diets should be more similar to each other than bison diets are to pronghorn, mule deer, or other native species. Thus, in some areas the loss of the smaller native ungulates such as pronghorn, or changes in their foraging patterns, may have had a larger ecological consequence for plant–herbivore interactions than the displacement of bison by cattle. Overall, differences in both diets and spatial and temporal patterns of grazing between native and introduced ungulates underlie ecological interactions of these herbivores and their respective effects on the composition and structure of vegetation.

Habitat Selection at Larger Scales

In addition to diet selection and small-scale patterns of herbivory, the distributions and forage utilization patterns of ungulates at intermediate scales are influenced by factors such as topography, slope, distance from water sources, vegetation composition, or patterns of insect herbivory (Hyder et al. 1966, Senft et al. 1985, Andrew 1988). At the larger landscape and regional scales, factors such as climate and fire are important. The movement and utilization patterns of ungulates at different scales determine the spatial and temporal patterns of grazing impacts on plant communities and habitats (Coughenour 1991).

Bison social behavior appears to interact with seasonal changes in forage conditions at the regional scale (Hansen 1984). Larger areas of high-quality mixed-prairie forage in a level to gently rolling landscape are preferred when bison congregate during the midsummer breeding season (Steuter et al. 1995). By contrast, as herd size declines during the dormant season, smaller patches of forage are selected, often in rough and broken topography. For example, small groups of nonbreeding bulls may select for cool-season grasses in wooded draws (Norland 1984), and small mixed groups may forage on windswept hilltops and ridges after snowstorms.

Our experience in the Nebraska Sandhills indicates that even the high-quality forage after a fire does not overcome the aversion bison have for hilly and wooded habitats when they are in large summer breeding groups. The open and productive grassland areas of the Flint Hills are preferred during the summer after a fire (Vinton et al. 1993). By contrast, cattle have a more variable group size and use both wooded and open habitats opportunistically (Van Vuren 1982). Livestock managers have generally matched the flexible group sizes of domestic ungulates with forage conditions, ownership and pasture configuration, and marketing needs rather than with consideration for regional habitat patterns.

Throughout much of the Great Plains, bison historically moved nomadically in response to vegetation changes associated with local and regional patterns of rainfall and fire. It has been suggested that these patterns would have imposed a deferred rotation in that it would be unlikely for any given area to be grazed at the same time every year. The lag time for return movements would provide a deferment during the regrowth period. During periods of severe or extended drought, mortality or emigration of ungulates likely would have reduced regional herbivore densities. During postdrought periods of adequate precipitation, the lag time associated with recovery of herbivore numbers would have provided a natural "rest" period. This view that large native herbivores move in a manner that conserves forage, prevents overgrazing, and results in regional equilibrium in plant-herbivore interactions was expressed a century ago in reference to bison on the Great Plains (Smith 1895). It is also consistent with contemporary theoretical model predictions that spatial heterogeneity may confer stability on plant-herbivore interactions (May and Beddington 1981, Coughenour 1991). Aggregations of ungulates rotating among patches in varying stages of postgrazing recovery in a shifting vegetation mosaic may have been an important process promoting vegetation-herbivore stability in the Great Plains grasslands (Coughenour 1991). This concept has been incorporated into some grazing management systems.

Influence of Large Ungulates on Great Plains Grasslands

Effects on Vegetation Composition and Diversity

Given the differences in the trophic ecology and habitat selection patterns between native and introduced ungulates, what are the consequences for Great Plains plant communities and habitats? Although effects of native ungulates have been studied in various grasslands (McNaughton 1985, Sala et al. 1986), no studies have directly compared plant community responses to native versus introduced species in a given habitat. The ability to draw inferences about vegetation responses to the historical displacement of native with domestic herbivores is limited.

Herbivores can modify vegetation in several ways. They can reduce plant canopy height and change plant morphology, creating "grazing lawns." They can alter horizontal structure and create patchiness or vegetation mosaics that differ from ungrazed landscapes. Also, modifications can reduce peak standing crops, change the proportion of biomass among various plant functional groups, increase live/dead biomass ratios, and increase bare ground (Dyer et al. 1982, Whicker and Detling 1988). This activity above ground can also impact belowground vegetation structure (e.g., reductions in root biomass), indirectly altering rates of energy and material flow through belowground consumers (Whicker and Detling 1988). Large grazers can also influence plant species diversity through a variety of mechanisms including selective or nonselective grazing or physical disturbance associated with nongrazing activities (Harper 1977, Huntly 1991, Pacala and Crawley 1992). Light-to-moderate grazing intensities may increase or decrease plant species diversity in grasslands. Theoretical models such as the intermediate disturbance hypothesis and predator-mediated coexistence models predict a nonlinear relationship between grazing intensity and plant species diversity, with maximum diversity occurring at intermediate grazing intensities. A graphical model proposed by Milchunas et al. (1988) predicts this relationship for subhumid grasslands such as tallgrass prairie but a gradual decline in diversity with increasing grazing intensity in the semiarid shortgrass steppe.

In addition to herbivory, physical disturbance generated by the grazing and nongrazing activities of large vertebrates plays an important role in the structure, composition, and spatial heterogeneity of Great Plains plant communities. All large ungulates generate soil disturbance and small-scale patchiness due to activities such as hoof action and the trampling of vegetation, bedding, trailing, and dung and urine deposition. The intensity or frequency of these small-scale disturbances may also impart a pattern at larger scales due to spatial gradients in animal activity associated with watering points or other habitat features. Plant species of the Great Plains grasslands have become as well adapted to the pervasive influence of animal-generated disturbance as they have to the selective pressures associated with ungulate herbivory.

The effects of ungulate grazers on plant community structure and composition have been studied in several types of grasslands within the Great Plains. Vegetation responses to cattle grazing in the shortgrass steppe include an increase in the relative cover of dominant grasses (e.g., *Bouteloua gracilis*) and succulents (e.g., *Opuntia*), reduced forb and shrub densities, and decreased total plant cover and species diversity (Milchunas et al. 1989). In tallgrass and mixed-grass prairie, studies of effects of bison or cattle grazing on floristic diversity generally support theoretical predictions. In the absence of grazing, a few tall warm-season grasses dominate the community, whereas moderate grazing results in a more species-rich mosaic pattern of short grasses, tall grasses, and forbs and a mosaic pattern of canopy structure. Preferential grazing of the dominant warm-season grasses reduces their relative abundances and competitive effects, which increases the survival of other subdominant species, thereby increasing species diversity (Collins and Barber 1985, Collins 1987, Hickman et al. 1996). In Kansas tallgrass prairie, plant species diversity increased with increasing light-to-moderate grazing intensities, as predicted by theoretical models (Milchunas et al. 1988, Hickman et al. 1996). Kucera (1956) found decreased plant species diversity in Missouri tallgrass prairie sites that were grazed intensively for many years and indicated a significant drought–grazing interaction affecting long-term trends in plant community composition. Plant diversity may be reduced and exotic invasion increased by large ungulate grazing of mesic tallgrass and meadow communities.

Tallgrass prairie plant communities were studied on the Konza Prairie in the Kansas Flint Hills after the reintroduction of bison in 1987 onto previously ungrazed watersheds. Cover and frequency of cool-season graminoids and some forbs increased, whereas dominant warm-season grasses and other forbs decreased in response to bison grazing (Hartnett et al. 1996a, b). The net result was that plant species diversity and community heterogeneity were significantly increased by bison (Table 4.2).

The plant community responses we observed suggested that bison influence tallgrass prairie diversity via multiple mechanisms. The preferential grazing of the dominant grasses and concomitant increases in subordinate species, increases the

Table 4.2. Plant Biodiversity Components in Tallgrass Prairie Sites[a,b]

	S		E		H′		Heterogeneity	
	U	G	U	G	U	G	U	G
Annually burned prairie								
Uplands	27	35	0.42	0.53[c]	1.40	1.87[c]	0.41	0.44
Lowlands	28	43[c]	0.51	0.49[c]	1.70	1.82[c]	0.48	0.59[c]
4-Year burn prairie								
Uplands	27	32[c]	0.53	0.59[c]	1.76	2.03[c]	0.33	0.38[c]
Lowlands	29	41[c]	0.38	0.50	1.29	1.85	0.40	0.39

[a]G, Sites grazed year-round by bison; U, ungrazed sites; S, species richness (mean number of species per sampling site); E, Evenness Index = H′/ln S); H′, Shannon Species Diversity Index; heterogeneity, mean percentage dissimilarity in species composition among plots within a site.
[b]From Hartnett et al. (1996b).
[c]Significant difference between grazed and ungrazed sites ($P \leq .05$).

equitability of species abundances (Hartnett et al. 1996b). Subordinate forb species in tallgrass prairies experience competitive release when bison graze grasses (Fahnestock and Knapp 1994). Grazing of the dominant matrix grasses results in an increase in light availability and increased water availability for ungrazed forbs due to the reduction in transpiring grass leaf area. Reduction of the grass canopy cover by bison also results in warmer soil temperatures, perhaps enabling subdominant ungrazed forbs to initiate growth earlier in the spring (Fahnestock and Knapp 1994). By contrast, increases in plant species richness and heterogeneity in grazed prairie were likely a result of greater microsite diversity generated by bison grazing and nongrazing activities. The mosaic of habitat patches generated by bison likely increase the diversity of establishment opportunities for several species that are otherwise excluded from the grassland community by the strong competitive dominance of the matrix grasses.

The patchy disturbances and heterogeneity generated by large ungulate activity play varying roles in the dynamics of other Great Plains grassland plant communities. In Sandhills prairie, for example, soil disturbance plays an important role in the dynamics of blowouts, unvegetated patches subjected to wind and soil movement in the expansive stabilized dune system. The blowout penstemon (*Penstemon haydeni*), which is highly adapted to these natural disturbances, has declined significantly in the Sandhills due to "improved" range management practices that have reduced the intensity of livestock-generated disturbance. As a result of reductions in these crucial dune disturbances, *P. haydeni* is now one of only a few federally endangered native plant species on the Great Plains.

In shortgrass plant communities, fecal pats are similar in size to established individuals of blue grama (*Bouteloua gracilis*), and their deposition causes adult plant mortality, resulting in reduced seedling-adult competition for limited soil water and increased successful seedling establishment (Coffin and Lauenroth 1988, Aguilera and Lauenroth 1995). Thus, fecal pats are an important patchy disturbance influencing the gap dynamics of these plant communities, and their influence on genet turnover rates may play a role in the maintenance of genetic diversity in populations of this dominant species. Urine deposition in grasslands creates patches with altered nutrient dynamics and plant species composition (Day and Detling 1990, Jaramillo and Detling 1992, Steinauer and Collins 1995), contributing to vegetation heterogeneity. Large vertebrate grazers also mediate a significant redistribution of nutrients through their dung and urine deposition.

Savory (1988) hypothesized that hoof action by large numbers of large grazers disturbs the surface soil structure, producing microsites with improved water infiltration and soil water availability, seed burial, and improved seedling germination and establishment. However, field studies have found no support for this hypothesis (Warren et al. 1986, Weigel et al. 1990). In general, all these disturbances and direct and indirect nonherbivory effects of large vertebrates contribute to the overall spatial and temporal patchiness of Great Plains grassland vegetation (Vinton and Collins, this volume).

Although direct empirical comparisons are lacking, differences in plant community responses to bison versus cattle grazing may be predicted based on their diets and foraging patterns. Because bison show lower dietary breadth and consume almost exclusively graminoids, they may reduce abundance of the dominant grasses, increase survival and growth of subordinate species, and increase species equitability and diversity to a greater degree than cattle. Greater microenvironmental heterogeneity generated by the wallowing, pawing, and other nongrazing activities of bison may also result in greater species richness and plant community heterogeneity in grasslands grazed by these native ungulates than those grazed at similar intensities by cattle. However, it is difficult to compare plant community responses with cattle versus bison, reported from other studies, due to confounding effects of site, weather, management, and other factors. For example, Collins (1987) reported that moderate cattle grazing increased local plant species richness by 2–15% in tallgrass prairie, whereas Hartnett et al. (1996b) reported that bison grazing increased tallgrass prairie plant species richness by 19–54%. This greater enhancement of plant species diversity by bison compared with cattle is consistent with previous predictions. However, the two studies were done in different years, at different locations (Oklahoma versus Kansas), and under different management regimes and temporal patterns of plant defoliation (seasonal cattle grazing versus year-round bison grazing). The lack of a significant increase in the abundance of cool-season C_3 graminoids in response to bison grazing compared with their increase under cattle grazing is also likely due to the greater cumulative grazing pressure experienced under year-round grazing than seasonal grazing rather than due to dietary differences between the two herbivores (Hartnett et al. 1996b). Furthermore, any plant community responses due to dietary differences between bison and cattle may be subtle at light-to-moderate grazing intensities but pronounced when animal densities are higher or grazing is prolonged.

There are also some important qualitative and quantitative differences between native and introduced ungulates in some of their nongrazing behaviors and their ecological consequences. For example, bison behaviors that differ from cattle, such as wallowing and pawing the ground (Reinhardt 1985), generate unique types of disturbance at certain scales. Wallows are saucerlike depressions created when bison trample the ground and roll in the exposed soil. Wallowing appears to be a sex-independent self-grooming behavior that helps relieve itching skin irritations caused by the process of molting (Reinhardt 1985). Rubbing on woody vegetation likely also has a similar function. The extensive dusting of the animal while wallowing may also serve to reduce or repel insects. An interesting contrast between bison and cattle is that the former engage in primarily self-grooming behavior (wallowing, rubbing) whereas cattle engage in social licking but do not typically wallow. Reinhardt (1985) suggested that this predominance of self-grooming rather than social grooming behavior may explain why bison are less associative than cattle.

Bison wallows were historically a common feature of the central grasslands (England and DeVos 1969) and are unique among the array of animal-generated disturbances. Unlike the mound disturbances generated by pocket gophers (*Geomys* spp.), prairie dogs (*Cynomys* spp.), and other animals, bison wallows are larger in size, are depressions rather than elevated mounds of loosely packed soil, and persist as distinct microhabitats for over a century rather than a few weeks, months, or years. Established wallows vary in depth from a few centimeters to 2–4 meters and in diameter from a few meters to more than 1 ha when individual wallows merge (Fig. 4.3). In Kansas tallgrass prairie, they occur most frequently on flat, upland soils. The compaction of soil results in the retention of standing water, creating localized habitats supporting species adapted to mesic conditions and ruderal species. On relict bison wallows in Kansas tallgrass prairie, prairie dropseed (*Sporobolus asper*), sedges (*Carex* spp.), and rushes (*Juncus* spp.) are the most abundant species (Gibson 1989). Many relict wallows in Colorado shortgrass prairie remain unvegetated. Compared with other animal disturbances, vegetation on bison wallows shows much lower similarity to undisturbed vegetation. Infrequently disturbed wallows tend to be dominated by sedges, forming dense rhizomatous mats that are well adapted to trampling and seasonal hydrologic fluctuations (Polley and Collins 1984). Inactive wallows are also largely dominated by mesic species but show seasonal shifts in relative abundances of cool- and warm-season grasses, similar to those of surrounding undisturbed vegetation.

Soil moisture appears to be the primary environmental factor influencing vegetation differences between wallows and their surrounding vegetation (Polley and Collins 1984), although other factors such as soil texture (higher clay content in wallows), available phosphorus (higher in wallows), and pH (wallow soils are more acidic) are also associated with these vegetation patterns. Because tallgrass prairie soils are inherently low in phosphorus and most of the dominant species are obligately mycorrhizal, vegetation responses in wallows may be mediated through effects on phosphorus availability coupled with differential plant species dependencies and responsiveness to mycorrhizae.

Together, these various studies of grazing and nongrazing activities indicate an important role of ungulate grazers in the regulation and maintenance of diversity in Great Plains plant communities. Effects of ungulates on plant species richness

Figure 4.3. (A) Bison wallows on shortgrass steppe, Weld County, Colorado. (B) Revege-tated wallow. (Photographs by F.L. Knopf.)

and spatial heterogeneity may have other important implications as well, in that these biodiversity components may strongly influence other structural and functional attributes of grassland ecosystems (Archer and Smeins 1991, West 1993). For example, empirical evidence for a relationship between diversity and stability has come from a variety of grasslands, where increased plant species diversity confers greater year-to-year stability in net primary production and species composition in response to drought or other stresses (Frank and McNaughton 1991, Tilman and Downing 1994, Hickman 1996).

Effects of Large Ungulates on Habitat Structure and Animal Populations

Concomitant with grazer-induced shifts in plant species composition and diversity are significant changes in the physical structure and nutritional dynamics of vegetation, both of which strongly influence other animal populations. Defoliation and disturbance generated by large herbivores alter aspects of habitat structure such as canopy density and heterogeneity, the relative abundances of woody vegetation and herbaceous cover, and amounts of standing dead vegetation and litter. These changes influence the breeding ecology and abundances of birds, and avifauna species richness declines monotonically with increasing grazing intensity (or fire frequency) (Zimmerman, this volume). Small mammal populations are similarly affected as grazers alter vegetation architecture, litter depth, and relative openness of the soil surface. Differential species responses to these habitat changes and the habitat mosaics of burned, unburned, grazed, and ungrazed patches result in shifts in small mammal species richness, dominance, and diversity (Kaufman and Kaufman, this volume). Consumers such as insects respond to grazer-induced changes in the taxonomic composition as well as the physical structure of the vegetation (Evans 1988).

Foraging by one herbivore may elicit vegetation patterns beneficial to other species using similar forage resources (e.g., removal of aging leaves and stimulation of new growth with greater nitrogen concentrations and digestibility) or cause a change in both the forage resource and vertebrate herbivore populations. For example, prairie dogs alter vegetation composition and nutritional quality on their colonies, producing patches of intense plant–herbivore activity and unique patterns of use by larger vertebrate herbivores such as elk, bison, and pronghorn (Coppock et al. 1983, Krueger 1986, Whicker and Detling 1988, Cid et al. 1991). The combined activities of these herbivores cause a shift in the competitive balance of plant species within the colony, a decrease in proportional contribution of graminoids to the plant community, increased graminoid equitability and overall plant species diversity, and increased plant shoot nitrogen concentrations (Whicker and Detling 1988). Increased habitat heterogeneity associated with the presence of these herbivores also results in increased small mammal and avian densities on colonies and increased diversity of other wildlife species (Whicker and Detling 1988). Historically, among native herbivores, bison and black-tailed prairie dogs probably had some of the greatest impacts on the structure and composition of grassland plant communities (Cid et al. 1991), and both the intensity and spatial pattern of their effects were significantly different than those of managed European cattle. Finally, effects of ungulates occur within the context of numerous interacting influences such as fire regimes, weather, topography, and management. In particular, variation in plant and animal populations among years associated with weather variability is typically greater than spatial or temporal variation associated with ungulate grazers or with fire.

Interactions with Fire, Climate, and Topography

Historically, grazing by large ungulate herbivores, periodic fire, and a variable continental climate all interacted to give rise to the grasslands of the Great Plains (Wells 1965, Axelrod 1985). Because plant community responses to grazing may be dependent on these other factors (Collins 1987, Vinton et al. 1993), attempts to understand general comparative effects of native and introduced ungulates must consider these interacting influences.

Fire in grasslands can modify plant–grazer interactions by influencing forage plant selectivity, habitat selection by grazers at the patch and larger scales, or the responses of plants to grazing. In tallgrass and Sandhills prairie, the occurrence of fire influences both bison forage plant selectivity and plant responses to grazing. For example, on unburned prairie, bison select the dominant rhizomatous grass big bluestem (*Andropogon gerardii*) but avoid grazing the bunchgrass little bluestem (*Schizachyruim scoparium*), consuming only 5% of the available stems (Pfeiffer and Steuter 1994, Pfeiffer and Hartnett 1995). On burned prairie, however, this discrimination disappeared, and the grazing frequency of little bluestem was more than threefold greater than on unburned habitats and equal to that of its dominant neighbor. This fire–grazing interaction is likely because little bluestem accumulates a persistent clump of standing dead tillers that serve as a physical deterrent to grazing, and burning removes this deterrent, exposing the plants to more frequent defoliation. Persistent grazing by bison reduces the biomass and shifts the population size distribution of little bluestem populations toward higher frequencies of smaller individuals. Because smaller individuals may be less drought tolerant than larger conspecifics (Butler and Briske 1988, Briske and Anderson 1990), declines in little bluestem populations under persistent grazing may be an indirect result of greater drought-induced mortality due to the altered population size structure (Pfeiffer and Hartnett 1995). Thus, fire, weather, population size structure, and growth form interact to influence bison grazing patterns and responses of little bluestem to grazing. The compensatory growth responses of dominant tallgrass prairie grasses such as big bluestem and switchgrass (*Panicum virgatum*) are also strongly influenced by the occurrence of fire and changes in plant resource availability in the postfire environment (Vinton and Hartnett 1992).

Patterns of bison and cattle grazing at larger scales are influenced by both the spatial and temporal patterns of fire and topography. During the growing season in Kansas tallgrass prairie, bison use areas burned in the spring much more frequently than unburned habitats and generate larger and more numerous intensively grazed patches in burns within the landscape mosaic (Vinton et al. 1993). During autumn and winter, bison graze burned and unburned areas more uniformly but graze most intensively in unburned portions of the landscape containing large stands of cool-season C_3 graminoids. These fire-induced patterns are further modified by topography, and many of the effects of bison on plant communities in the Flint Hills are more pronounced on uplands and level terraces than on lowland sites or slopes. Thus, fire and topographic interactions influence habitat selection by ungulates at scales ranging from that of individual plants up to the entire landscape.

Cattle and bison impacts interact with fire and topography at the landscape level in habitats such as south Texas thorn scrub (Archer et al. 1988), central Great Plains grasslands (Gartner et al. 1993), and Canadian parkland (Campbell et al. 1994). A consistent pattern in these habitats is the change from scattered individual trees, often located on topographically protected sites, to woodlands expanding into grassland (Steinauer and Bragg 1987). This increase in tree cover has been attributed primarily to a reduction in fire frequency (Bragg and Hulbert 1976), along with less physical damage caused to young trees by cattle as compared with bison. Probably as important is the reduced competitive ability of herbaceous plants versus woody plants. Grass vigor is reduced by season-long grazing as compared with the intense but intermittent grazing under which Great Plains plant communities evolved. Because cattle seek out wooded sites for protection from heat and cold and for preferred forage, the grasses will be suppressed unless access to these sites is controlled. As noted earlier, bison tend to avoid wooded areas and rough topographies, especially during the growing season. These differences in selection of woodlands by bison and cattle can lead to significant differences in grazing distribution and thus fine-fuel loads associated with woodlands. High fine-fuel loads under scattered trees and woodland patches would have increased the chances that summer lightning strikes would have spread as ground fires, causing significant damage to woody plants (Streng and Harcombe 1982, Platt et al. 1988). Thus, Great Plains woodlands were more restricted to sites protected by topography from both bison and the regional fire regime (Steuter et al. 1990). At this larger scale, the interaction between topography, fire, and large grazers has changed significantly with the change to European cattle management.

Ecological Consequences of Ungulate Management

Historical landscape changes and management practices that coincide with domestic herbivore introductions included significant subdivision of the landscape at the hectare-to-kilometer scale, a shift resulting in increasing similarity to the European landscape. Fencing, additions of fixed watering points, and the implementation of various grazing systems all altered spatial and temporal patterns of herbivory and its impact on vegetation. Grassland responses to the replacement of native with introduced domestic herbivores is likely explained by the changes in spatial patterns of animal distribution and herbivory associated with these management practices and landscape changes.

Fencing disrupted the spatial patterns and temporal dynamics of ungulate–plant interactions by restricting animal movements in response to forage and other environmental factors. The shift from large landscape mosaics to smaller fenced units influences the density and distributions of herbivore populations. Restricted dispersal can lead to high local or regional herbivore densities and overgrazing (Coughenour 1991). The development of additional watering points for domestic livestock also generates spatial patterns of habitat use and impacts that are different from those associated with geomorphic water sources for ungulates (Fig. 4.2C). Zones of attenuating animal impact away from watering points are typical of domestic ungulates. These "piospheres" represent potentially significant gradients in defoliation intensity, feces accumulation, soil nutrients and compaction, density of livestock trails, and cover of soil cryptogamic crusts and bare ground (Walker and Heitschmidt 1986, Andrew 1988). Modern water developments for livestock are located with consideration for many social and economic concerns not part of the long-term evolutionary background of the landscape. Other spatial gradients in animal activity may be associated with wooded habitats (shaded cover), feed or salt/mineral supplementation sites, or other habitats that are foci of animal activity (prairie dog towns).

Differences between native and introduced ungulates in their relationships to these habitat factors and landscape changes may also contribute to significant differences in the pattern of their impact on plant communities and habitat structure. For example, the time large vertebrate herbivores spend in or near streams varies seasonally (Siekert et al. 1985), and these seasonal patterns may differ considerably between native and introduced species. Cattle and bison in tallgrass prairie differ considerably in their seasonal patterns and intensity of use and impact on streams and riparian vegetation corridors. Managed cattle herds grazing during the growing season in Kansas tallgrass prairie use wooded areas principally for cover during the hot periods of the summer. Confined herds of bison grazing year-round in the same habitat have greatest impact on riparian areas during the winter months when permanent natural water sources are reduced and intermittent streams are dry. These differences in habitat use and foraging patterns between cattle and bison may have important consequences for the dynamics of prairie-riparian woodland ecotones, potentially extending the distribution of some plant species into grass-dominated areas and broadening wooded riparian zones. In general, patterns of habitat use by European cattle are more constrained by proximity to water than are those of native herbivores. Thus, confined cattle with limited water distribution may result in impacts on areas close to watering points. These "sacrifice areas" may have no native perennial vegetation remaining and provide highly disturbed habitat patches for the colonization and establishment of ruderal plant species.

Additionally, the development and implementation of various livestock grazing systems also brought changes to the spatial and temporal patterns of ungulate herbivory in the Great Plains. One rationale for the early development of cattle grazing systems was to manipulate the spatio-temporal pattern of grazing in a way that simulated the migratory grazing behavior of native ungulates. Grazing systems were viewed as a means to compensate for the continuous grazing pressure placed on the vegetation under a season-long or year-long grazing schedule (Smith 1895, Coughenour 1991). Specialized grazing systems control the length, intensity, and temporal and spatial patterns of herbivory during the growing season (see reviews by Gammon 1978, Holechek et al. 1989, Vallentine 1990). There are many rationales for the use of different grazing systems, including increasing vegetation productivity, improving pasture condition, controlling impacts on key forage species, or delaying grazing until after seed production to maintain plant population vigor (Hart et al. 1988, Holechek et al. 1989). Various grazing systems were implemented to provide a period of rest (no grazing) to allow adequate recovery of grazed vegetation and adequate allocation of stored metabolic reserves in belowground tissues (Kothmann 1984). Many grazing systems, such as intensive-early stocking or high intensity-short duration, were developed partially to reduce spatial heterogeneity in grazing intensity and increase the uniformity of forage use (Smith and Owensby 1978). Thus, these management practices have profound effects on the spatial components of plant-herbivore interactions in prairie communities.

Effects of various grazing systems such as continuous, season-long, intensive-early stocking, and short-duration grazing have been studied in several grasslands (Trlica et al. 1977, Heitschmidt et al. 1987, Pitts and Bryant 1987, Brummer et al. 1988, Hart et al. 1988, Gillen et al. 1990, 1991, Olson et al. 1993, Taylor et al. 1993). Both empirical studies and theoretical models (Noy-Meir 1976) suggest that the consequences of changing spatial and temporal herbivory patterns through different grazing systems are complex, and effects are dependent on other factors such as animal densities or initial range conditions. Much recent interest and controversy concerning grazing systems have been generated by claimed ecological and animal production benefits of time-controlled grazing systems advocated by Savory (1983, 1988). Studies of various grazing systems in Great Plains grasslands indicate that grazing systems can influence grazing distribution and defoliation patterns, vegetation standing crop, amount of litter, spatial patterns of trails and other nonherbivory effects, and plant species composition, although effects on vegetation composition are often minor (Herbel and Anderson 1959, Ring et al. 1985, Walker and Heitschmidt 1986, Owensby et al. 1988, Gillen at al. 1991, Olson et al. 1993). In most cases, however, the effects of different grazing systems on plant communities may be dependent on grazing animal density (stocking rates) or may be small and overridden by effects of animal stocking density (Gammon and Roberts 1978, Hart 1978, Heitschmidt et al. 1987, Brummer et al. 1988, Hart et al. 1988, 1993, Gillen et al. 1990). All these parameters are a function of management rather than the grazing animals per se. Most studies of vegetation responses to grazing systems have focused on forage plant and animal responses relevant to livestock production. Grazing system studies have often excluded infrequent species or lumped species into broad groups and thus have not adequately assessed the consequences of these management practices on the

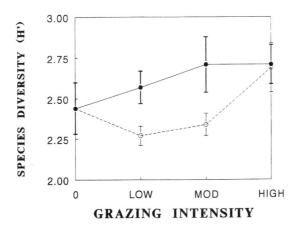

Figure 4.4. Mean (± 1 SE) plant species diversity (Shannon's index) at varying cattle stocking rates in tallgrass prairie. Solid line and closed circles, season-long stocking. Dotted line and open symbols, late-season rest rotation. Stocking rates: 0, ungrazed, LOW, 3.8; MOD, 2.8; and HIGH, 1.8 ha per cow-calf pair, respectively. (From Hickman et al. 1996.)

composition and diversity of plants or other life-forms or other ecologically important aspects of community structure. A comparison of the effects of two grazing systems at varying cattle stocking densities in Kansas tallgrass prairie showed that grazing systems had a significant effect on plant species diversity but that the effect diminished at high stocking rates (Fig. 4.4). More studies assessing effects of grazing management on community structure and diversity of plants and other life-forms are needed.

Although perennial grasses dominate the primary production of native Great Plains grasslands at the larger scales, the number of forb species outnumbers grasses by an order of magnitude. Distribution and abundance of forbs are largely responsible for the pattern of small- and mid-scale herbaceous diversity. Forbs are also more diagnostic of environmental conditions and disturbance patterns (Biondini et al. 1989). The forb community is highly dynamic in the presence of large grazers. Significant differences in the forb community result from the inter-action of grazers, slope, and soil associations (Steuter et al. 1995). Because forbs provide a resource base for a diverse group of Great Plains invertebrates, amphibians and reptiles, birds, and small mammals, their response to grazing management deserves additional attention.

Regionally, the perennial grass matrix is stable under a wide range of bison or cattle grazing intensities. Grass vigor and vertical and horizontal structure can be manipulated with grazing management, however. These changes in grass cover directly affect the habitat quality of grassland amphibians, reptiles, birds, and mammals, which are individually adapted to the full range of short or sparse or tall or dense grassland. For example, the Upland Sandpiper (*Bartramia longicauda*) requires relatively large areas of grasslands of short stature or that are closely cropped, whereas the Grasshopper Sparrow (*Ammodramus savannarum*) selects the taller and denser areas associated with light-to-moderate grazing (Kantrud 1981). In the Nebraska Sandhills, the northern earless lizard (*Holbrookia maculata*) and the prairie lizard (*Sceloporus undulatus*) select habitats with distinctly different levels of open sand, which can be increased or decreased through grazing management (Ballinger and Jones 1985, Jones and Ballinger 1987). In the Kansas Flint Hills, changes in litter depth and other habitat features resulting from grazing management similarly influence small mammal distributions and abundances (Kaufman and Kaufman, this volume). Given the landscape heterogeneity required to support the array of native Great Plains species, a large herbivore with a grass diet is fundamental.

Summary

Patterns of ungulate herbivory and habitat selection at several scales determine the spatial and temporal pattern impacts on plant communities and habitats of the Great Plains grasslands. The settlement of the Great Plains by immigrants from Europe in the mid- to late 1800s brought not only a change in ungulate species, but a myriad of changes in landscape spatial and temporal patterns associated with their management. These historical changes in plant–herbivore interactions occurred within the context of the complex and interacting influences of fire, topography, and a variable continental climate.

We have described significant differences between bison and cattle from small-scale diet selection to the larger scales of landscape use. These differences would be expected based on the differences in evolutionary history of the two species and human management of their respective landscapes. However, to date, the direct comparisons between bison and cattle that control for management differences in a rangeland setting have not been done. For example, if a bison herd was managed with the same year-long nutrient supplementation strategies used to support cattle herds on native grasslands, their selection for high-quality forbs and woodland habitats may change as seasonal nutrient deficits are relieved. Similarly, if bison on native grasslands were managed by season-long stocking or other grazing system rather than year-round grazing, their impacts on plant community structure may be significantly different. The specific morphological and physiological differences (heavier winter pelage and head and chest coverage, taller and narrower frame, lower water requirement) between bison and cattle result in different environmental tolerances. Yet, domestic cattle have a wide variation in these traits, which might also be more effectively exploited in land management.

In addition to significant differences between bison and cattle, we have identified corresponding changes in the landscape during the past 500 years of native and introduced ungulate use. Native ungulates remain well adapted to the undeveloped portions of this landscape. With proper management, we expect that native and introduced ungulates can continue to adapt to moderate rates of landscape change. Factors such as land area, management, surrounding land use, and others influence the relative suitability of native and domestic ungulates for Great Plains

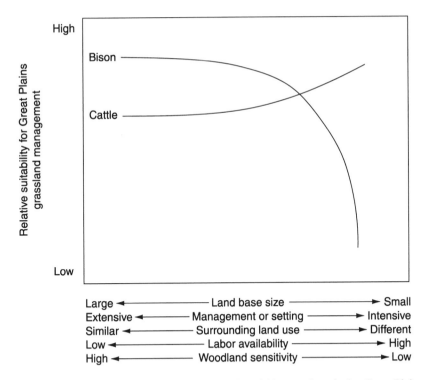

Figure 4.5. Simplified depiction of the suitability of bison and cattle for Great Plains grassland management. (Adapted from Plumb and Dodd 1993.)

grasslands management (Fig. 4.5). Local species diversity should be maintained where large ungulate grazing remains the dominant land use. However, many species migrate within and outside of the Great Plains. Their survival cannot be predicted based solely on suitable habitat within the Plains. Given the significant effects of ungulate grazers on plant community structure and the recent demonstration of clear relationships between diversity and stability in Great Plains grasslands, future changes in management will have implications for further long-term changes in patterns of biodiversity and ecosystem stability.

The Great Plains has responded to a complex combination of changes in the species of ungulates, the influence of humans and their management practices, and the interacting influences of numerous biotic and abiotic influences. In total, the change in dominant ungulate from a native to introduced species represents only one of many events and factors influencing the plant communities of the Great Plains. The consequences of intensive ungulate management by the human inhabitants of the plains were likely much larger than the effects of a change in the species of animal. The question of whether bison and cattle are analogous herbivores may be of less importance than asking how Great Plains plant and animal communities are influenced by landscape subdivision, grazing systems, large ungulate densities, fire–grazer interactions, or other management activities. Each of these factors is under human control, and each has important influences on the biodiversity and stability of Great Plains communities. Thus, given an understanding of their specific adaptations, both bison and cattle provide opportunities for managing plant-herbivore interactions in a way that maintains the long-term productivity, diversity, and biotic integrity of the Great Plains.

References

Aguilera, M.O., and W.K. Lauenroth. 1995. Influence of gap disturbances and type of microsites on seedling establishment in *Bouteloua gracilis*. J. Ecol. 83:87-97.

Andrew, M.H. 1988. Grazing impact in relation to livestock watering points. Trends Ecol. Evol. 3:336-339.

Archer, S., C.J. Scifres, S.R. Bassham, and R. Maggio. 1988. Autogenic succession in a subtropical savanna: conversion of grassland to thorn scrub. Ecol. Monogr. 58:111-127.

Archer, S., and F.E. Smeins. 1991. Ecosystem-level processes. Pp. 109-139 *in* R.K. Heitschmidt and J.W. Stuth, eds. Grazing management: an ecological perspective. Timber Press, Portland, OR.

Axelrod, D.I. 1985. Rise of the grassland biome, central North America. Bot. Rev. 51:164-201.

Ballinger, R.E., and S.M. Jones. 1985. Ecological disturbance in a sandhills prairie: impact and importance to the lizard community on Arapaho Prairie in western Nebraska. Prairie Nat. 17:91-100.

Belovsky, G.E. 1986. Optimal foraging and community structure: implications for a guild of generalist grassland herbivores. Oecologia 70:35-52.

Bennett, J.P., and R.B. Dahlgren. 1982. Seasonal food habits of bison on mixed grass prairie. Pp. 143-146 *in* D.D. Briske and M.M. Kothman, eds. Proceedings, a national conference on grazing management technology. Texas A&M Univ., College Station.

Biondini, M.E., A.A. Steuter, and C.E. Grygiel. 1989. Seasonal fire effects on the diversity patterns, spatial distribution and community structure of forbs in the northern mixed prairie, USA. Vegetatio 85: 21-31.

Bragg, T.B., and L.C. Hulbert. 1976. Woody plant invasion of unburned Kansas bluestem prairie. J. Range Manage. 29:19-24.

Briske, D.D., and V.J. Anderson. 1990. Tiller dispersion in populations of the bunchgrass *Schizachyrium scoparium:* implications for herbivory tolerance. Oikos 59:50-56.

Brummer, J.E., R.I. Gillen, and F.T. McCollum. 1988. Herbage dynamics of tallgrass prairie under short duration grazing. J. Range Manage. 41:264-266.

Butler, J.L., and D.D. Briske. 1988. Population structure and tiller demography of the bunchgrass *Schizachyrium scoparium* in response to herbivory. Oikos 51:306-312.

Campbell, C., I.D. Campbell, C.B. Blyth, and J.H. McAndrews. 1994. Bison extirpation may have caused aspen expansion in western Canada. Ecography 17:360-362.

Caughley, G. 1979. What is this thing called carrying capacity? Pp. 2-8 *in* M.S. Boyce and L.D. Hayden-Wing, eds. North American elk: ecology, behavior, and management. Univ. Wyoming, Laramie.

Cid, M.S., J.K. Detling, A.D. Whickler, and M.A. Brizuela. 1991. Vegetational responses of a mixed-grass prairie site following exclusion of prairie dogs and bison. J. Range Manage. 44:100-105.

Coffin, D.P., and W.K. Lauenroth. 1988. The effects of disturbance size and frequency on a shortgrass plant community. Ecology 69:1609-1617.

Collins, S.L. 1987. Interaction of disturbances in tallgrass prairie: a field experiment. Ecology 68:1243-1250.

Collins, S.L., and S.C. Barber. 1985. Effects of disturbance on diversity in mixed-grass prairie. Vegetatio 64:87-94.

Coppock, D.L., J.K. Detling, J.E. Ellis, and M.I. Dyer. 1983. Plant–herbivore interactions in a North American mixed-grass prairie. I. Effects of black-tailed prairie dogs on intraseasonal aboveground plant biomass and nutrient dynamics and plant species diversity. Oecologia 56:1–9.

Coughenour, M.B. 1991. Spatial components of plant-herbivore interactions in pastoral, ranching, and native ungulate ecosystems. J. Range Manage. 44:530–542.

Day, T.A., and J.K. Detling. 1990. Grassland patch dynamics and herbivore grazing preference following urine deposition. Ecology 71:180–188.

Dyer, M.I., J.K. Detling, D.C. Coleman, and D.W. Hilbert. 1982. The role of herbivores in grasslands. Pp. 255–295 in J.R. Estes, R.J. Tyrl, and J.N. Brunken, eds. Grasses and grasslands: systematics and ecology. Univ. Oklahoma Press, Norman.

England, R.C., and A. DeVos. 1969. Influence of animals on pristine conditions on the Canadian grasslands. J. Range Manage. 22:87–94.

Evans, E.E. 1988. Grasshopper (Insecta: Orthoptera: Acrididae) assemblages of tallgrass prairie: influences of fire frequency, topography, and vegetation. Can. J. Zool. 66:1495–1501.

Fahnestock, J.T., and A.K. Knapp. 1994. Plant responses to selective grazing by bison: interactions between light, herbivory and water stress. Vegetatio 115:123–131.

Frank, D.A., and S.J. McNaughton. 1991. Stability increases with diversity in plant communities: empirical evidence from the 1988 Yellowstone drought. Oikos 62:360–362.

Fuller, W.A. 1960. Behavior and social organization of the wood bison of Wood Buffalo National Park, Canada. Arctic 13:3–19.

Gammon, D.M. 1978. A review of experiments comparing systems of grazing management on natural pastures. Proc. Grassland Soc. S. Africa 13:75–82.

Gammon, D.M., and R.B. Roberts. 1978. Patterns of defoliation during continuous and rotational grazing of the Matapos Sandveld in Rhodesia. 1. Selectivity of grazing. Rhodesia J. Agric. Res. 16:117–131.

Gartner, F.R., W.W. Thompson, and K.J. Wrage. 1993. Bison, pine and environmental stress. Pp. 85–90 in N.H. Granholm, ed. Biostress: mechanisms, responses, management. Proc. Stress Symp. Coll. of Agric. and Biol. Sci., South Dakota State Univ., Brookings.

Geist, V. 1991. Phantom subspecies: the wood bison Bison "athabascae" Rhoads 1897 is not a valid taxon, but an ecotype. Arctic 44:283–300.

Gibson, D.J. 1989. Effects of animal disturbances on tallgrass prairie vegetation. Am. Midl. Nat. 121:144–154.

Gillen, R.L., F.T. McCollum, and J.E. Brummer. 1990. Tiller defoliation patterns under short duration grazing in tallgrass prairie. J. Range Manage. 43:95–99.

Gillen, R.L., F.T. McCollum, M.E. Hodges, J.E. Brummer, and K.W. Tate. 1991. Plant community responses to short duration grazing in tallgrass prairie. J. Range Manage. 44:124–128.

Hansen, J.R. 1984. Bison ecology in the northern Great Plains and a reconstruction of bison patterns for the North Dakota region. Plains Anthropol. 29:93–113.

Harper, J.L. 1977. The population biology of plants. Academic Press, New York.

Hart, R.H. 1978. Stocking rate theory and its application to grazing on rangelands. Pp. 547–550 in D.N. Hyder, ed. Proc. First Intl. Rangelands Congr. Soc. Range Manage., Denver, CO.

Hart, R.H., S. Clapp, and P.S. Test. 1993. Grazing strategies, stocking rates, and frequency and intensity of grazing on western wheatgrass and blue grama. J. Range Manage. 46:122–126.

Hart, R.H., M.J. Samuel, P.S. Test, and M.A. Smith. 1988. Cattle, vegetation, and economic responses to grazing systems and grazing pressure. J. Range Manage. 41:282–286.

Hartnett, D.C., K.R. Hickman, and L.E. Fischer-Walter. 1996a. Effects of bison on plant species diversity in tallgrass prairie. Pp. 215–216 in N.E. West, ed. Proc. Fifth Int. Rangeland Congr. Soc. Range Manage., Denver, CO.

Hartnett, D.C., K.R. Hickman, and L.E. Fischer-Walter. 1996b. Effects of bison grazing, fire, and topography on floristic diversity in tallgrass prairie. J. Range Manage. 49:413–420.

Hayes, W.P. 1927. Prairie insects. Ecology 8:238–250.

Heitschmidt, R.K., S.L. Dowhower, and J.W. Walker. 1987. Some effects of rotational grazing treatment on quantity and quality of available forage and amount of ground litter. J. Range Manage. 40:318–321.

Herbal, C.H., and K.L. Anderson. 1959. Response of true prairie vegetation on major Flint Hills range sites to grazing treatment. Ecol. Monogr. 29:171–186.

Hickman, K.R. 1996. Effects of large ungulate herbivory on tallgrass prairie plant populations and community structure. PhD dissertation. Kansas State Univ., Manhattan.

Hickman, K.R., D.C. Hartnett, and R.C. Cochran. 1996. Effects of grazing systems and stocking rates on plant species diversity in Kansas tallgrass prairie. Pp. 228–229 in N.E. West, ed. Proc. Fifth Int. Rangelands Congr. Soc. Range Manage., Denver, CO.

Hobbs, N.T., and D.M. Swift. 1988. Grazing in herds: when are nutritional benefits realized? Am. Nat. 131:760–764.

Holechek, J.L., R.D. Pieper, and C.H. Herbel. 1989. Range management: principles and practices. Prentice-Hall, Englewood Cliffs, NJ.

Hudson, R.J., and S. Frank. 1987. Foraging ecology of bison in Aspen boreal habitats. J. Range Manage. 40:71–75.

Huntly, N. 1991. Herbivores and the dynamics of communities and ecosystems. Annu. Rev. Ecol. Systematics 22:477–503.

Hyder, D.N., R.E. Bement, E.E. Remmenga, and C. Terwilliger, Jr. 1966. Vegetation—soils and vegetation-grazing relations from frequency data. J. Range Manage. 19:11–17.

Jaramillo, V.J., and J.K. Detling. 1992. Small-scale grazing in a semi-arid North American grassland. I. Tillering, N uptake, and retranslocation in simulated urine patches. J. App. Ecol. 29:1–8.

Jones, J.K., Jr., D.C. Carter, and H.H. Genoways. 1973. Checklist of North American mammals north of Mexico. Texas Tech Univ. Mus. Occas. Pap. 12.

Jones, S.M., and R.E. Ballinger. 1987. Comparative life histories of Holbrookia maculata and Sceloporus undulatus in western Nebraska. Ecology 68:1828–1838.

Kantrud, H.A. 1981. Grazing intensity effects on the breeding avifauna of North Dakota native grasslands. Can. Field Nat. 95:404–417.

Kautz, J.E., and G.M. Van Dyne. 1978. Comparative analyses of diets of bison, cattle, sheep and pronghorn antelope on shortgrass prairie in northeastern Colorado, USA. Pp. 438–442 in D.N. Hyder, ed. Proc. First Int. Rangeland Congr. Soc. Range Manage., Denver, CO.

Kothmann, M.M. 1984. Concepts and principles underlying grazing systems: a discussant paper. Pp. 903–916 in Developing strategies for rangeland management. Nat. Res. Counc.-Nat. Acad. Sci., Westview Press, Boulder, CO.

Krasinska, M., and Z.A. Krasinski. 1995. Composition, group size, and spatial distribution of European bison bulls in Balowieza Forest. Acta Theriologica 40:1–21.

Krueger, K. 1986. Feeding relationships among bison, pronghorn, and prairie dogs: an experimental analysis. Ecology 67:760–770.

Kucera, C.L. 1956. Grazing effects on composition of virgin prairie in north-central Missouri. Ecology 37:389–391.

Lewis, M., and W. Clark. 1893. The history of the Lewis and Clark expedition. E. Coues, ed., originally published by F. P. Harper, New York. Republished in 1964 by Dover, New York.

Lott, D.F. 1979. Dominance relations and breeding rate in mature male American bison. Z. Tierpsychologie 49:418–432.

Lott, D.F. 1981. Sexual behavior and intersexual strategies in American bison. Z. Teirpsychologie 56:97–114.

Mack, R.N., and J.N. Thompson. 1982. Evolution in steppe with few large, hooved mammals. Am. Nat. 119:757-773.

May R.M. and J.R. Beddington. 1981. Notes on some topics in theoretical ecology in relation to management of locally abundant populations of mammals. Pp. 205-215 *in* P.A. Jewell, S. Holt, and D. Hart, eds. Problems in management of locally abundant wild mammals. Academic Press, New York.

McHugh, T. 1979. The time of the buffalo. Univ. Nebraska Press, Lincoln.

McNaughton, S.J. 1979. Grazing as an optimization process: grass-ungulate relationships in the Serengeti. Am. Nat. 113:691-703.

McNaughton, S.J. 1985. Ecology of a grazing ecosystem: the Serengeti. Ecol. Monogr. 55:259-294.

Milchunas, D.G., W.K. Lauenroth, P.L. Chapman, and M.K. Kazempour. 1989. Effects of grazing, topography, and precipitation on the structure of a semiarid grassland. Vegetatio 80:11-23.

Milchunas, D.G., O.E. Sala, and W.K. Lauenroth. 1988. A generalized model of the effects of grazing by large herbivores on grassland community structure. Am. Nat. 132:87-106.

Morey, D.F. 1994. The early evolution of the domestic dog. Am. Sci. 82:336-347.

Norland, J. 1984. Habitat use and distribution of bison in Theodore Roosevelt National Park. M.S. thesis. Montana State Univ., Bozeman.

Noy-Meir, I. 1976. Rotational grazing in a continuously growing pasture: a simple model. Agric. Syst. 1:87-112.

Olson, K.C., J.R. Brethour, and J.L. Launchbaugh. 1993. Shortgrass range vegetation and steer growth response to intensive-early stocking. J. Range Manage. 46:127-132.

Owensby, C.E., R.C. Cochran, and E.F. Smith. 1988. Stocking rate effects on intensive-early stocked Flint Hills bluestem range. J. Range Manage. 41:483-487.

Pacala, S.W., and M.J. Crawley. 1992. Herbivores and plant diversity. Am. Nat. 140:243-260.

Peden, D.G. 1976. Botanical composition of bison diets on shortgrass plains. Am. Midl. Nat. 96:225-229.

Peden, D.G., G.M. Van Dyne, R.W. Rice and R.M. Hansen. 1974. The trophic ecology of *Bison bison* L. on shortgrass plains. J. Appl. Ecol. 11:489-497.

Pfeiffer, K.E., and D.C. Hartnett. 1995. Bison selectivity and grazing response of little bluestem in tallgrass prairie. J. Range Manage. 48:26-31.

Pfeiffer, K.E., and A.A. Steuter. 1994. Preliminary response of Sandhills Prairie to fire and bison grazing. J. Range Manage. 47:395-397.

Pitts, J.S., and F.C. Bryant. 1987. Steer and vegetation response to short duration continuous grazing. J. Range Manage. 40:386-389.

Platt, W.J., G.W. Evans, and S.L. Rathbun. 1988. The population dynamics of a long-lived conifer (*Pinus palustuis*). Am. Nat. 131:491-525.

Plumb, G.E., and J.L. Dodd. 1993. Foraging ecology of bison and cattle on a mixed prairie: implications for natural area management. Ecol. Appl. 3:631-643.

Polley, H.W., and S.L. Collins. 1984. Relationships of vegetation and environment in buffalo wallows. Am. Midl. Nat. 112:178-186.

Reiger, J.F. 1972. Passing of the Great West. Scribner & Sons, New York.

Reinhardt, V. 1985. Quantitative analysis of wallowing in a confined bison herd. Acta Theriologica 30:149-156.

Ring, C.B., R.A. Nicholson, and J.L. Launchbaugh. 1985. Vegetational traits of patch-grazed rangeland in west-central Kansas. J. Range Manage. 38:51-55.

Roe, F.G. 1970. The North American buffalo: a critical study of the species in its wild state. Second ed. Univ. Toronto Press, Toronto, Ontario.

Sala, O.E., M. Oesterheld, R.J.C. Leon, and A. Soriano. 1986. Grazing effects upon plant community structure in subhumid grasslands of Argentina. Vegetatio 67:27-32.

Savory, A. 1983. The Savory grazing method or holistic resource management. Rangelands 5:155-159.

Savory, A. 1988. Holistic resource management. Island Press, Covelo, CA.

Schaefer, A.L., B.A. Young, and A.M. Chimwano. 1978. Ration digestion and retention times of digesta in domestic cattle (*Bos taurus*), American bison (*Bison bison*), and Tibetan yak (*Bos grunniens*). Can. J. Zool. 56:2355-2358.

Schonewald, C. 1994. *Cervus canadensis* and *C. elaphus:* North American subspecies and evaluation of clinical extremes. Acta Theriologica 39:431-452.

Schwartz, C.C., and J.E. Ellis. 1981. Feeding ecology and niche separation in some native and domestic ungulates on the shortgrass prairie. J. Appl. Ecol. 18:343-353.

Senft, R.L., L.R. Rittenhouse, and R.G. Woodmansee. 1985. Factors influencing patterns of cattle grazing behavior on shortgrass steppe. J. Range Manage. 38:82-87.

Siekert, R.E., Q.D. Skinner, M.A. Smith, J.L. Dodd, and J.D. Rodgers. 1985. Channel response of an ephemeral stream in Wyoming to selected grazing treatments. Pp. 276-278 *in* R. Roy Johnson, C.D. Ziebell, D.R. Patton, P.F. Efollian, and R.H. Hamre, eds. Riparian ecosystems and their management: reconciling conflicting uses. Proc. First N. Am. Riparian Conf., Tucson, AZ.

Smith, E.F., and C.E. Owensby. 1978. Intensive-early stocking and season-long stocking of Kansas Flint Hills range. J. Range Manage. 31:14-17.

Smith, J.G. 1895. Forage conditions of the prairie region. Pp. 309-324 *in* U.S. Dept. Agric. Yearbook of Agric.—1895. U.S. Dept. Agric.

Steinauer, E.M., and T.B. Bragg. 1987. Ponderosa pine (*Pinus ponderosa*) invasion of Nebraska Sandhills Prairie. Am. Midl. Nat. 118:358-365.

Steinauer, E.M., and S.L. Collins. 1995. Effects of urine deposition on small-scale patch structure in prairie vegetation. Ecology 76:1195-1205.

Steuter, A.A., B. Jasch, J. Ihnen, and L.L. Tieszen. 1990. Woodland Boundary changes in the middle Niobrara Valley of Nebraska identified by C values of soil organic matter. Am. Midl. Nat. 124:301-308.

Steuter, A.A., E.M. Steinauer, G.L. Hill, P.A. Bowers, and L.L. Tieszen. 1995. Distribution and diet of bison and pocket gophers in a Sandhills prairie. Ecol. Appl. 5:756-766.

Streng, D.R., and P.A. Harcombe. 1982. Why don't east Texas savannas grow up to forests? Am. Midl. Nat. 108:278-294.

Taylor, C.A., T.D. Brooks, and N.E. Garza. 1993. Effects of short duration and high-intensity, low-frequency grazing systems on forage production and composition. J. Range Manage. 46:118-121.

Telfer, E.S., and A. Cairns. 1986. Resource use by moose versus sympatric deer, wapiti, and bison. Alces 22:114-137.

Telfer, E.S., and J.P. Kelsall. 1984. Adaptations of some large North American mammals for survival in snow. Ecology 65:1828-1934.

Tieszen, L.L., D.J. Ode, P.W. Barnes, and P.M. Bultsma. 1980. Seasonal variation in C3 and C4 biomass at the Ordway Prairie and selectivity by bison and cattle. Pp. 165-174 *in* D.C. Hartnett, ed. Proc. 7th N. Am. Prairie Conf. Kansas State University, Manhattan.

Tilman, D., and J.A. Downing. 1994. Biodiversity and stability in grasslands. Nature 367:363-365.

Trlica, M.J., M. Buwai, and J.W. Menke. 1977. Effects of rest following defoliations on the recovery of several range species. J. Range Manage. 30:21-27.

Vallentine, J.F. 1990. Grazing management. Academic Press, New York.

Van Vuren, D. 1982. Comparative ecology of bison and cattle in the Henry Mountains, Utah. Pp. 449-457 *in* L. Nelson and J.M. Peek, eds. Proc. Wildlife-Livestock Relationships Symp. Univ. Idaho, Coeur d'Alene.

Van Vuren, D. 1984. Summer diets of bison and cattle in southern Utah. J. Range Manage. 37:260-261.

Van Zyll de Jong, C.G., C. Gates, H. Reynolds, and W. Olson. 1995. Phenotypic variation in remnant populations of North American bison. J. Mammal. 76:391-405.

Vinton, M.A., and S.L. Collins. In press. Landscape gradients and habitat structure in native grasslands of the central Great Plains. Ecol. Stud.

Vinton, M.A., and D.C. Hartnett. 1992. Effects of bison grazing on *Andropogon gerardii* and *Panicum virgatum* in burned and unburned tallgrass prairie. Oecologia 90:374–382.

Voisin, A. 1988. Grass productivity. Island Press, Covela, CA.

Walker, J.W., and R.K. Heitschmidt. 1986. Effect of various grazing systems on type and density of cattle trails. J. Range Manage. 39:428–431.

Warren, S.D., T.L. Thurow, W.H. Blackburn, and N.E. Garza. 1986. The influence of livestock trampling under intensive rotation grazing on soil hydrologic characteristics. J. Range Manage. 39:491–495.

Weigel, J.R.C.M. Britton, and G.R. McPherson. 1990. Trampling effects from short-duration grazing on tobosa-grass range. J. Range Manage. 43:92–95.

Wells, P.V. 1965. Scarp woodlands, transported grassland soils, and the concept of grassland climate in the Great Plains region. Science 148:246–249.

West, N.E. 1993. Biodiversity of rangelands. J. Range Manage. 46:2–13.

Whicker, A.D., and J.K. Detling. 1988. Modification of vegetation structure and ecosystem processes by North American grassland mammals. Pp. 301–316 *in* M.J.A. Werger, H.J. During, and P.J.M. Van Der Aart, eds. Plant form and vegetation structure. SPB Acad. Publ., The Hague.

Wydeven, A.P., and R.B. Dahlgren. 1985. Ungulate habitat relationships in Wind Cave National Park. J. Wildl. Manage. 49:805–813.

2. Ecology of Vertebrate Assemblages Within Grassland Landscapes

5. Historical Changes in the Landscape and Vertebrate Diversity of North Central Nebraska

Michael A. Bogan

Introduction

In the United States of the mid-nineteenth century, change was the order of the day. In the East, the Industrial Revolution was affecting everyday life in cities and farms. Clearing of forests for a familiar, European form of agriculture was continuing, and factories were producing goods needed by an expanding nation. In much of the West, change was a function of Manifest Destiny; exploration for precious minerals was beginning and exploitation of water resources to sustain agriculture and livestock in often arid settings was underway. Transcontinental railway links were literally gathering steam as government exploring parties searched for the best pathways to the west coast. Settlement of the American West beyond the Mississippi was to bring major changes, but nowhere did settlement of the land bring such major and irreversible change to a place, its flora and fauna, and its people as it did in the Great Plains. In an American landscape that has been as unappreciated and undervalued as any on the continent, the Great Plains suffered grieviously. Settlement of the prairies and plains brought not just settlers, agriculture, livestock, and their support systems such as merchants and railways, but drastic changes to a land and people largely dependent on roving herds of ungulates—and one in particular, the bison (*Bison bison*). Present-day citizens of the United States marvel at the immense herds of ungulates in Africa, often not realizing that similar sights were common as little as 120 years ago on the Great Plains. What must it have been like to be witness to herds of animals that stretched

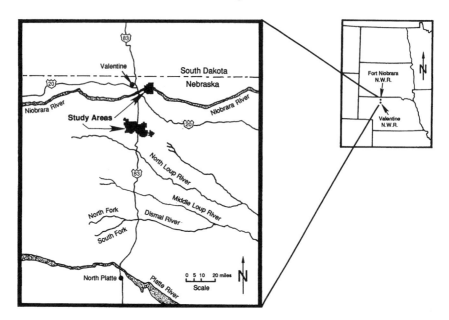

Figure 5.1. Area map of the Sandhills showing the locations of Fort Niobrara and Valentine National Wildlife Refuges.

to the horizon and the subsequent changes as millions of these same animals were eliminated?

Many of the most important changes wrought by settlement of the prairie are obvious. Prairie lands are now fenced and no longer stretch unconfined to the horizon. Land within the fences has been broken for agriculture or grazed to varying degrees by domestic livestock. Some areas, often on sites used for centuries by humans, are paved and developed for an urban population increasingly out of touch with the natural world, its inhabitants, and their cycles of life. Although bison are increasingly popular with some ranchers (as an alternative to domestic cattle), they are no longer a functional part of prairie ecosystems. Conversely, ungulates such as deer (*Odocoileus* spp.) and pronghorn (*Antilocapra americana*), recovering from population lows suffered near the turn of the century, are again integral to natural processes in many areas.

Does this mean that the prairie has weathered the storm of settlement successfully? What do recovering populations of large ungulates tell us about insults to the prairie ecosystem? What has been the effect of settlement and change on other (less charismatic) species of prairie animals? Are there species that are lost forever, locally extirpated, or that exhibit downward population trends? And have other species, with or without the aid of humans, invaded the prairie, and what has been their impact? Perhaps as always, the prairie is enigmatic. In many respects, we know more about bison and pronghorn and less of the meadow jumping mouse (*Zapus hudsonius*) and the prairie vole (*Microtus ochrogaster*). Do these smaller

Table 5.1. Species of Mammals of the Northern Sandhills[a]

Extant species
Sorex cinereus—masked shrew
Blarina brevicauda—northern short-tailed shrew
Cryptotis parva—least shrew
Scalopus aquaticus—eastern mole
Myotis ciliolabrum—western small-footed myotis
Myotis septentrionalis—northern myotis
Lasiurus borealis—eastern red bat
Lasiurus cinereus—hoary bat
Lasionycteris noctivagans—silver-haired bat
Eptesicus fuscus—big brown bat
Tadarida brasiliensis—brazilian free-tailed bat
Sylvilagus audubonii—desert cottontail
Sylvilagus floridanus—eastern cottontail
Lepus townsendii—white-tailed jackrabbit
Spermophilus spilosoma—spotted ground squirrel
Spermophilus tridecemlineatus—thirteen-lined
    ground squirrel
Cynomys ludovicianus—black-tailed prairie dog
Geomys bursarius—plains pocket gopher
Perognathus fasciatus—olive-backed pocket mouse
Perognathus flavescens—plains pocket mouse
Chaetodipus hispidus—hispid pocket mouse
Dipodomys ordii—Ord's kangaroo rat
Castor canadensis—beaver
Reithrodontomys megalotis—western harvest mouse
Reithrodontomys montanus—plains harvest mouse
Peromyscus leucopus—white-footed mouse
Peromyscus maniculatus—deer mouse
Onychomys leucogaster—northern grasshopper mouse
Neotoma floridana—eastern woodrat
Microtus ochrogaster—prairie vole
Microtus pennsylvanicus—meadow vole
Ondatra zibethicus—muskrat
Synaptomys cooperi—southern bog lemming
Zapus hudsonius—meadow jumping mouse
Erethizon dorsatum—porcupine
Canis latrans—coyote
Vulpes velox—swift fox
Mustela frenata—long-tailed weasel
Mustela nivalis—least weasel
Mustela vison—mink

Extant species (*continued*)
Taxidea taxus—badger
Mephitis mephitis—striped skunk
Lutra canadensis—northern river otter
Lynx rufus—bobcat
Odocoileus hemionus—mule deer
Odocoileus virginianus—white-tailed deer
Antilocapra americana—pronghorn

Extirpated species
Canis lupus—gray wolf
Ursus americanus—black bear
Ursus arctos—grizzly bear
Mustela nigripes—black-footed ferret
Gulo gulo—wolverine
Felis concolor—mountain lion
Lynx lynx—lynx
Cervus elaphus—wapiti or elk
Bison bison—bison
Ovis canadensis—bighorn sheep

Invading species
Didelphis virginiana—Virginia opossum
Lepus californicus—black-tailed
    jackrabbit
Marmota monax—woodchuck
Sciurus niger—eastern fox squirrel
Procyon lotor—raccoon
Spilogale putorius—eastern spotted skunk
Vulpes vulpes—red fox
Urocyon cinereoargenteus—grey fox

Introduced species
Sciurus carolinensis—eastern grey squirrel
Mus musculus—house mouse
Rattus norvegicus—Norway rat

Species of possible occurrence
Sorex merriami—Merriam's shrew
Spermophilus franklinii—Franklin's ground
    squirrel
Perognathus flavus—silky pocket mouse

[a] Scientific and vernacular names follow Jones et al. (1992) or Bogan and Ramotnik (1995).

animals survive and, if so, in what relative kind of security? How have they responded to change on the prairies?

During 1991 and 1992, research personnel from the U.S. Fish and Wildlife Service were presented with an opportunity to examine these and other questions in one of the most unique ecological areas of the Great Plains, the Sandhills along the Niobrara River valley of north central Nebraska (Fig. 5.1). These studies, centered on Valentine and Fort Niobrara National Wildlife Refuges (NWR) and reported in detail in Bogan (1995), provided new information on vertebrates of this area and how native prairie flora and fauna have changed over the past 150 years. My goal in this chapter is to provide an overview of historic changes in flora and fauna in the northern Sandhills near Valentine, Nebraska. Because my own specialty is mammals and the historical record for mammals tends to better than for other vertebrate groups, my emphasis is on that group (see Table 5.1 for names of mammals). However, some information is provided on other vertebrate groups as well.

Historical Sandhills Landscape (1800–1900)

Our understanding of the historical landscape and its vertebrates is admittedly incomplete. Early travelers to the West tended to comment on large and impressive forms of wildlife (e.g., elk and bison) and often showed little interest in smaller or more secretive forms. Thus, descriptions of the early landscape and its fauna tend to be dominated by accounts of those large ungulates that, along with climate, geology, and fire, were ecological forces responsible for shaping the Great Plains. At the least, these large animals were the essence of the Plains to European eyes unaccustomed to animals of such size and numbers.

The incomplete accounting of early flora and fauna of north central Nebraska also is due to other more practical reasons. Most of the early exploring expeditions tended to follow larger rivers, such as the Missouri, which allowed overwater travel, and the Platte, whose level floodplain allowed easier overland movement, rather than the smaller, more challenging Niobrara River valley. Explorers also tended to avoid areas such as the Sandhills that appeared desolate, inhospitable, and difficult to traverse. In addition, early collections and museum records sometimes demonstrate a tendency toward opportunistic rather than systematic collecting and emphasize new or unique specimens, to the omission of more common or widespread, but often dominant, species. Finally, place names, boundaries, and landscapes have changed, often making it difficult to reconstruct routes of early travelers and their activities. Nonetheless, useful information can be gleaned from observations of explorers and naturalists who ventured in or near the Sandhills.

The first group to record observations of what is now Nebraska was early explorers, fur traders, and missionaries who visited the area in the 1700s and early 1800s. Diller (1955) names several early explorers but says that their routes are not known with certainty. One route that is more certain is that of J. Mackay, who was the leader of the second expedition sent out by the Upper Missouri Company of St. Louis in 1795. According to Diller, Mackay traveled westward north of the Niobrara River, crossed that river a little below where Valentine now stands, and continued southward past the lakes of present-day Cherry and Thomas counties, and then on south and east. Mackay observed that white bears (perhaps "silvertip" grizzlies, although Smith [1958] attributes the observation to bighorn sheep) were common (Todd County, South Dakota); described a "great desert of drifting sand without trees, soil, rocks, water, or animals of any kind, excepting some little varicolored turtles, of which there are vast numbers" (Cherry County); and said there are many wild oxen (buffalo) in summer and autumn (Loup County) and many beaver in the Elkhorn River. Smith (1958) and Miller (1990) provide other details of Mackay's travels.

Captains Lewis and Clark passed through Nebraska in 1804 and 1806 followed by Major Stephen H. Long in 1819–1820; neither expedition passed through the Sandhills. It was Long's map of 1823 that labeled the area east of the Rocky Mountains as the "Great [American] Desert" and dampened official enthusiasm about the West (Viola 1987). After the discovery of gold in California in 1849 and the subsequent rush of settlers to the West, the U.S. War Department was assigned tasks of surveying the area for wagon and railway routes and protecting those who traveled west (Jones 1964). Of the many reports that emanated from these military surveys, only one discusses the Sandhills. During the period 1855–57, Lieutenant G.K. Warren passed through the Sandhills on two of his three northern Plains expeditions. In 1855, Warren traveled from Fort Pierre (South Dakota) to Fort Kearney (Nebraska), which took him directly through the Sandhills. In 1857, Warren returned to Nebraska and traveled up the "Loup Fork" (of the Platte River), through the Sandhills north to the Niobrara River, then west to Fort Laramie. Later that year, his party returned east along the Niobrara River to its junction with the Missouri (Warren 1875).

Warren (1875, p. 25) described the Sandhills as an area comprising no less than 20,000 square miles (19,300; Bleed and Flowerday 1990) "covered with barren sand, which, blown by the wind into high hills, renders this section not only barren, but in a measure impracticable for travel." The hills were said (p. 26) to vary in height from 10 to 200 ft, often arranged in ridges running east and west. Near the sources of the Loup Fork (North and Middle Loup rivers), he found many of the lakes impregnated with salts and unfit to drink (although he notes on p. 18 that some of their animals drank without injury). Warren (1875, p. 26) noted that the present form of the hills is mainly due to wind and that where grass protects the surface the sand did not drift, but where grass is removed, the wind "whirls the sand in the air, and often excavates deep holes."

Less seems to have been written of the interspersed lakes within the Sandhills, most of which are recharged from below rather than filled by runoff. Warren (1856) noted that some lakes contained fishes, and this suggests they were permanently watered. Unlike the river bottom of the Niobrara, the lakes were sandy bottomed with shorelines vegetated with sedge (*Carex* spp.); woody vegetation was restricted primarily to scattered willows (*Salix* spp.) (Warren 1856, Smith and Pound 1893, Pound and Clements 1900; plant names in text follow Kaul 1990). Grasses were relatively thick around the lakes and provided nesting habitats for many species of waterbirds (Wolcott 1906).

Warren (1875, p. 40) described the Niobrara River valley as destitute of wood, with not even enough to build a campfire. Except for occasional pines (*Pinus ponderosa*) on distant hills to the north, the valley was described as treeless until near longitude 102° 30′ (near the present town of Rushville), at which point the river entered between high steep banks. Here, trees became more continuously distributed, with pine on the bluffs and small clusters of cottonwood (*Populus deltoides*), elm (*Ulmus americana*), and ash (*Fraxinus pennsylvanica*) along the river. Where the river was walled in by high bluffs, Warren (1875, p. 40) described small, deep valleys with luxuriant grass, abundant pine, ash, and oak (*Quercus macrocarpa*) and gushing springs of clear cold water. Warren suggested that agriculture would be more productive to the east of 99°20′ (western Boyd County), where the valley bottoms are wider with abundant timber. He described (p. 41) Minnechaduza (or Rapid) Creek at its mouth near Valentine as being about "eight yards" wide, with a fertile valley about "a quarter to half a mile" wide. The banks were said to be "scantily fringed with small trees." The Niobrara River is spring fed and, unlike the Platte, does not arise in high-elevation mountains to the west. Thus, it was probably not subject to annual, extreme spring flooding from melting snow, a fact that may have allowed the development of riparian forests.

Wildfire frequency also likely played an important role in determining vegetative components in the riparian habitats around lakes, streams, and rivers. Steinauer and Bragg (1987) provide data supporting an average fire "return" interval of 3.5 years in ponderosa pine in the middle Niobrara valley before settlement (1851–1900). Frequent, historic fires surely precluded significant establishment of deciduous trees, such as cottonwood, around the lakes while promoting growth of root-sprouting willows. In the river valley, deciduous trees were present where protected by cliffs and canyons from wildfires.

Higgins (1986) compiled historical accounts of fires in the northern Great Plains pre-1900 and states that lightning-set fires were recorded less frequently than were fires set by Native Americans. The most prevalent type of fires were scattered, single events of short duration and small areal extent. Fires occurred mainly in two periods: March–May with a peak in April, and July–November with an October peak. Higgins believes that Native Americans purposely used fire as a tool to aid hunting and gathering of food and other materials and that seasonal patterns of fires were timed with movements of bison. Westover (no date), who believes that most prairie fires were the result of lightning strikes, provides first-hand accounts of early settlers' experiences with prairie fires in the Great Plains. His accounts span a time period from 1873 to 1925.

Early photographs (Fig. 5.2), including some from drainages just northwest and southeast of old Fort Niobrara, demonstrate the presence of woody riparian vegetation, dominated by cottonwood, along the Niobrara River. The presence of timber, especially pine, was likely important in the decision to establish a military fort in the vicinity (Buecker 1982). Fort Niobrara was established in 1880 to protect settlers, gold seekers, and fur traders from the Rosebud Sioux (Beed 1936); by 1907, the fort was abandoned but used as a supply depot for 5 years. Tolstead (1942) discusses the historical presence, cutting, and regeneration of riparian vegetation in the Valentine area.

Figure 5.2. (Top) Minnechaduza Falls on Minnechaduza Creek near Valentine, Nebraska, in 1887. Note the deciduous woody vegetation in the photograph. The man with the heavy beard is Jules Sandoz. (Bottom) Fort Niobrara from a hill north of the Niobrara River looking south in 1886. Note extent of riparian vegetation on south side of the river and only scattered trees on the warmer, north side. (Photographs courtesy of the John A. Anderson Collection, Nebraska State Historical Society, Lincoln.)

In the late 1800s, several biologists with the Division of Economic Ornithology and Mammalogy (later the Bureau of Biological Survey) worked in the Valentine, Nebraska, area. Their journals (indexed in Cox 1986) contain considerable information on the northern Sandhills. Vernon Bailey worked near Kennedy, just west of the present-day headquarters of Valentine NWR, from June 9–14, 1888. Bailey (unpublished field notes) described the immediate area as "high, very sandy hills, grassy valleys, creeks and a few sloughs and lakes, no timber or brush for miles. . . . The hills are of clear, white sand, (no pebbles) but scantily covered with vegetation. . . . The tops of some hills are being scooped out into great holes and the sand piled up in some other place. . . . The hills seem to me to be formed almost wholly by the wind." On a subsequent trip in April 1890, he elaborated on the nature of the country, noting that "the Gordon Valley has dark fertile soil and produces a fine growth of grass, that of the hills is light sand and the grass is thin and in bunches." He commented that the water in the shallow ponds is strongly alkaline. He apparently found trees and brush only along the Niobrara and its branches; away from the streams, "there is neither tree nor brush."

Bailey then moved north to "Clarks Canon (16 km south of Valentine)," through which Gordon Creek flows. This area was characterized by "very sandy, hilly, deep canons (or ravines), river valleys, and no trees or brush except in vallies." Bailey worked in this canyon, the valley of the Niobrara, and along a "few plowed fields." Along the river and in the canyons, he found a scattering of timber including "yellow pine, red cedar [*Juniperus virginiana*], bur oak, elm, ash, cottonwood, and of bushes, chokecherry [*Prunus virginiana*], raspberry [*Rubus* sp.], willow, plum [*Prunus americana*]" and others. He noted that in the canyons and along the river, pine and bur oak are the "most important" trees. At the time he was there, corn was the principal crop in the area.

In April 1890, Bailey visited the "village" of Valentine and described it as "situated on a level table land forming a wedge shaped point between the Niobrara River and . . . the Minnechaduza Creek . . . the streams flowing through narrow deep valleys with cut and broken sides. In each valley there is considerable brush and timber of . . . oak, elm, cottonwood, boxelder (*Acer negundo*), green ash," redcedar, and yellow pine near the top of the valley banks. Describing the top of the tablelands, he said that they were "sandy, grassy, and treeless."

A.B. Baker of the division traveled from Trego County, Kansas, to Valentine and back between May and July 1889. He commented that south of the Middle Loup the hills are "composed entirely of yellowish-white sand and are covered very sparsely with grass. The valleys between are . . . covered more thickly with grass than are the hills. There is no timber except a very little on the banks of the Dismal River." He stated that north from the Middle Loup the character of the country changed rapidly and that for 10 miles "there are Sandhills with fertile table-lands between: then the valleys become lower and more wet, till, at the North Loup they expand into broad marshes. For twenty miles northward from the North Loup River, the marshes and numerous small lakes occupy about one half of the territory, the remaining being isolated areas of Sandhills."

Early Fauna

As he passed through the Sandhills in 1855, Warren (1856) found only scattered buffalo chips (to build fires with) but noted that pronghorn were abundant, as were mosquitoes and flies. On his 1857 trip, Warren (1875, p. 18) noted that he had been "very much disappointed in the amount of game," and although there was evidence of the country having recently been occupied by large herds of buffalo, "only a few bulls were seen." Although it is unclear if Warren actually saw them, he (1875, p. 28) also referred to clouds of grasshoppers that "occasionally destroy the vegetation." His comments on big game may have been influenced by the wildlife he had seen the previous year (1856) on the Yellowstone River, about which he said (1875, p. 16) they "enjoyed the greatest abundance of large game of all kinds."

Observations on the status of large animals in the Upper Missouri region in 1855-57 were made by Hayden (Warren 1875, pp. 90-91), who noted that many of them were "fast passing away, and in a few years must become extinct." He said buffalo were gathering into a smaller area and were annually decreasing at a rapid rate but that they were still abundant in the Platte River valley. Warren (1875), in a footnote, noted the propensity of buffalo to move annually and seasonally. Hayden believed that all larger animals (buffalo, elk, deer, pronghorn, bighorn, and beaver) were more abundant in the Yellowstone valley than in any other portion of the Upper Missouri. He said that beaver were increasing rapidly now that major trapping had ceased and that many mountain streams literally swarmed with beaver. Regarding the distribution of raccoons, he said that they were seldom seen beyond the edge of the frontier (i.e., the Missouri River in eastern Nebraska). Hayden saw skins of bobcats, wolves ("most abundant where the buffalo range"), coyote ("exceedingly abundant"), and foxes.

Collections of Nebraska mammals were made as early as 1856. Hayden's report (Warren 1875) lists 47 species of recent mammals collected during the Warren expeditions in 1856 and 1857, 39 of which came from in or near the Sandhills (Table 5.2). Baird's (1857) report on mammals of the Pacific railway routes includes records from Warren's 1856 trip up the Missouri (Viola 1987) but none from 1857. Hayden's report mentions specimens of river otter, Franklin's and thirteen-lined ground squirrels, plains pocket gopher, silky pocket mouse, and kangaroo rat from localities visited in 1857 (Niobrara and Loup rivers). The records of the National Museum of Natural History list the following species collected (localities in parentheses) by Warren or Hayden in what is now Nebraska: hoary bat (Loup Fork, 1857), plains pocket mouse (Loup Fork, 1857; perhaps the one identified as a silky pocket mouse by Hayden), beaver (Loup Fork, no date; Platte River, 1857), prairie vole (no locality or date but with a catalog number close to others), coyote (Platte River, 1857), river otter (Platte River, no date), striped skunk (Loup Fork, no date), badger (Platte River, no date), wapiti (no date or locality), and antelope (Platte River, 1857). Most, but not all, of these specimens are listed by Jones (1964).

Table 5.2. Thirty-Nine Species of Mammals Occurring in the Sandhills and Collected by Hayden and Warren During Their Trips[a]

Sorex haydeni (*S. cinereus*)
Blarina brevicauda
Scalops argentatus (*Scalopus aquaticus*)
Lynx rufus
Canis occidentalis (*C. lupus*)
Canis latrans
Vulpes macrourus (*V. vulpes*)
Vulpes velox
Putorius longicauda (*Mustela frenata*)
Lutra canadensis
Mephitis mephitica (*M. mephitis*)
Taxidea americana (*T. taxus*)
Procyon lotor
Ursus horribilis (*U. arctos*)
Sciurus ludovicianus (*S. niger*)
Sciurus carolinensis
Spermophilus franklini (*S. franklinii*)
Spermophilus tridecemlineatus
Cynomys ludovicianus
Castor canadensis
Geomys bursarius
Jaculus hudsonius (*Zapus hudsonius*)
Perognathus flavus
Dipodomys ordii
Mus musculus
Hesperomys sonoriensis (*Peromyscus maniculatus*)
Hesperomys leucogaster (*Onychomys leucogaster*)
Arvicolo haydeni (*Microtus ochrogaster*)
Lepus campestris (*L. townsendii*)
Lepus sylvaticus (*Sylvilagus floridanus*)
Cervus canadensis (*C. elaphus*)
Cervus leucurus (*Odocoileus virginianus*)
Cervus macrotis (*O. hemionus*)
Antilocapra americana
Ovis montana (*O. canadensis*)
Bos americanus (*Bison bison*)
Vespertilio pruinosus (*Lasiurus cinereus*)
Vespertilio noctivagans (*Lasionycteris noctivagans*)
Vespertilio novaboracensis (*Lasiurus borealis*)

[a]They collected an additional eight species from localities outside the sandhills for a total of 47 species. Taxa are as listed in Warren (1875). Currently accepted scientific names are in parentheses; see Table 5.1 for Common Names.

The paleontologist O.C. Marsh made a series of annual trips to the West in search of fossils from 1870 to 1873; on the first of these, one of his assistants was G.B. Grinnell, whose papers have been edited by Reiger (1972). The first expedition went north from Fort McPherson, near the present town of Maxwell, through the Sandhills past Cody Lake, crossed the Dismal River, and on to the middle fork of the Loup. From here they traveled west past the forks of the Middle Loup and then turned south and proceeded past Dry Lake and into Birdwood Creek. Grinnell saw big game everywhere along this trip. One herd of elk on the north bank of the Loup numbered from 200 to 250 animals. By far the most abundant big game was pronghorn, and Grinnell (Reiger 1972, p. 37) characterized them as being "still nearly in their primitive numbers."

In 1873, Marsh made another paleontological expedition through the Sandhills and reached what is now Cherry County, Nebraska. This trip is recounted in the diary of Dr. Maghee (or McGhee) (Lindsey 1929, Smith 1958). The 1873 party left North Platte and traveled up the North Platte River to Birdwood Creek, where they turned north, crossed the head of the Dismal River, and proceeded across the Sandhills to the Niobrara River south of Gordon. They then traveled down the Niobrara; near Valentine they turned south and took a route close to the present route of Highway 83 to North Platte.

Marsh's party saw large numbers of pronghorn, elk, and (mule) deer throughout their journey. They also commented on seeing beaver (or sign), coyote, and "red deer" (white-tailed deer) (Jones 1964). Maghee described the valley of the Niobrara as winding and fertile, with cedar trees of great size, and large pines in the canyons. They apparently spent several days near the confluence of the Minnechaduza and Niobrara, an area described as a magnificent camp with a swift clear stream, wooded at its mouth, with a beautiful grove of timber to the north. Bear hunting was mentioned, but there is no record that they were successful. Fish were caught occasionally, as was a beaver kitten. The party must have traveled near Valentine NWR (Lindsey's map depicts their route as being just east of Marsh Lake). Maghee commented in general on this most "ellegant [sic] scope of country" with a well-watered "lake and pond valley" of "surpassing beauty and fertility" as they passed through the area.

In 1873, General D.S. Stanley led the North Pacific Railroad Expedition to the Dakota and Montana territories and was accompanied by the naturalist J.A. Allen. Although this expedition was north of the Sandhills, his observations shed light on the general nature of the northern Plains at this time. Allen (1874) noted that although this region was little frequented by white men, the relative abundance of the larger mammals had already changed considerably. He believed that bison had disappeared east of the Yellowstone, as far up as the Tongue River, and with it the coyote and wolf. The grizzly bear and beaver were sparsely distributed, the bighorn occurred sparingly in the Badlands, and the pronghorn was the most numerous large mammal and the only herbivore still generally distributed. Allen's narrative suggests that several animals were showing population decreases, including coyote, wolf, elk, mule deer, bighorn sheep, prairie dog, and "striped gopher" (probably the thirteen-lined ground squirrel). The commensal house mouse was already common in a 1-year-old fort.

That the fauna of the Sandhills was changing by the 1870s and 1880s is shown in recollections of early settlers assembled and published by Beel (1986). A cowboy named Rowley saw a buffalo bull south of the Snake River in the early 1880s and claimed that the next spring hunters killed two buffalo 75 miles south. These were the "last wild buffalo he heard of." About 1882 near Bull Lake (west of Valentine NWR), Rowley shot a "spike-horn buck elk, the only one he ever saw there." Other recollections state that in the early 1870s gray wolves were scattered thickly through the Sandhills but became scarce as their prey, deer and pronghorn, were killed by advancing settlers. At that time, wolves apparently turned to livestock and were said to prey on yearling heifers, steers, horses, and mules. In the spring of 1894, ranchers organized a wolf hunt in the area between the North Loup and Snake rivers where wolves were thought to be most abundant; the country around Hackberry Lake (now on Valentine NWR) marked the eastern boundary of this area. After several days on the hunt, more than two dozen wolf "scalps" were secured. County records show that in the spring of 1894 a trapper named Pratt collected bounty on 94 wolves. He was required to swear on the Bible that the wolves all came from Cherry County. In 1908, the *Valentine Reporter* said that wolf numbers were increasing and a hunt was being organized. A wolf kill as late as 1915 in Cherry County is included in the recollections (Beel 1986).

Interactions of settlers and wild animals on the Great Plains between 1865 and 1879 were reviewed by Fleharty (1995). He points out that interactions with all wildlife were important, not just those with buffalo, and that smaller, less obvious animals may have been more important to survival of the average settler. Fleharty provides convincing evidence that for much of the Great Plains the "bison/settler coaction" was completed by 1879 and that other species of wildlife also had been (negatively) affected by that time. Choate (1987) provides data on declines of large wildlife in western Kansas that generally mirror the situation in northern Nebraska (coyote, declines by 1880s; gray wolf, nearly gone by 1895; black bear, gone from northern Kansas by 1875; grizzly bear, gone by 1885; mountain lion, disappearing by 1890; elk, declining markedly by 1870, extirpated by 1890; deer, gone by 1896; pronghorn, less numerous after 1870 and eventually disappeared entirely by 1910; and bison, greatly reduced by 1870s, extirpated by 1900).

W.W. Granger collected mammals during travels in New Mexico, Utah, Wyoming, and Nebraska in 1895 and 1896 and Allen (1896, p. 244) discussed the 132 specimens of 12 species, taken from near Perch and Bassett in Rock County, Nebraska. Allen stated that these localities were in typical Sandhills and that the specimens of voles and shrews were taken at a large marshy lake near Perch post office. This collection of small mammals is one of the earliest from the Sandhills and included kangaroo rat, plains pocket mouse, meadow and prairie voles, grasshopper mouse, western harvest mouse, deer mouse, spotted and thirteen-lined ground squirrels, eastern mole, northern short-tailed shrew, and masked shrew. Allen said that the prairie voles all came from dry sandy ground bordering a cornfield and that grasshopper mice came from old fields; both species were said to be common. Two species were noted as rare, the spotted ground squirrel and northern short-tailed shrew. The masked shrew was said to be much more common.

Apparently, beaver were trapped along the Niobrara for years, but by 1901 they were protected in the state. Other wildlife was becoming scarce, as was shown by a Game and Fish Commission report that indicated that there were only 50 deer present in the entire state in 1901–02 (Beel 1986). By the early 1900s, reports of large numbers of animals killed by hunters were of waterfowl, not large ungulates.

According to Jones (1964), the first published list of Nebraska mammals was that of Samuel Aughey (1880) in "Sketches of the Physical Geography and Geology of Nebraska." Aughey enumerated 82 species in his list. As Jones noted, it is primarily a recapitulation of previous compendia by Baird, Coues, and others.

An early work containing more original data on Nebraska mammals is Swenk's (1908) "A Preliminary Review of the Mammals of Nebraska, with Synopses." Swenk summarized the current knowledge of the status and distribution of 95 species and subspecies of mammals in the state. Swenk (1908, p. 68) discussed what were then ongoing changes in the mammal fauna of Nebraska, including the extirpation of large forms such as elk, deer, bison, pronghorn, bighorn, gray wolf, and mountain lion, and the gradual decline or retreat westward of other species such as badger, beaver, plains (= swift?) fox, otter, and porcupine as human settlement encroached. He noted that some mammals such as the opossum, spotted skunk, gray squirrel, and red fox originally found only in extreme southeastern Nebraska had followed streams into parts of the prairie region.

In summary, declines of large wildlife in the Sandhills likely followed a familiar pattern, with relatively dense populations in the late 1700s and early 1800s, before the major push of westward expansion. By the 1860s, populations of some species were declining in Nebraska while remaining common in less settled areas (e.g., Yellowstone). By 1870, buffalo, never abundant, were extirpated, and major herds were south and west of northern Nebraska. Perhaps only pronghorn were still truly abundant. By the 1880s, large carnivores as well as ungulates were disappearing from the Plains. Increasing settlement and breaking of the prairie to cultivation and grazing continued the decline, often dramatically, up to the turn of the century, when game laws began to be enacted and enforced.

The Contemporary Landscape

Compared with other areas on the Great Plains, the landscape of the northern Sandhills has changed relatively little since settlement by Europeans. Probably the most significant alteration to this area since 1850 has been in vegetative coverage and composition due to fire suppression and all-season grazing of domestic live-stock. In the Niobrara valley, pines, bur oak, and eastern redcedar have expanded up the south canyon walls and into the surrounding prairie. Steuter et al. (1990) suggest that the increasing canopy cover, reduced fuel load, higher fuel moisture, and less frequent fire return will protect the expanding woody vegetation and ultimately result in further expansion of woodland patches. Secondary growths of hardwood including ironwood (*Ostyra virginiana*), linden (*Tilia americana*), and green ash have sprouted from much of the old growth cut by early settlers and are now dominants in the downslope community. Excepting advancement of the pine woodland up the southern canyon walls, much of the flora of the Niobrara valley likely resembles that of 1850.

Of all habitats near the Niobrara River valley, the Sandhills are the most diverse and possibly least understood. The Sandhills have historically supported grasslands, although the sands may be more stabilized and vegetated at present. Hayden (Warren 1875) and Wolcott (1906) describe the hills as bare shifting sands subjected to much wind and only scantily covered by grasses with little to no woody vegetation except yucca (*Yucca* spp.). Smith (1892) stated that grass species composition on the Sandhills did not change appreciably from 1839 to 1892, but Frolik and Shepherd (1940) and Bragg (1978) believe that bluestems (*Andropogon* spp.) have decreased in coverage since 1900, possibly due to fire control (Burzlaff 1962). Anecdotal information gathered from long-term residents also suggests that vegetation was less dense after early settlement (M. Beel, personal communication); however, changes, if any, in species composition are not known. Some residents attribute these changes to dune stabilization efforts and fire suppression.

Generalizations regarding landscape changes are difficult at best, and few recent comprehensive studies have been conducted. Bragg (1978) points out that many investigators have described the vegetative composition of the Sandhills but because of variations in location, season, successional stage, and aspect, each has reported different results. Kantak (1995) recognized seven major community types in the Niobrara River valley: sandhill, mixed prairie, tallgrass meadow, upland pine, juniper, hardwood, and streambank. Kantak's analysis may be the first to deal convincingly with this biogeographically unique region and the influence of glacial and dispersal events as well as the role of geomorphological and hydrologic conditions in producing the Sandhills landscape. At least 581 species of vascular plants are known from the Niobrara valley, one-third of the known total for Nebraska (Kantak 1995). Kaul et al. (1988) discussed the occurrence of many of these species as a result of postglacial changes and noted the steep decline in total number of vascular plant species westward through the Niobrara valley.

At present, Sandhills dunes are stabilized by prairie sandreed (*Calamovilfa longifolia*), sand dropseed (*Sporobolus cryptandrus*), and sand bluestem (*Andropogon hallii*) (Harrison 1980), with forbs, yucca, and cacti, in order of decreasing coverage, constituting the remaining vegetation (Bragg 1978). Species that dominate the dry valleys include cool-season grasses such as wheatgrass (*Agropyron* spp.) and needlegrass (*Stipa* spp.), whereas rushes (*Juncus* spp.) and spikerushes (*Eleocharis* spp.) are most prevalent in the wet meadows (Kaul 1990).

The establishment of introduced woody vegetation in the Sandhills and, to a greater extent, along the shorelines of major lakes suggests contemporary fire regimes are significantly altered from historical ones. In the absence of fire, eastern redcedar has spread from planted windbreaks into the surrounding Sandhills, whereas plantings of cottonwood, Siberian elm (*Ulmus pumila*), hackberry (*Celtis occidentalis*), and black locust (*Robinia pseudoacacia*) are found along the shorelines of major lakes. Native willow still persists on the shoreline of many lakes. Pines also have moved out of the canyons and encroached into the grasslands with fire suppression in recent times (Steinauer and Bragg 1987, Steuter et al. 1990). Some floral changes including continued encroachment of woody vegetation and possible changes in grass species composition may be directly linked to existing management and agricultural practices.

Present-Day Mammals

The first modern study of mammals of north central Nebraska is Beed's (1936) "A Preliminary Study of the Animal Ecology of the Niobrara Game Preserve" (now known as Fort Niobrara NWR). The study, an example of the "new" quantitative science of ecology, was conducted in the summer of 1934 on the Crookston Table, north of the Niobrara River. Beed (1936, p. 4) provides some early information on land use, noting that there was little grazing by domestic livestock on this tract of land between 1880 and 1907, although much of the native timber along the valley was cut for firewood. From 1907 to 1912, the area was heavily grazed by horses and cattle of homesteaders. In 1912, the preserve was established, and from then until 1925, little grazing by cattle and horses was permitted. Bison and wapiti were introduced in 1913, and herds grew to 126 bison and 37 wapiti by 1936. Longhorn cattle were introduced in 1936. Beed noted that his study area was recovering from a period of intense grazing pressure. In addition, older trees showed burn scars of fires occurring at regular intervals; however, between 1912 and 1936 only one fire (in 1927) occurred. Beed's observations and trapping efforts allowed him to record the presence of 36 species of mammals on the preserve. Although he saved some specimens, their existence today is in doubt.

The most recent overall compilation of mammals in the northern Great Plains is that of Jones et al. (1983); Armstrong et al. (1986) provide additional information on habitat affinities and biogeography of mammals of this area. The current status of 55 species of mammals in the Sandhills was summarized by Freeman (1990), who noted an additional three species were associated only with the Niobrara River. Habitat selection and movement patterns in several species of Sandhills rodents in Arthur County, including plains pocket mouse, hispid pocket mouse, kangaroo rat, and prairie vole, were reported by Lemen and Freeman (1986). A 10-year study of mammals in the Bessey Division of the Nebraska National Forest was conducted by Manning and Geluso (1989). This forest was planted during the 1930s and is an example of how pine can survive on the Great Plains when protected from fire. Manning and Geluso examined the effects that this artificial forest has had on relative abundance and species composition of mammals in both native and non-native (forested) habitats and documented the presence of 47 species.

The native mammalian fauna of the northern Sandhills probably comprised 57 species (Jones 1964, Hall 1981, Jones et al. 1983, Bogan and Ramotnik 1995)

(Table 5.1). We have no information on historically dominant small mammals, but in 1991 and 1992, the most abundant mammal at several localities in the Sandhills (on Valentine NWR) was the deer mouse, followed by Ord's kangaroo rat, western harvest mouse, white-footed mouse, and prairie vole (Bogan and Ramotnik 1995). Species richness was greatest in wet meadows and ungrazed Sandhills and lowest in cedar plantings. At Fort Niobrara NWR, the most abundant mammal was the white-footed mouse, followed by deer mouse, prairie vole, Ord's kangaroo rat, and western harvest mouse. Species richness was greatest in riparian and Sandhills habitats and lowest in cedar plantings (Bogan and Ramotnik 1995).

Of the 57 presumed native species, 10 (17.5%, Table 5.1) are now extirpated from north central Nebraska (Jones et al. 1983). These 10 species (gray wolf, black and grizzly bears, black-footed ferret, wolverine, mountain lion, lynx, wapiti, bison, and bighorn sheep) were probably all extirpated before establishment of the Niobrara Game Preserve and certainly gone from the Sandhills by the time Valentine NWR was established (1935). These species were extirpated by a combination of overhunting (black and grizzly bears, bison, wapiti, bighorn sheep), predator and rodent control (gray wolf, black-footed ferret, mountain lion), intolerance of humans and their settlement of the prairies (lynx and wolverine), or perhaps a combination of the three.

Just as the settlement of the prairies resulted in the extirpation of some species, other species have accompanied humans westward, accidentally or otherwise. Bogan and Ramotnik (1995) cite evidence that up to eight species have invaded the Sandhills. These species are Virginia opossum, black-tailed jackrabbit, wood-chuck, eastern fox squirrel, red and gray foxes, raccoon, and eastern spotted skunk. These species may be more tolerant of human activities (opossum, rac-coon), increase as a result of breaking of the prairie and subsequent agricultural practices (black-tailed jackrabbit, woodchuck, foxes, eastern spotted skunk), or benefit by the desire of humans to have familiar species nearby (perhaps the fox squirrel, although it likely has spread naturally as well). Specifically, it appears that black-tailed jackrabbits have spread northward at the expense of white-tailed jackrabbits, although in Cherry County white-tailed jackrabbits appear to be holding their own (Bogan and Ramotnik 1995). Some of these species (opossum, foxes, raccoons, fox squirrel) also have found the prairies more habitable with the spread of woody vegetation and a less frequent fire regime.

Finally, humans have knowingly or unknowingly directly introduced three species to the general area of the refuges. Two of these species are human commensals (house mouse and Norway rat) that have accompanied humans on travels around the globe. One introduced species, known from the town of Valen-tine but not from the refuges yet, is the gray squirrel. Humans also maintain a variety of domestic animals on farms and ranches in the Sandhills, and a herd of about 350 Texas longhorn is kept at Fort Niobrara NWR.

Thus, the original native fauna consisted of 57 species, including 10 carnivores and ungulates that have since been extirpated. The 47 remaining native species have been augmented by up to 11 additional species that we either know or suspect were not members of the native Sandhills mammalian fauna; thus the current total of 58 species includes a sizable proportion (20%) of invading or introduced species. Three additional peripheral species (Merriam's shrew, Franklin's ground squirrel, and silky pocket mouse) may eventually be found in the area, making a total of 61 extant kinds.

Even though changes in plant composition and density may have occurred, the basic floral components of the 1850 landscape are still present. Contemporary assemblages of large mammals in the Sandhills prairie are not representative of what the current landscape can support. Bison, historically a seasonal wanderer in the Sandhills, are now restricted to pastures within Fort Niobrara NWR. Wapiti, once abundant along riparian habitats, are now also limited to the refuge. The historically abundant pronghorn now seems to be scarce in north central Nebraska.

Only deer are likely as numerous and widespread within the Sandhills as they were in 1850. Their recovery is due to more stringent regulation of harvest, increase from persisting populations, lower populations of large predators, and changing landscapes on the prairie. In the northern Sandhills, and specifically on the two NWRs, white-tailed deer slightly outnumber mule deer (Bogan and Ramotnik 1995) and the zone of white-tailed deer dominance (as determined from hunter kill statistics) has shifted west. Between 1962 and 1968, this zone shifted westward from around the 98th to the 100th meridian (Menzel 1984). Increased woody vegetation may favor white-tailed over mule deer; additionally, the two species hybridize, and the nature and extent of such introgression also may be involved in numerical dominance at a given site (e.g., Carr and Hughes 1993).

Other Vertebrates

Fishes

Jennings (1995) summarized historical data for fishes of the northern Sandhills but noted that occurrence and distributional information of fishes from historic sources are incomplete because many early travelers were unfamiliar with western fishes or they failed to record and report catches. Information that was recorded was likely biased because early fishing devices were crude and techniques may have selected for larger, edible species. Additionally, identification of species was difficult; many common names were interchangeable and often colloquial. Hrabik (1990) listed fishes of the Sandhills and identified 61 species from 14 families that were endemic to headwaters, streams, and rivers of this region, although drainage-specific localities were not identified. Names of fishes follow Hrabik (1990).

According to Jennings (1995), the clear, cool waters of the Niobrara and its tributaries near Valentine likely maintained fish assemblages dominated by cyprinids, catostomids, and percids, totaling about 32 species. More specifically, small stream order tributaries may have contained associations that included the central stoneroller (*Campostoma anomalum*), brassy minnow (*Hybognathus hankinsoni*), bigmouth shiner (*Notropis dorsalis*), blacknose shiner (*N. heterolepis*), sand shiner (*N. stramineus*), fathead minnow (*Pimephales promelas*), longnose dace (*Rhinichthys cataractae*), finescale dace (*Phoxinus neogaeus*), northern redbelly dace

(*P. eos*), plains topminnow (*Fundulus sciadicus*), white sucker (*Catastomus commersoni*), Iowa darter (*Etheostoma exile*), and brook stickleback (*Culaea inconstans*). Near the mouth of tributaries and in the river, associations likely included the flathead chub (*Hybopsis gracilis*), sand shiner, red shiner (*N. lutrensis*), bigmouth shiner, river shiner (*N. blennius*), emerald shiner (*N. atherinoides*), suckermouth minnow (*Phenacobius mirabilis*), river carpsucker (*Carpiodes carpio*), white sucker (*Catostomus commersoni*), shorthead redhorse (*Moxostoma macrolepidotum*), green sunfish (*Lepomis cyanellus*), and stonecat (*Noturus flavus*). Other less common species that likely occurred in the river were the plains minnow (*Hybognathus placitus*), silver chub (*Hybopsis storeriana*), creek chub (*Semotilus atromaculatus*), plains killifish (*F. zebrinus*), black bullhead (*Ameiurus melas*), and flathead (*Pylodictus olivaris*) and channel catfish (*I. punctatus*).

Historical fish associations of the Sandhills lakes south of Valentine are more difficult to establish, but grass pickerel (*Esox americanus*) and fathead minnow were likely numerous in open drainage lakes because of their tolerance to alkaline waters. Brook stickleback and Iowa darter were previously reported from streams near Valentine NWR and may historically have entered the headwaters of some low-alkalinity Sandhills lakes during periods of high water. Although available information does not indicate distributions including Sandhills lakes, red shiner, river carpsucker, quillback (*C. cyprinus*), black bullhead, flathead catfish, and green sunfish have affinities for a wide variety of aquatic habitats and therefore may have occasionally found their way into Sandhills lakes (Jennings 1995).

Numerous accidental (Jones 1963) and intentional (Johnson 1942) releases of nonendemic species in larger rivers and lakes of north central Nebraska confound distributional data. The impacts these stockings had on abundance and distribution of native fish assemblages are not clear, but the predatory nature of most introduced fish likely affected endemic fish communities where they co-occurred.

Amphibians and Reptiles

Corn et al. (1995) reviewed historical records of amphibians and reptiles from the Sandhills. In general, these species are poorly documented before the early 1900s. Most amphibians and reptiles are diminutive, secretive, and cryptically colored and were likely overlooked by early explorers. Because amphibians and reptiles provided neither food nor sport, they were rarely mentioned during the early exploration of the West. Indeed, the apparent absence of some species today may simply be the result of short-term studies and weather conditions. In turn, lack of evidence of historical occurrence is not positive proof that a species did not occur historically. Scientific and common names are those used by Corn et al. (1995).

While with Warren's expeditions, Hayden (1863) collected 593 species of plants, 65 molluscs, 291 birds, 47 mammals, and only 33 amphibian and reptile species. Of the 33 species, 9 were collected in or near the Sandhills of Nebraska (Hayden 1863, pp. 177-178): bullsnake (*Pituophis catenifer*), plains hognose snake (*Heterodon nasicus*), garter snake (*Thamnophis radix* and *T. sirtalis*), prairie racerunner (*Cnemidophorus sexlineatus*), many-lined skink (*Eumeces multivirga-*

tus), lesser earless lizard (*Holbrookia maculata*), northern prairie lizard (*Sceloporus undulatus*), and tiger salamander (*Ambystoma tigrinum*). Corn et al. (1995) found the current species richness of amphibians and reptiles (22 species) in north central Nebraska to be similar to historic accounts. They failed to find eight species previously recorded from Fort Niobrara NWR (tiger salamander, Blanding's turtle [*Emydoidea blandingii*], yellow mud turtle [*Kinosternon flavescens*], lesser earless lizard, northern prairie lizard, eastern hognose snake [*H. platyrhinos*], northern water snake [*Nerodia sipedon*], and prairie rattlesnake [*Crotalus viridis*]), and two species known from Valentine NWR (many-lined skink and prairie rattlesnake). Most of these species likely still exist in the area. Snakes, particularly, are secretive, and two (eastern hognose and northern water snake) are near the western edge of their distribution. Many-lined skinks and prairie rattlesnakes have been found since 1983 by staff at Valentine NWR. Corn et al. (1995) added the plains spadefoot (*Spea bombifrons*) to the known fauna of Valentine NWR, and Stuart and Scott (1992) added the western spiny softshell turtle (*Apalone spinifera*) from Minnechaduza Creek.

Establishment of ponds has likely augmented populations of some amphibians and turtles, whereas introductions of sport fish have probably reduced populations of leopard frogs (*Rana pipiens*) and tiger salamanders. Corn et al. (1995) noted that many terrestrial reptiles favor open sandy habitats created by blowouts and that protection of the Sandhills from fire and grazing may have lowered populations of lizards and snakes that prefer some amount of open habitat.

Birds

Sedgwick (1995) has summarized available information on birds in the northern Sandhills; most species that he recorded in north central Nebraska also occurred historically. All grassland species were present, except perhaps the Field Sparrow (*Spizella pusilla*). All forest species also were present, except for Mallard (*Anas platyrhynchos*), Eastern Phoebe (*Sayornis phoebe*), Great Crested Flycatcher (*Myiarchus crinitus*), Brown Thrasher (*Toxostoma rufum*), Cedar Waxwing (*Bombycilla cedrorum*), Ovenbird (*Seiurus aurocapillus*), and Swamp Sparrow (*Melospiza georgiana*). All of the dominant grassland species occurring today were present before 1920, and of the 33 dominant forest species occurring today, only 2 were not noted historically (Ovenbird and Great Crested Flycatcher). Those species for which historical records are lacking are either difficult to identify or inconspicuous (e.g., Alder Flycatcher [*Empidonax alnorum*], Least Flycatcher [*E. minimus*], Willow Flycatcher [*E. traillii*], Ruby-crowned Kinglet [*Regulus calendula*]), uncommon migrants in the Sandhills (e.g., Blackburnian Warbler [*Dendroica fusca*], Mourning Warbler [*Oporonus philadelphia*], Orange-crowned Warbler [*Vermivora celata*], White-throated Sparrow [*Zonotricha albicollis*]), or have only recently expanded their range and probably did not occur there historically (e.g., Northern Cardinal [*Cardinalis cardinalis*]).

Available information suggests that much of the escarpment north of the Niobrara River supported a vegetational landscape similar to that occurring there today. Likewise, the perennial flow of the Niobrara River suggests that a treed riparian corridor was present. These facts, in conjunction with museum and other historical records, imply that most of the species, including the dominants occurring today in treed habitats (e.g., Great Crested Flycatcher, Eastern Kingbird [*Tyrannus tyrannus*], Black-capped Chickadee [*Parus atricapillus*], House Wren [*Troglodytes aedon*], Red-eyed Vireo [*Vireo olivaceous*], Black-and-white Warbler [*Mniotilta varia*], Common Yellowthroat [*Geothlypis trichas*], Ovenbird, Orchard Oriole [*Icterius spurius*]), also were present historically (Sedgwick 1995). In fact, these species may have benefited from expansion of woodlands since European settlement.

Sandhills habitats may have been less vegetated historically than at present and may have supported fewer species and in lower densities than today. Grasshopper Sparrows (*Ammodramus savannarum*), the dominant species in Sandhills grasslands, were probably always present but perhaps in reduced densities. The occurrence of more frequent fires before European settlement (Steinauer and Bragg 1987) also suggests that there was far less tree and shrub development around Sandhills lakes than at present; early photographs (Fig. 5.3) support this premise. Wooded areas near the Sandhills lakes are typically plantings, often for windbreaks, and fire suppression and exclusion from grazing have allowed woodlands to become a permanent part of today's landscape. Again, that portion of the avifauna dependent on trees today has benefited from this change. Eastern Kingbirds, Black-capped Chickadees, House Wrens, Bell's Vireos (*Vireo bellii*), Warbling Vireos (*V. gilvus*), Yellow Warblers (*Dendroica petechia*), Northern Orioles (*Icterus galbula*), Orchard Orioles, and American Goldfinches (*Carduelis tristis*) all either did not occur as breeders in the Sandhills or occurred at much lower densities. The wetland avifauna, including the avifauna associated with marshes (Common Yellowthroat, Swamp Sparrow, Red-winged [*Agelaius phoeniceus*] and Yellow-headed [*Xanthocephalus xanthocephalus*] blackbirds), was probably much the same historically as it is today (Sedgwick 1995).

Some species with historical records are known to occur presently in the northern Sandhills but were not found by Sedgwick. These species included winter residents or migrants occurring outside the time frame of his surveys (e.g., Trumpeter Swan [*Cygnus buccinator*], Rough-legged Hawk [*Buteo lagopus*], Bald Eagle [*Haliaeetus leucocephalus*], Sandhill Crane [*Grus canadensis*], and Townsend's Solitaire [*Myadestes townsendi*]). Other species recorded historically and occurring commonly today were not recorded due to survey technique, plot locations, or survey emphasis (i.e., Neotropical migrants), including American Wigeon, American Avocet, Bufflehead (*Bucephala albeola*), Burrowing Owl (*Athene cunicularia*), Canvasback (*Aythya americana*), Black-billed Magpie, Green-winged Teal, Ring-necked Duck, and Marbled Godwit.

Notably absent in historic Sandhills landscapes were European Starling (*Sturnus vulgaris*, not introduced to North America until 1890 and still uncommon today on the refuges) and Ring-necked Pheasant (*Phasianus colchicus*, not introduced to Nebraska until 1909). The Wild Turkey (*Meleagris gallopavo*) occurred but was extirpated, reintroduced in the 1920s and 1930s unsuccessfully, then successfully reintroduced along the Niobrara River in 1959-1960. The House

Figure 5.3. Two views of shoreline of Hackberry Lake, now on Valentine National Wildlife Refuge, about 1911. Note absence of woody vegetation. (Photographs courtesy of the Frank Shoemaker Collection, the University Libraries, University of Nebraska, Lincoln.)

Sparrow (*Passer domesticus*), introduced to North America about 1850, was recorded as early as about 1899 near Valentine.

Thirty-seven species were recorded pre-1920 but were not found by Sedgwick (1995). More intensive work in appropriate habitats at selected times would likely reveal the presence of these species. The apparent absence of these species reflects, in part, the disadvantages of short-term studies. However, seven of these species (Ferruginous Hawk [*Buteo regalis*], Marbled Godwit [*Lemosa fedoa*], Burrowing Owl [*Athene cunicularia*], Dickcissel [*Spiza americana*], McCown's Longspur [*Calcarius mccownii*], Chestnut-collared Longspur [*C. ornatus*], Savannah Sparrow [*Passerculus sandwichensis*]) are endemic to the Great Plains. As a group, endemic grassland bird species have declined more than other groups of birds, including Neotropical migrants. Of 12 primary and 25 secondary grassland bird species, breeding ranges have shifted for 9 species, breeding habitats are disappearing for 6 species, and populations are known to be declining for 3 species (Knopf 1994). Current efforts that focus on groups such as Neotropical migrants, to the exclusion of species endemic to unique habitats, may be misguided. Continued expansion of woodlands in the northern Sandhills may result in increased species richness, including Neotropical migrants. But such gains may come at the cost of a diminished pool of grassland endemic species.

Conclusions

The vastness of the northern Sandhills, in comparison with the relatively small human settlements, at first suggests that here, perhaps, humans have not had a major impact on the area. Much of the landscape seems little-changed and, in places, quite wild. And to a large extent, this is true. The Niobrara River runs free (and is popular with present-day river runners), and in the Sandhills themselves, the wind still "whirls the sand in the air and excavates deep holes." The studies in Bogan (1995) suggest that the great majority of vertebrate species are still present and that populations of most species are healthy. Many of the species that were not documented (Bogan 1995) are known to be there and were missed in that short-term study. It is often difficult to prove that a species does not occur in an area and somewhat easier to demonstrate that one does.

As we have seen, the reality is that some species (at least of vertebrates and about 17.5% of the original mammal species) have been lost from the area, especially the larger carnivores and some ungulates. These losses stem from activities related to settlement in the late 1800s and predate establishment of preserves and refuges. At present, about 20% of the mammal species in the northern Sandhills can be categorized as either invading or introduced species. The presence of these species is due, directly or indirectly, to human agency. For birds, it appears that woodland-dwelling species are perhaps more common than they were historically. The picture for endemic grassland species of birds in the northern Sandhills is not completely clear, but current summaries (e.g., Knopf 1994) provide evidence that they are declining.

Ongoing vegetational changes in the northern Sandhills are somewhat more insidious, as they, like the long-ago disappearance of bison, are perhaps less obvious to the casual observer. Decreased fire frequency, perhaps partly due to overgrazing at some sites, allowed trees and shrubs to become established, often when they escaped from plantings. Fire suppression allows this process to continue so that woodlands are spreading out from river valleys and, elsewhere, slowly encircling Sandhills lakes and ponds. The trend of increasing woody vegetation, although perhaps good for Neotropical migrants favoring such vegetation and generalized mammals inhabitating eastern forests, does not bode well for native Sandhill species. Attempts should be initiated to adopt a more flexible and liberal policy on lightning-caused fires, and thought should be given to prescription burning in some areas to reduce encroachment of woody vegetation, as well as to reduce fuel loads that may create larger fires than occurred historically.

Much has been written recently of grazing. In the Sandhills, much of the range appears to be in good shape (although surveys and rigorous experimental studies are needed here as elsewhere). Grazing can and should be used as a tool to encourage growth of either cool- or warm-season grasses on specific sites as desired and to open up areas and reduce litter accumulation as part of a well-designed program that is integrated with the use of fire for similar purposes. For those interested in a comprehensive review of grazing, Fleischner (1994) summarizes considerable information on grazing in the western United States. Not all data or conclusions in Fleischner are applicable to the northern Sandhills. Readers also should examine comments by Noss (1994), Brussard et al. (1994), and Brown and McDonald (1995), who present additional thoughts and opinions on the role of grazing in the West.

It is probably axiomatic that a researcher would recommend more research, and this report is no different. For many species in the Sandhills, there is still much we do not know about their basic natural history and their response to changing environments. Long-term survey and monitoring efforts are needed to provide current information on status and on population trends. Continued spread of woodland-adapted species and potential loss of habitat for grassland dwellers is a concern; there is a need to focus on species characteristic of the prairies in general and the Sandhills in particular. Without such data, it will be impossible to take management actions that will help retain the fundamental character of this unique region of the Great Plains.

References

Allen, J.A. 1874. Notes on the natural history of portions of Dakota and Montana territories, being the substance of a report to the Secretary of War on the collections made by the North Pacific Railroad Expedition of 1873, Gen. D.S. Stanley, Commander. Proc. Boston Soc. Nat. Hist. 17:33–86.

Allen, J.A. 1896. List of mammals collected by Mr. Walter W. Granger, in New Mexico, Utah, Wyoming and Nebraska, 1895–96, with field notes by the collector. Bull. Am. Mus. Nat. Hist. 8:241–258.

Armstrong, D.M., J.R. Choate, and J.K. Jones, Jr. 1986. Distributional patterns of mammals in the Plains states. Occas. Pap., Mus., Texas Tech Univ. 105:1–27.

Aughey, S. 1880. Sketches of the physical geography and geology of Nebraska. Daily Republican Book and Job Office, Omaha, NE.

Baird, S.F. 1857. Mammals: general report upon the zoology of the several Pacific railroad routes. Reports of explorations and surveys to ascertain the most practicable and economical route for a railroad from the Mississippi River to the Pacific Ocean. Senate Executive Document 78, vol. 8, pt. 1. Washington, DC.

Beed, W.E. 1936. A preliminary study of the animal ecology of the Niobrara Game Preserve. Bull. Conserv. Surv. Div., Univ. Nebraska, Lincoln 10:1-33.

Beel, M.B. 1986. A sandhill century. Book 1. The land. A history of Cherry County, Nebraska. Cherry County Centennial Comm., Valentine, NE.

Bleed, A., and C. Flowerday, eds. 1990. An atlas of the Sandhills. Resource Atlas 5a. Conserv. Surv. Div., Inst. Agric. Nat. Resour., Univ. Nebraska, Lincoln.

Bogan, M.A., ed. 1995. A biological survey of Fort Niobrara and Valentine National Wildlife Refuges. Unpubl. final rep. U. S. Fish and Wildl. Serv., Denver, CO.

Bogan, M.A., and C.A. Ramotnik. 1995. The mammals. Pp. 140-186 in M.A. Bogan, ed. A biological survey of Fort Niobrara and Valentine National Wildlife Refuges. Unpubl. final rep. U. S. Fish and Wildl. Serv., Denver, CO.

Bragg, T.B. 1978. Effects of burning, cattle grazing, and topography on vegetation of the choppy sand range site in the Nebraska Sandhills prairie. Proc. First Int. Rangeland Cong. 1:248-253.

Brown, J.H., and W. McDonald. 1995. Livestock grazing and conservation on southwestern rangelands. Conserv. Biol. 9:1644-1647.

Brussard, P.F., D.D. Murphy, and C.R. Tracy. 1994. Cattle and conservation biology—another view. Conserv. Biol. 8:919-921.

Buecker, T.R. 1982. Fort Niobrara, Nebraska. Nebraska State Hist. Soc., Lincoln, NE.

Burzlaff, D.F. 1962. A soil and vegetation inventory and analysis of three Nebraska Sandhills range sites. Univ. Nebraska Agric. Exp. Stat. Res. Bull. 206. Lincoln.

Carr, S.M., and G.A. Hughes. 1993. Direction of introgressive hybridization between species of North American deer (Odocoileus) as inferred from mitochondrial-cytochrome-b sequences. J. Mammal. 74:331-342.

Choate, J.R. 1987. Post-settlement history of mammals in western Kansas. Southwestern Nat. 32:157-168.

Corn, P.S., M.L. Jennings, and R.B. Bury. 1995. Amphibians and reptiles. Pp. 32-59 in M.A. Bogan, ed. A biological survey of Fort Niobrara and Valentine National Wildlife Refuges. Unpubl. final rep. U. S. Fish and Wildl. Serv., Denver, CO.

Cox, W.E. 1986. Guide to the field reports of the United States Fish and Wildlife Service circa 1860-1961. Archives and special collections 4. Smithsonian Institution, Washington, DC.

Diller, A. 1955. James MacKay's journey in Nebraska in 1796. Nebraska Hist. 36(2):123-128.

Fleharty, E.D. 1995. Wild animals and settlers on the Great Plains. Univ. Oklahoma Press, Norman.

Fleischner, T.L. 1994. Ecological costs of livestock grazing in western North America. Conserv. Biol. 8:629-644.

Freeman, P. 1990. Mammals. Pp. 181-188 in A. Bleed and C. Flowerday, eds. An atlas of the Sandhills. Resour. Atlas 5a. Conserv. and Surv. Div., Inst. Agric. Nat. Resour., Univ. Nebraska, Lincoln.

Frolik, A.L., and W.O. Shepherd. 1940. Vegetative composition and grazing capacity of a typical area of Nebraska Sandhill range land. Univ. Nebraska Agric. Exp. Stat. Res. Bull. 117. Lincoln.

Hall, E.R. 1981. The mammals of North America. Second ed. 2 vols. John Wiley & Sons, New York.

Harrison, A.T. 1980. The Niobrara Valley Preserve: its biogeographic importance and description of its biotic communities. An unpublished working report to The Nature Conservancy, Minneapolis, MN.

Hayden, F.V. 1863. On the geology and natural history of the upper Missouri. Trans. Am. Philos. Soc. 12:1-218.

Higgins, K.F. 1986. Interpretation and compendium of historical fire accounts in the northern Great Plains. Resour. Publ. 161. U. S. Fish and Wildl. Serv., Washington, DC.

Hrabik, R.A. 1990. Fishes. Pp. 143-154 in A. Bleed and C. Flowerday, eds. An Atlas of the Sandhills. Resour. Atlas 5a. Conserv. Surv. Div., Inst. Agric. Nat. Resour., Univ. Nebraska, Lincoln.

Jennings, M.L. 1995. Historical distribution of fishes in northcentral Nebraska. Pp. 25-31 in M.A. Bogan, ed. A biological survey of Fort Niobrara and Valentine National Wildlife Refuges. Unpubl. final rep. U. S. Fish and Wildl. Serv., Denver, CO.

Johnson, R.E. 1942. The distribution of Nebraska fishes. Unpubl. PhD dissertation. Univ. Michigan, Ann Arbor.

Jones, D.J. 1963. A History of Nebraska's Fishery Resources. Proj. F-4-R. Nebraska Game, Forestation, Parks Comm.

Jones, J.K., Jr. 1964. Distribution and taxonomy of mammals of Nebraska. Univ. Kansas Publ. 16. Mus. Nat. Hist..

Jones, J.K., Jr., D.M. Armstrong, R.S. Hoffman, and C. Jones. 1983. Mammals of the northern Great Plains. Univ. Nebraska Press, Lincoln.

Jones, J.K., Jr., R.S. Hoffman, D.W. Rice, C. Jones, and M.D. Engstrom. 1992. Revised checklist of North American mammals north of Mexico, 1991. Occas. Pap., Mus., Texas Tech Univ. 146: 1-23.

Kantak, G.E. 1995. Terrestrial plant communities of the middle Niobrara valley, Nebraska. Southwestern Nat. 40:129-138.

Kaul, R. 1990. Plants. Pp. 127-142 in A. Bleed and C. Flowerday, eds. An atlas of the Sandhills. Resour. Atlas 5a. Conserv. and Survey Div., Inst. Agric. Nat. Resour., Univ. Nebraska, Lincoln.

Kaul, R.B., G.E. Kantak, and S.P. Churchill. 1988. The Niobrara River Valley, a postglacial migration corridor and refugium of forest plants and animals in the grasslands of central North America. Bot. Rev. 54:44-81.

Knopf, F.L. 1994. Avian assemblages on altered grasslands. Studies Avian Biol. 15:247-257.

Lemen, C.A., and P.W. Freeman. 1986. Habitat selection and movement patterns in Sandhills rodents. Prairie Nat. 18:129-141.

Lindsey, C.L. 1929. The diary of Dr. Thomas G. Maghee. Nebraska Hist. Mag. 12(3):249-304.

Manning, R.W., and K.N. Geluso. 1989. Habitat utilization of mammals in a man-made forest in the sandhill region of Nebraska. Occas. Pap., Mus., Texas Tech Univ. 131:1-34.

Menzel, K.E. 1984. Central and southern plains. Pp. 449-456 in L.K. Halls, ed. White-tailed deer ecology and management. Wildlife Management Inst., Stackpole Books, Harrisburg, PA.

Miller, S.M. 1990. Land development and use. Pp. 207-226 in A. Bleed and C. Flowerday, eds. An Atlas of the Sandhills. Resour. Atlas 5a. Conserv. and Surv. Div., Inst. Agric. Nat. Resour., Univ. Nebraska, Lincoln.

Noss, R.F. 1994. Cows and conservation biology. Conserv. Biol. 8:613-616.

Pound, R., and R.E. Clements. 1900. The phytogeography of Nebraska. Univ. Nebraska, Bot. Surv. I, General Surv. Jacob North, Lincoln, NE.

Reiger, J.F. 1972. The passing of the Great West—selected papers of George Bird Grinnell. Winchester Press, New York.

Sedgwick, J.A. 1995. Occurrence, diversity, and habitat relationships of birds. Pp. 60-139 in M.A. Bogan, ed. A biological survey of Fort Niobrara and Valentine National Wildlife Refuges. Unpubl. final rep. U. S. Fish and Wildl. Serv., Denver, CO.

Smith, C.E. 1958. Natural history of Thomas County Nebraska. Privately published by author.

Smith, J.G. 1892. The grasses of the sand hills of northern Nebraska. Nebraska State Board Agric., Rep. Botanist:1892:280-291.

Smith, J.G., and R. Pound. 1893. Flora of the sand hill region of Sheridan and Cherry counties, and a list of plants collected on a journey through the sand hills in July and August, 1892. Univ. Nebraska, Bot. Surv. II:5-30.

Steinauer, E.M., and T.B. Bragg. 1987. Ponderosa pine (*Pinus ponderosa*) invasion of Nebraska Sandhills prairie. Am. Midl. Nat. 118:358-365.

Steuter, A.A., C.E. Grygiel, and M.E. Biondini. 1990. A synthesis approach to research and management planning: the conceptual development and implementation. Nat. Areas J. 10:61-68.

Stuart, J.N., and N.J. Scott. 1992. Geographic distribution. *Apalone spinifera hartwegi.* Herp. Rev. 23:87.

Swenk, M.H. 1908. A preliminary review of the mammals of Nebraska, with synopses. Proc. Nebraska Acad. Sci. 8:61-144.

Tolstead, W.L. 1942. Vegetation of the northern part of Cherry County, Nebraska. Ecol. Monogr. 12:255-292.

Viola, H.J. 1987. Exploring the West. Smithsonian Books, Smithsonian Institution, Washington, DC.

Warren, G.K. 1856. Explorations in the Dacota country in the year 1855. Senate Exec. Doc. 76., 34th U.S. Congress, 1st Session, Washington, DC.

Warren, G.K. 1875. Preliminary report of explorations in Nebraska and Dakota in the years 1855-'56-'57. Engineer Dept., U.S. Army, Gov. Printing Office, Washington, DC. (Reprint: Schubert, F. N. 1981. Explorer on the northern plains: Lieutenant Gouverneur K. Warren's preliminary report of explorations in Nebraska and Dakota, in the years 1855-'56-'57. Engineer Hist. Ser. 2. Gov. Printing Office, Washington, DC.)

Westover, D.E. No date. Prairie fires and the Nebraska pioneer. EC 78-1744. Coop. Ext. Serv., Inst. Agric. Nat. Resour., Univ. Nebraska, Lincoln.

Wolcott, R.H. 1906. Biological conditions in Nebraska. Nebraska Acad. Sci. 2:23-34.

6. Ecology of Fishes Indigenous to the Central and Southwestern Great Plains

Kurt D. Fausch and Kevin R. Bestgen

Introduction

The Great Plains of western North America are a harsh environment for fishes. Due to the increasing aridity and general lack of permanent water one finds traveling westward across the Plains, even trained biologists rarely consider fishes or their habitats when traversing the region. The same was true of early explorers. For example, none of the four scientific expeditions that ascended the Platte River in the first half of the 1800s described Plains fishes, whereas flora and other fauna were collected and described extensively (e.g., Frémont expedition, 1842–1843; Jackson and Spence 1970). Indeed, no collections of fish from the Plains region of the South Platte River Basin in Colorado were preserved until 1873 (Cope and Yarrow 1875).

It is therefore not surprising that relatively little is known about indigenous fishes of the Great Plains, especially considering the extensive modification of aquatic environments that began about 1860 (Eschner et al. 1983). Distributions of some species were likely altered by transfers among waters, whereas others were probably locally extirpated before any specimens were curated (cf. Jordan 1891). Moreover, little is known about the basic life history of many fishes that range widely throughout the Plains, and even less is understood about the ecology of other biota in Plains streams and riparian zones (Matthews 1988, Zale et al. 1989, Brown and Matthews 1995). This information is urgently needed, due to the continued acceleration of human degradation and loss of species' populations from these aquatic ecosystems (Cross and Moss 1987).

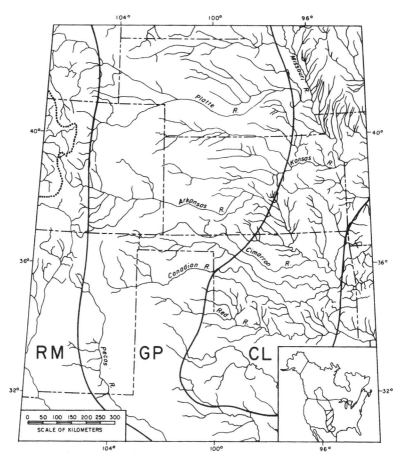

Figure 6.1. Major river basins of the central and southern Great Plains, a region shown by the hatched area on the inset. Solid lines show boundaries of three physiographic regions, Rocky Mountains (RM), Great Plains (GP), and Central Lowlands (CL). (After Cross et al. 1986.)

Our goal in this chapter is to draw together information about the distribution and ecology of fishes indigenous to the central and southern Great Plains, a physiographic region encompassed largely by upstream reaches of the Platte, Kansas, Arkansas, Canadian, and Pecos rivers (Cross et al. 1986) (Fig. 6.1). We emphasize the Platte and Arkansas rivers in eastern Colorado and the Pecos River in New Mexico, because we have firsthand knowledge and experience in these basins. However, we incorporate literature from other Plains rivers and from similar rivers in the adjacent Central Lowlands to the east, where appropriate.

Because one can interpret little about the ecology of organisms without understanding the physical template to which they are adapted (cf. Southwood 1988), we begin by describing the physical characteristics of Plains streams. We then discuss what is known of the original fish fauna and its characteristics, reproductive ecology of Plains fishes, and factors influencing variability of populations and assemblages. We end by attempting a synthesis of ecological mechanisms with broad implications for conservation of plains fishes of the region and by summarizing needs for further research.

Physical Habitat of Plains Streams

Most rivers of the central and southern Great Plains are unique in that their headwaters drain the eastern slope of the Rocky Mountains (Fig. 6.1), whereas downstream reaches traverse plains of Quaternary sediments underlain by the Ogallala aquifer (Eschner et al. 1983), a vast bed of sediments eroded from the mountains during the Miocene (Cross et al. 1986). Thus, most flow in the mainstems of rivers during early summer was originally derived from snowmelt runoff, which also percolated into the highly transmissive alluvium and underlying aquifer. Effluents from this aquifer supplied fall and winter baseflow (Eschner et al. 1983, Cross and Moss 1987). In contrast, high flows in wholly Plains tributaries of these rivers are primarily from spring rains or summer thunderstorms, which may produce flash floods in basins with relatively impermeable soils that produce high runoff (Fausch and Bramblett 1991). Other tributaries drain highly permeable sandy soils that yield little runoff and little flooding, such as in south central Kansas (Cross and Collins 1995). The Kansas River is unique because its headwaters are fed by Ogallala groundwater rather than melting snow and originally had cool, clear, stable baseflow (Cross and Moss 1987, Sanders et al. 1993). Thus, diversity in geology and geomorphology, especially among Plains tributaries, produced different flow regimes in different streams and different responses among systems to human alteration of habitat.

Another important set of lotic habitats for fishes are transition zone stream reaches, intermediate between mountains and plains (Ellis 1914, Propst 1982). These reaches are characterized by cooler water, cobble-gravel substrate, and single-thread channels shaded by riparian trees, which likely contributed large woody debris to the channels. They also harbor unique fish assemblages, due to these unusual characteristics (see below, and Rahel and Hubert 1991). Most early settlements, and present-day cities, in this transition zone are located where rivers emerge from the mountains. Natural lakes are nearly absent throughout the Plains and transition zone, although both main channel and off-channel reservoirs now abound.

Journals from early scientific expeditions, such as John C. Frémont's two trips up the Platte River system in 1842 and 1843 (Jackson and Spence 1970), are assumed to be accurate representations of undisturbed Plains ecosystems (Eschner et al. 1983), although the cumulative effects of indigenous peoples and early beaver trappers are unknown. Accounts of these explorers and drawings and photographs before 1910 indicate that during the middle and late 1800s, mainstem Plains rivers were wide and shallow with braided sand beds and little riparian vegetation (Gregg 1952, Eschner et al. 1983, Cross et al. 1985). Timber was scarce along the upper Platte and Arkansas rivers (Mead 1896, Nadler and Schumm 1981) except on larger islands and in scattered groves, often located in hollows and ravines formed by tributary mouths. However, Scott et al. (1996, in press) suggested that abundance of cottonwoods (*Populus deltoides*), the dominant riparian tree, may cycle on ca. 50-150-year intervals because successful reproduction is primarily limited to elevated floodplain deposits from large floods. In contrast to the large rivers, transition zone streams and Plains tributaries with more stable baseflows and lower flood peaks likely had fringing riparian forest that contributed large woody debris to channels (Brown and Matthews 1995; F.B. Cross, personal communication).

Main river flow regimes were dominated by snowmelt runoff in early summer, after which flow declined to low levels, leaving some channels intermittent for long reaches during late summer (Jordan 1891, Mead 1896, Eschner et al. 1983, Cross et al. 1985). As a result, physicochemical variables such as water temperature, dissolved oxygen, turbidity, and salinity fluctuated drastically among seasons and often reached extreme values in late summer (Matthews and Zimmerman 1990). Runoff magnitude also varied year to year (cf. reports for Frémont's 1842 and 1843 expeditions, Jackson and Spence 1970), but there were few reports of overbank flooding during 1800-1870, probably because river banks were either high or nearly nonexistent and channels were wide. Some Plains tributaries were dry when explorers crossed them in early summer, whereas others had permanent flow and riparian vegetation. Various springs, fed by artesian or unconstrained aquifers, provided permanent pools and clear water in some locations (Matthews et al. 1985, Matthews 1988, Hubbs 1994). Channel beds of mainstem rivers were largely shifting sand, which scoured and filled rapidly forming innumerable bars and islands, and the water was turbid during summer due to high suspended load (Cross and Moss 1987, Bramblett and Fausch 1991a). However, Frémont (Jackson and Spence 1970) reported, while ascending the South Platte in July 1842, that the river became clear and the bed changed to coarse gravel about 100 km east of the mountains and that the transition zones of mountain tributaries were clear and cold with cobble and gravel beds.

Water development in the Platte and Arkansas basins of Colorado's eastern Plains began in earnest after gold was discovered in the mountains west of Denver in 1858. Development in the South Platte Basin proceeded in four stages (Eschner et al. 1983), starting from small ditches (1840-1860) followed by larger canals used to irrigate terraces above the valley floor (1860-1885). Because even perennial rivers ran dry by late summer due to overappropriation of flow, reservoirs were then built to store snowmelt runoff (1885-1930). After these supplies were again overexploited, groundwater was pumped from the Ogallala aquifer, and water was diverted from basins on the western slope of the Rocky Mountains across the Continental Divide (1930-present).

Changes in flow regimes caused by this water development had drastic effects on river channels. The North, South, and upper mainstem Platte river channels narrowed to about 15% (range, 5-40%) of their 1860 width and became more sinuous, due to encroachment of riparian vegetation (Eschner et al. 1983, Kircher and Karlinger 1983). Nadler and Schumm (1981) reported similar changes for two reaches of the Arkansas River in southeastern Colorado. Reduced snowmelt runoff peaks due to reservoir storage and increased baseflows caused by irrigation allowed seedlings of woody vegetation to stabilize the shifting sand bars. These vegetated bars trapped sediment and eventually aggraded and attached to the floodplain, changing the straight, wide, braided channels to single narrow sinuous ones. Channels narrowed most during the 1930s, apparently because the prolonged drought during that decade provided favorable conditions for establishing vegetation in the channel.

These changes in channel morphology were similar throughout the Platte River Basin, despite different changes in hydrology. For example, four large reservoirs on the mainstem North Platte River in Wyoming each decreased snowmelt runoff flood flows, which were further altered by construction of Lake McConaughy in Nebraska starting in 1935 (Eschner et al. 1983, Kircher and Karlinger 1983). In contrast, there were no significant changes through 1980 in peak flows of the South Platte River from Julesburg (near Colorado's eastern border) upstream, which lacks such mainstem reservoirs. As a result, floods in the Platte River downstream from the confluence of the two main tributaries are largely produced by South Platte snowmelt runoff, whereas Platte River baseflows are influenced by both tributaries (Kircher and Karlinger 1983). Summer baseflows have increased in both streams due to raised water tables and return flows from irrigation. Water for irrigation is supplied primarily by reservoir storage in the North Platte and groundwater pumping and transbasin diversion in the South Platte. However, summer baseflows also fluctuate markedly compared with historic conditions, due to daily variation in irrigation needs.

Changes in flow regime were even more drastic in the Arkansas and Kansas river basins in western Kansas (Cross et al. 1985, Cross and Moss 1987, Sanders et al. 1993). The closing of John Martin Reservoir on the mainstem Arkansas River in Colorado in 1942 initially moderated discharge by reducing peak flows and increasing baseflows. However, subsequent groundwater pumping in Colorado and western Kansas completely eliminated flow in 160 km of the Arkansas, except for discharge from municipal wastewater treatment plants. Similarly, in the western reaches of the Kansas River Basin, groundwater pumping and terraces and retention ponds that increase infiltration and evapotranspiration have reduced flows or dried up streams altogether.

In transition zone streams, water clarity and channel substrates have also changed drastically, based on early reports, although riparian vegetation may have changed little. For example, on July 12, 1842, Frémont crossed the Cache la Poudre River, a major tributary of the South Platte River, about 16 km above its mouth near what is now Greeley, Colorado. His description (Jackson and Spence 1970) reads, "This is a very beautiful mountain stream, about one hundred feet wide, flowing with a full swift current over a rocky bed. We halted under the shade of some cottonwoods, with which the stream is wooded scatteringly." Although cottonwoods are still sparsely distributed along the riparian, cobble-gravel substrates in this reach are now almost entirely overlain by silt and organic muck, due to runoff from crop lands and enrichment from fertilizer and domestic wastewater. The water is highly turbid, even during snowmelt runoff. Substrates in mainstem rivers with modified flows, such as the Arkansas, have also changed from loose quicksands to homogenous mixtures of sand and fine clay, which are firmer, probably because floods no longer deposit fine substrates on the floodplain nor sort coarser substrates by size and texture (F.B. Cross, personal communication).

In summary, the physical template of habitat for Plains stream fishes has changed markedly from the original condition due to human modification, but the response varied among streams due to differences in geomorphology and human development (Eschner et al. 1983). Some mainstem rivers that were once wide, braided, and shallow with little riparian vegetation are now narrow and sinuous, have moderately deep pools, and are bordered by trees. Large mainstem reservoirs damp flood maxima from snowmelt runoff in some rivers, but not others that lack such dams. Reservoir releases, transbasin diversions, and seepage from irrigation increase summer baseflow in some reaches, but others are dewatered by diversion, groundwater withdrawal, and retention ponds. Channel substrates have changed due to human development in transition zone streams and flow modification in mainstem rivers. The increase in cottonwood riparian forests along main rivers undoubtedly contributes more woody debris to these channels, but humans remove most debris from plains streams of all sizes (Brown and Matthews 1995; K.D. Fausch, personal observation). Where it occurs, such debris is likely to have important effects on fish and aquatic invertebrate assemblages, as reported for other sand-bed streams (Benke et al. 1985).

Historical Distribution of Fishes in Great Plains Streams

Understanding distribution and abundance patterns of biota, and the historic and current mechanisms that produce them, is central to much of ecology and conservation biology. Here we review historical records of fish collections for transition zone and Plains streams of eastern Colorado to show the challenges encountered when attempting to reconstruct native fish assemblages.

Early Collections in Colorado

The history of collections in the South Platte and Arkansas river basins in Colorado exemplifies the poorly known distribution and status of Plains stream fishes (Fig. 6.2). The first collections, from the Wheeler Survey (1872–1874) (Cope and Yarrow 1875), provided material for the original description of *Haplochilus floripinnis,* a species later synonymized with *Fundulus sciadicus* (see Table 6.1 for common names of native fishes). However, fishes were collected at only 12 sites before 1900, and most were in the transition zone near current cities (Fig. 6.2A).

In the most comprehensive early work, Ellis (1914) summarized distribution of Colorado fishes compiled from his own collections and those of others (Cope and

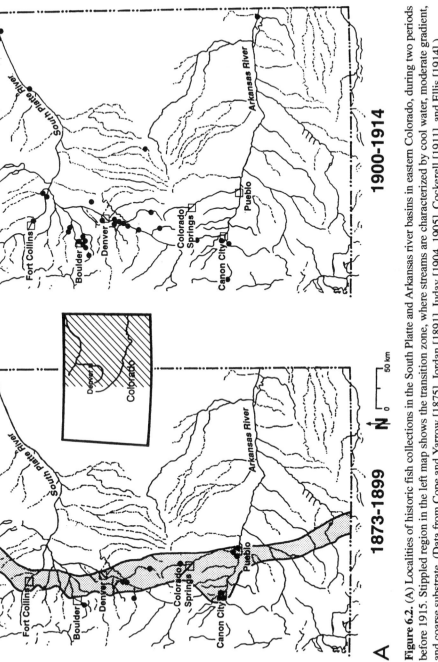

Figure 6.2. (A) Localities of historic fish collections in the South Platte and Arkansas river basins in eastern Colorado, during two periods before 1915. Stippled region in the left map shows the transition zone, where streams are characterized by cool water, moderate gradient, and coarse substrate. (Data from Cope and Yarrow [1875], Jordan [1891], Juday [1904, 1905], Cockerell [1911], and Ellis [1914].)

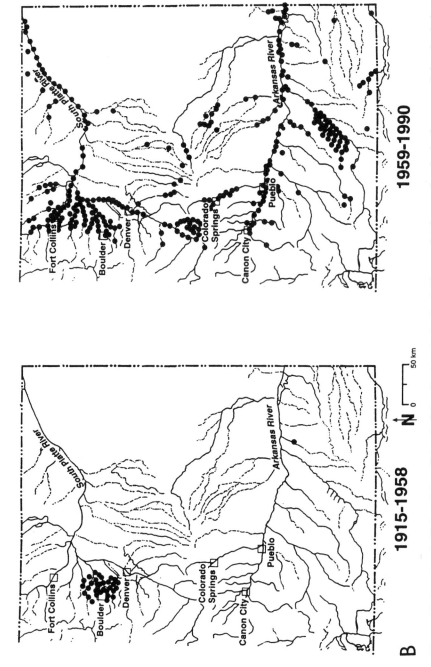

Figure 6.2. (*continued*) (B) Localities of historic fish collections in the South Platte and Arkansas river basins in eastern Colorado, during two periods after 1914. (Data from Hendricks [1950], unpublished collections records from F. Cross and A. Metcalf [University of Kansas Museum of Natural History, Lawrence], Li [1968], Propst [1982], and Loeffler et al. [1982].)

Table 6.1. Native Fishes of the South Platte and Arkansas River Basins, Colorado[a]

Scientific Name	Common Name	Occurrence and Date of First Collection	
		South Platte	Arkansas
Salmonidae			
Oncorhynchus clarki stomias	Greenback cutthroat trout	X (1856)	X (1856)
Cyprinidae			
Campostoma anomalum	Central stoneroller	X (1903)	X (1875)
Couesius plumbeus	Lake chub	X (1903)	
Cyprinella lutrensis	Red shiner	X (1889)	X (1875)
Hybognathus hankinsoni	Brassy minnow	X (1889)	
Hybognathus placitus	Plains minnow	X (1900?)	X (1889)
Luxilus cornutus	Common shiner	X (1889)	
Macrhybopsis aestivalis	Speckled chub		X (1889)
Nocomis biguttatus	Hornyhead chub	X (1903)	
Notropis dorsalis	Bigmouth shiner	X (1889)	
Notropis girardi	Arkansas River shiner		X (none)[b]
Notropis heterolepis	Blacknose shiner	X (1889)	
Notropis stramineus	Sand shiner	X (1889)	X (1875)
Phenacobius mirabilis	Suckermouth minnow	X (1903)	X (1979?)
Phoxinus eos	Northern redbelly dace	X (1903)	
Phoxinus erythrogaster	Southern redbelly dace		X (1979?)[c]
Pimephales promelas	Fathead minnow	X (1903)	X (1875)
Hybopsis gracilis	Flathead chub		X (1875)
Rhinichthys cataractae	Longnose dace	X (1889)	X (1875)
Semotilus atromaculatus	Creek chub	X (1889)	
Catostomidae			
Carpiodes carpio	River carpsucker	X (1903)	X (1980?)[d]
Catostomus catostomus	Longnose sucker	X (1889)	
Catostomus commersoni	White sucker	X (1889)	X (1875)
Ictaluridae			
Ameiurus melas	Black bullhead	X (1900)	X (1914)
Ictalurus punctatus	Channel catfish	X (1914)	X (1914)
Noturus flavus	Stonecat	X (1984)	
Cyprinodontidae			
Fundulus sciadicus	Plains topminnow	X (1875)	
Fundulus zebrinus	Plains killifish	X (1903)	X (1875)
Centrarchidae			
Lepomis cyanellus	Green sunfish	X (1903)	X (1913)
Lepomis humilis	Orangespotted sunfish	X (1959?)	X (1979?)
Percidae			
Etheostoma cragini	Arkansas darter		X (1903)
Etheostoma exile	Iowa darter	X (1903)	
Etheostoma nigrum	Johnny darter	X (1889)	
Etheostoma spectabile	Orangethroat darter[e]	X (1969)	

[a]Native occurrence (X) and date of first collection were determined from literature accounts, museum records, and unpublished data. Questionable or uncertain data are indicated by question marks (?). Common and scientific names are from Robins et al. (1991).
[b]Based on reports from the Arkansas River in Kansas and both upstream and downstream from Colorado in the Cimarron River (see text).
[c]Taxonomy of *Phoxinus* sp. in the Arkansas River Basin, Colorado, needs clarification because both *P. eos* and *P. erythrogaster* have apparently been found there (Loeffler et al. 1982, see Fig. 11.5 in Cross et al. 1986). We suspect that *P. eos* collected in the Arkansas River upstream from Pueblo, Colorado, by F.B. Cross in 1959 (University of Kansas Museum of Natural History Catalog No. 4760) stemmed from release of imported bait minnows.
[d]Based on records in Lee et al. (1980) of unknown date before 1980.
[e]Based on a single Wyoming specimen record (Baxter and Simon 1970) for Lodgepole Creek, a tributary to the South Platte River in eastern Colorado.

Yarrow 1875, Jordan 1891, Juday 1904, 1905, Cockerell 1911) and described many initial occurrences of Plains stream fishes. Collections during 1900-1914 were typically close to population centers along Colorado's Front Range mountains or in cold mountain streams with salmonids. Few sites were sampled on the eastern Plains. Sampling during 1915-1958 was limited to Boulder County (Hendricks 1950), except for one site on an Arkansas River Plains tributary (Fig. 6.2B), further emphasizing the importance of Ellis's work.

Only starting in 1959 were collections made throughout the two basins (Fig. 6.2B), mostly by university fish ecologists (F. Cross and A. Metcalf, University of Kansas, unpublished collections records; Li 1968, Loeffler et al. 1982, Propst 1982, Miller 1984). These recent comprehensive surveys documented the distribution and status of South Platte and Arkansas river basin fishes and provided critical comparative data for ongoing work. Colorado Division of Wildlife biologists are currently sampling all stream segments in both the Arkansas and South Platte basins in Colorado (J. Melby and R. Van Buren, personal communication).

Native Fish Assemblages

Despite the modest early sampling effort, the taxonomic composition of Colorado Plains fish assemblages was relatively well documented from the few collections made by 1914 (Ellis 1914) because most species eventually considered native to the Platte and Arkansas river basins (other than those perhaps extirpated early) had been collected by then (Table 6.1). All but 4 of 20 species thought native to the Arkansas Basin, including the only *Macrhybopsis aestivalis* specimen known from Colorado (Jordan 1891), had been recorded by 1914, and all but 3 of 29 South Platte native species had been collected.

Relatively late dates of first collection for some species attests to the lack of early work in Plains reaches. For example, *Lepomis humilis* was not collected until about 1959 in the South Platte and 1979 in the Arkansas, although it is considered native by most researchers because of reports lacking corroborative specimens (Beckman 1952) and because it is considered native in downstream reaches of the Platte in Nebraska and the Arkansas in Kansas (Johnson 1942, Cross and Collins 1995). Similarly, *Carpiodes carpio* (cf. Lee et al. 1980) and *Phenacobius mirabilis* (Loeffler et al. 1982) were only recently collected in the Arkansas River in Colorado, both about 1980 (Table 6.1).

Native status of some Colorado species will likely always be difficult to decipher. For example, some researchers considered either *Stizostedion vitreum* (walleye) or *S. canadense* (sauger) native to the South Platte River, Colorado (Wiltzius 1985, Propst and Carlson 1986) based on early popular reports, although no specimens were preserved. Relatively early (1879) stocking records (Wiltzius 1985) of "wall-eyed pike" in the South Platte River and its larger tributaries, and the fact that original plains streams were not typical habitat for either species, led us to initially discount a *Stizostedion* sp. as native. However, Wiltzius (personal communication) informed us of newspaper accounts describing catches of a *Stizostedion* sp. from the South Platte River near Greeley, Colorado, that predate stocking records, which strongly suggests native status for this taxon. We believe (like Wiltzius 1985) that the species involved was *S. canadense,* because it is considered native to the North Platte River in Wyoming (Baxter and Simon 1970) and the lower mainstem Platte River in Nebraska, whereas the status of *S. vitreum* is uncertain (Cross et al. 1986).

Sparse early collection records and taxonomic confusion also confound efforts to understand distribution of small, inconspicuous species. For example, both *Hybognathus placitus* and *H. hankinsoni* are known from recent South Platte River collections (Propst and Carlson 1986). However, the early distribution of these species (e.g., Ellis 1914) is unknown because *H. hankinsoni* was not described until 1929 (Bailey 1954). Ellis described large-eyed (presumably *H. hankinsoni*) and small-eyed (*H. placitus*) forms of *H. nuchalis* (which never occurred in Colorado) throughout the warm water reaches of the South Platte River and reported the former most abundant. Ellis's (1914) records of *Hybognathus* specimens greater than 100 mm total length (TL) also suggest that *H. placitus* was present, because *H. hankinsoni* are typically less than 100 mm TL (Bestgen and Propst 1996). Historic distributions of these species will remain elusive until museum specimens, if available, are examined.

Some researchers also believe that *Culaea inconstans* is native to the South Platte River in Colorado, because it is presently widespread where bait transfer would be unlikely (Propst and Carlson 1986). However, lack of *C. inconstans* records in the basin in Colorado before 1975 (Propst and Carlson 1986), coupled with evidence that it is native only in the lowermost Platte River, Nebraska (Lee and Gilbert 1980), suggest that it was introduced. In contrast, although *Notropis girardi* has never been collected in the state, it is likely native to Colorado based on records from the Arkansas River just downstream of the Colorado-Kansas state line (see Fig. 6.1), and on records in the Cimarron River just downstream of Colorado in Kansas and just upstream in Oklahoma (the type locality, Gilbert 1980). *Pylodictis olivaris* (flathead catfish) and *Dorosoma cepedianum* (gizzard shad) are unlikely natives in Colorado because they typically inhabit larger rivers and the nearest records are far downstream in Nebraska and Kansas. *Notropis dorsalis, Semotilus atromaculatus,* and *Catostomus catostomus* are also unlikely native in the Arkansas Basin because early collections did not document these species where they now occur (e.g., Fountain Creek, a transition zone tributary of the Arkansas River) and because of their restricted distribution. If native, we would expect these habitat generalist species to be widespread.

Despite recent surveys of Plains stream fishes in Colorado (Loeffler et al. 1982, Propst 1982), new distributional records continue to surface. We first collected *Noturus flavus,* a species previously only suspected native (Beckman 1952), in the transition zone of St. Vrain Creek, a South Platte tributary, in 1984 (cf. Platania et al. 1986). *Phoxinus eos,* a rare transition zone species in Colorado, was relatively common in off-channel habitat of West Plum Creek near Denver in 1985 (Bestgen 1989). We also discovered *Couesius plumbeus* in the lower montane zone of South St. Vrain Creek in 1989, after an 85-year hiatus of occurrence in Colorado (Juday 1905, Bestgen et al. 1991).

Knowledge of the ichthyofauna in Great Plains streams of western Kansas and Nebraska is probably also somewhat incomplete because relatively few collections were made before 1900 (Johnson 1942, Metcalf 1966, Cross 1967). Similarly, fish assemblages of New Mexico Plains streams were little known until W. Koster began collections in 1939 (Koster 1957, Bestgen and Platania 1990, 1991). These early collections likely yield an incomplete picture of fish assemblages in Plains streams because substantial land-use, hydrologic, and biotic changes that began before 1900 may have eliminated species before they could be detected. For example, Ellis (1914) reported that fishes in several miles of Boulder Creek near Boulder, Colorado, were decimated in 1907 by "introduction of mine and mill waste," presumably from nearby gold mining operations. Comparison of collections made before (1903) and after (1912) the fish kill indicated local extirpation of five of eight species, including two (*Nocomis biguttatus, Notropis heterolepis*) that were apparently extirpated from the state. He also reported extensive damage to fish assemblages from irrigation diversion and water depletion (see also Jordan 1891) and introduction of non-native fishes, all before 1914.

In summary, defining historic and recent distributions of Plains stream fishes is difficult, especially at the western edge of the Great Plains in Colorado and New Mexico. Sparse records and taxonomic confusion point to the importance of museum collections and published literature for defining the components of ecological communities. Lacking such sources, inferences about historic distribution and abundance must be based on zoogeography, habitat preferences, and life histories of the potential species involved. For example, lack of early collections in the Arkansas River in Colorado frustrates efforts to define historic distribution of *M. aestivalis,* although it likely occurred throughout the mainstem Arkansas, given its known upstream locality, semibuoyant egg reproductive strategy (see below), and presence of suitable habitat downstream. Alternatively, it is unlikely that distribution of *Luxilus cornutus* was continuous in the South Platte River from transition zone streams to the lower mainstem near the Colorado-Nebraska border. The warm, shallow, sandy, turbid habitat of the mainstem contrasts sharply with their preferred pool habitat in cool, clear streams with sand and gravel substrate. Instead, early records from the lower mainstem probably represent dispersers from Lodgepole Creek (see Table 6.1 footnote), where it was widespread. Unfortunately, few collections were made there after 1914 (Fig. 6.2), and downstream reaches of Lodgepole Creek are now normally dry (K.R. Bestgen, personal observation).

Characteristics of Great Plains Fish Assemblages

The ichthyofauna of Great Plains streams is relatively depauperate compared with more mesic basins to the east (Cross et al. 1986, Rabeni 1996) due to interactions among habitat size, environmental stability, and evolutionary history. The relative influence of these factors also likely varies longitudinally, as Schlosser (1990) reported for the Illinois River, a major tributary of the Upper Mississippi River in

Illinois. Here we make a similar longitudinal analysis of taxonomic and life-history attributes for the fish fauna of the Platte River Basin, which has fluctuating environmental conditions and relatively simple fish habitat, and compare it with the Illinois River Basin (Burr and Page 1986, Schlosser 1990), which has more diverse and stable habitat. This comparison is used to highlight unique ecological attributes of Plains stream fish assemblages.

Taxonomic Comparisons

Like Schlosser (1990), we reconstructed fish assemblages based on literature (mainly Cross et al. 1986) for the entire Platte River Basin (including the Niobrara River in Nebraska) and determined habitats and three life history attributes (maximum body length [TL], length at first reproduction, and maximum lifespan) from regional fish guides (Brown 1971, Scott and Crossman 1973, Lee et al. 1980, Becker 1983) and other published or unpublished data. Missing data for a few species' attributes were estimated from congeners. Species were classified as occurring in four habitat types: transition zone or highlands streams, Plains tributaries, the mainstem North and South Platte rivers (mid-Platte), and the mainstem Platte River downstream from their confluence (lower Platte). Many species occurred in two or more habitats, so our habitat classifications are not mutually exclusive, unlike Schlosser's (1990).

A striking attribute is the relative paucity of fishes in the Platte Basin, with only 78 species in 19 families presumed native (Table 6.2), compared with 129 species (Schlosser 1990) in 25 families (Burr and Page 1986) for the Illinois River. Although the Platte River Basin is about three times the area of the Illinois, its lower species richness can be ascribed to the simpler habitat and harsh physicochemical environment of Plains streams (Matthews 1987) and the greater distance from postglacial centers of dispersal (Cross 1970, Fausch et al. 1984).

Lower species richness in the Platte Basin was due mainly to lower diversity in five families (7 fewer Cyprinidae, 4 fewer Catostomidae, 4 fewer Ictaluridae [mostly *Noturus* spp.], 10 fewer Centrarchidae [mostly *Lepomis* spp.], 12 fewer Percidae [mostly *Etheostoma* spp.]), and 13 missing species in families not identified by Schlosser (1990). Many fish families underrepresented in the Platte River Basin, and in Plains streams in general, have relatively specialized habitat requirements or are intolerant of harsh, fluctuating physicochemical conditions such as high temperature and low oxygen (Matthews 1987, Smale and Rabeni 1995b). Representatives of these more specialized groups in Plains streams, such as the madtoms (*Noturus flavus* and *N. gyrinus*) and the darters (*Etheostoma exile* and *E. nigrum*), are among the most widespread, tolerant, nonspecialized members of these taxa. The lack of centrarchids in Plains streams is probably due to lack of deep pool habitat, shifting substrate that would reduce reproductive success of these mostly nest-building taxa, and turbid water that may limit foraging efficiency of these sight-feeders. Schlosser (1982, 1987, 1990) highlights the importance of deep pools to large-bodied stream fishes, because they provide refugia from floods, droughts, extreme physicochemical conditions, and terrestrial predators.

Table 6.2. Taxonomic Composition of Fishes from Four Habitat Types in the Platte River Basin[a]

	Number of Species				
Family	Transition	Tributaries	Mid-Platte	Lower Platte	Total Species
Petromyzontidae				1	1
Acipenseridae				3	3
Polyodontidae				1	1
Lepisosteidae				2	2
Anguillidae				1	1
Clupeidae			1	2	2
Hiodontidae				2	2
Esocidae			1	2	2
Cyprinidae	17	20	14	17	30
Catostomidae	3	3	6	9	11
Ictaluridae	2	2	3	7	7
Percopsidae				1	1
Gadidae				1	1
Cyprinodontidae	1	2	2	2	2
Gasterosteidae	1	1	1	1	1
Centrarchidae	1	2	3	4	4
Percidae	4	4	3	2	5
Sciaenidae				1	1

[a]Transition zone streams are intermediate in elevation between mountains and plains (see Fig. 6.2) and are generally small, cool, and clear. Tributaries are intermittent or perennial Plains streams less than 10 m wide. Mid-Platte rivers are large tributaries downstream of the transition zone and the North and South Platte river mainstems, and the Lower Platte is downstream from their confluence. Species were assigned to stream habitats based on individual accounts in Lee et al. (1980), Cross et al. (1986), the general literature, and the junior author's personal experience.

In the Platte Basin, Cyprinidae were the dominant taxa in each habitat type, particularly in smaller transition zone and Plains tributaries (Table 6.2). There were more species of Catostomidae, Ictaluridae, and other typically large-bodied taxa in the mid- and lower Platte, whereas Percidae declined downstream. Although our classification techniques differed, Schlosser (1990) also found that Cyprinidae were dominant in most habitat types, but that taxa were more evenly distributed among families in larger downstream reaches.

Longitudinal Differences in Life History Attributes

Fishes with maximum TL less than 100 mm dominated transition zone and especially Plains tributaries of the Platte Basin, probably due to the small size of potential habitats (Fig. 6.3A). Average TL is undoubtedly much smaller than maximum TL for all species, especially for large-bodied fishes (e.g., catostomids)

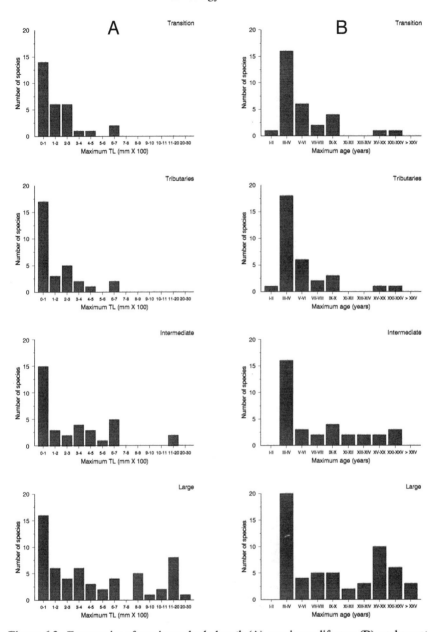

Figure 6.3. Frequencies of maximum body length (A), maximum lifespan (B), and age at first reproduction (C) for fish species found in four habitat categories of the Platte River Basin. Some fish species were classified in more than one category (see Table 6.2 and text).

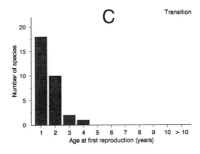

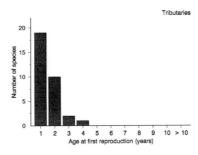

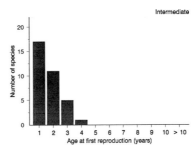

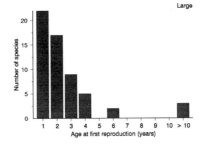

in these habitats because most individuals are juveniles less than 200 mm. Mid-and lower Platte assemblages include many larger-bodied species, indicative of the larger habitats. Schlosser (1990) also reported that Illinois River headwaters were dominated by species less than 200 mm TL, whereas downstream reaches supported taxa from small to large body size.

Patterns of maximum lifespan were similar to those for maximum body size. Maximum lifespans for most native fishes in transition zone and Plains tributaries of the Platte Basin are 6 years or less (Fig. 6.3B), except for a few longer-lived Catostomidae. However, average lifespan of most species is probably only 1 or 2 years. In contrast, mid- and lower Platte assemblages support increasing numbers of taxa with longer lifespans. These patterns were similar to the Illinois River, where short-lived species dominated headwater reaches but longer-lived taxa dominated intermediate and large river assemblages (Schlosser 1990).

Distributions of age at first reproduction changed less from upstream to downstream (Fig. 6.3C). Fishes reproducing at 2 years or less predominated throughout the basin, with some that reproduce at 3 or 4 years appearing in the mid- and lower Platte. Headwater reaches of the Illinois River were also dominated by early-maturing species, whereas downstream reaches had more taxa that mature later (Schlosser 1990).

Thus, larger downstream habitats for fishes in both the Platte and Illinois river basins support more taxa with larger maximum body size, longer lifespans, and older age at first reproduction than smaller headwater habitats. Because Schlosser (1990) used mutually exclusive habitat classes, many short-lived, small-bodied taxa not reported from downstream reaches of the Illinois Basin nevertheless probably occurred there in relatively low numbers. However, the Platte River Basin had fewer late-maturing species than the Illinois, suggesting that many Plains fishes evolved reproductive life-history traits in response to the harsh, fluctuating environment.

Unique Fish Fauna

Some constituents of the Great Plains fish fauna inhabit primarily cool, clear, spring-fed streams (often with macrophytes) or occur only in streams of the transition zone or other upland areas. These reaches contrast sharply with typical warm, turbid fluctuating Plains streams. Many of these unique taxa are glacial relict populations of Cyprinidae, which have centers of distribution to the north and east, although several taxa are distributed to the south or are endemic (Table 6.3). Low endemism is probably due to the rigors of Plains stream environments, because many species found in the fossil record during cooler and wetter preglacial and interglacial periods were later naturally extirpated as the region became warmer and more arid (e.g., *Esox, Perca;* Cross 1970, Cross et al. 1986). Thus, recent colonization from centers of endemism in the Mississippi Basin was more important than speciation in shaping the fauna.

Most of the unique taxa we list occur in Colorado, probably because the Arkansas and South Platte rivers have more extensive transition zone and spring-

Table 6.3. Selected Fishes with Isolated or Disjunct Distributions That Inhabit Spring-Stream, Transition Zone, or Upland Stream Habitats of the Great Plains Physiographic Region[a]

Species	States with Disjunct Populations[b]	Glacial Relict, Satellite, or Endemic	Present Center of Distribution
Couesius plumbeus	CO, NE	GR	N, UMW
Dionda episcopa	NM	SAT	S
Hybognathus hankinsoni	CO, KS	GR?, SAT	UMW
Luxilus cornutus	CO, KS, NE, WY	GR, SAT	UMW, E
Margariscus margarita	NE, WY	GR	N, NE
Nocomis biguttatus	CO, KS, WY	GR, SAT	UMW, E
Notropis heterolepis	CO, KS, ND, NE	GR, SAT	UMW
Phoxinus eos	CO, NE	GR	N, UMW
Phoxinus erythrogaster	CO, KS, NM, OK	SAT	E, UMW
Phoxinus neogaeus	CO[c],NE,SD, WY	GR	N, UMW
Ictalurus lupus	NM	SAT	S
Cyprinodon elegans	TX	END	
Cyprinodon pecosensis	NM, TX	END	
Gambusia nobilis	NM, TX	END	
Culaea inconstans	NM?	SAT	N, NE
Etheostoma exile	CO, WY	GR, SAT	UMW, N
Etheostoma lepidum	NM	SAT	S
Etheostoma cragini	AR, CO, KS, MO, OK	END	

[a]These unique fauna are classified as glacial relicts (GR), satellite populations (SAT) disjunct from a main population center but not glacially distributed (some species occur as both types of populations), or endemic (END). Their probable zoogeographic affinities are classified as northern (N), upper midwest (UMW), northeast (NE), east (E), or south (S) based on the present center of their distribution. Distributional data were derived from species accounts in Lee et al. (1980), regional fish guides, and unpublished data.

[b]Species were distributed among Arkansas (AR), Colorado (CO), Kansas (KS), Missouri (MO), Nebraska (NE), New Mexico (NM), North Dakota (ND), Oklahoma (OK), South Dakota (SD), Texas (TX), and Wyoming (WY).

[c]Present in Colorado only as *Phoxinus neogaeus* × *P. eos* hybrids (Bestgen 1989).

stream habitats than other Plains basins, and these reaches have been sampled extensively. The Platte Basin was also closer to the centers of distribution for more northern species that were displaced southward during glacial maxima (Metcalf 1966), resulting in more glacial relicts. Thus, of the 33 fish taxa considered native to the Arkansas and Platte basins in Colorado (Table 6.1, excluding *O. clarki stomias*), 9 (27%) are unique elements of Great Plains aquatic ecosystems, testimony to their importance in enriching regional biodiversity.

Because these unique taxa have relatively narrow habitat requirements and typically live in habitats of restricted size, populations are susceptible to groundwater depletions (spring systems), urban development (transition zone streams), and stochastic events. As a result, many are already extirpated from various states (e.g., *Nocomis biguttatus* and *Notropis heterolepis* in Colorado) or are listed by state agencies as at risk of extirpation. Although most are not in danger of extinction throughout their ranges, such isolated populations of unique species are often good indicators of more widespread habitat degradation and loss of ecological integrity (Fausch et al. 1990, Karr 1991) and perhaps may be effective indicators of global climate change (cf. Cross 1970, Matthews and Zimmerman 1990). Natural isolated populations in marginal habitats may also represent opportunities for divergence and evolution of distinctive taxa (Scudder 1989).

Non-native Species

Non-native fishes introduced into Great Plains streams have met with varied success, depending on whether the receiving habitats were similar to those in their native range. For example, at least 18 of the approximately 50 species that now occur in the South Platte Basin are non-natives (Schrader 1989), thus making up a large percentage of the current fauna. However, most non-natives from the eastern United States play minor ecological roles except in unique habitats. Non-native centrarchids, percids, and moronids are stocked by fishery management agencies in western Plains reservoirs but occur in the connecting streams primarily as juveniles and rarely reproduce there. However, *P. eos* was absent from its preferred cool, off-channel pond habitat in West Plum Creek when large numbers of non-native centrarchids were present (Bestgen 1989), suggesting that invasion of stable lentic habitats by non-native species may reduce unique native taxa. Similarly, non-native species thrive in regulated rivers and reservoirs of the mainstem Arkansas, Kansas, and Missouri rivers in the Central Lowlands to the east, and may partly account for the demise of native fauna there (Cross and Moss 1987; F.B. Cross, personal communication).

Harsh environmental conditions were likely responsible for excluding potential invaders from the Central Lowland Mississippi fauna, even without barriers to dispersal. For example, until recently no non-native species were found in a canyon reach of the Purgatoire River, a tributary of the Arkansas River that dissects the southeastern Colorado Plains, despite potential sources both upstream and downstream (Bramblett and Fausch 1991a). This situation was attributed to the flash floods that characterize the flow regime, to which only native species are presumably adapted (cf. Minckley and Meffe 1987). However, *Cyprinus carpio,* which is relatively successful even in Plains rivers, recently invaded the reach (S.C. Lohr and K.D. Fausch, unpublished data).

Limited dispersal and low abundance of most introduced sport fishes in Plains streams contrast sharply with the success of certain Plains taxa introduced beyond their native range. Rapid dispersal and reproduction of *H. placitus* and *N. girardi* introduced into the Pecos River, New Mexico, are discussed below. Noteworthy, however, is that introduced *H. placitus* hybridized with (Cook et al. 1992) and replaced native *H. amarus* throughout 260 km of the Pecos River before the introduction was even detected (Hatch et al. 1985, Bestgen and Platania 1991). Similarly, introduced *Cyprinodon variegatus* hybridized with native *C. pecosensis* in the Pecos River, Texas (Echelle and Connor 1989, Wilde and Echelle 1992). Hybrids apparently dispersed from a single introduction site there and replaced *C. pecosensis* throughout 430 km of habitat in less than 5 years. Genetically pure populations are secure only in upstream portions of the Pecos River drainage in New Mexico and at two off-channel localities in Texas.

Other cases also suggest that Plains fishes transplanted into other Plains streams tend to disperse and become widely established. *Pimephales vigilax* was introduced into the Kansas River system around 1976, and by 1981 was widespread and the fourth most abundant species in the lower Kansas River (Cross and Haslouer 1984). *Notropis bairdi* (Red River shiner), a species native to the Red River, Texas and Oklahoma, was introduced into the Cimarron River drainage (Arkansas River Basin) and is now common (Felley and Cothran 1981, Cross et al. 1983). *N. bairdi* has been implicated in the decline of the closely related native *N. girardi* via competition (Felley and Cothran 1981, Cross and Collins 1995), although streamflow alteration was probably a more important cause (Bestgen et al. 1989). *N. girardi* is now extirpated from its original extensive range in Kansas (Cross and Collins 1995) and reduced elsewhere, prompting the U.S. Fish and Wildlife Service to consider it for listing as an endangered species. *Gambusia affinis* was widely introduced into Plains streams and is now found throughout much of the Platte River, Nebraska (Lynch 1988), and in much of Kansas (Brown 1987, Cross and Collins 1995). Although *G. affinis* prey on native fishes in Arizona desert streams (Meffe 1985), effects in Plains streams are unknown. Testing effects of interspecific competition or predation between Plains stream fishes will require controlled field and laboratory studies (Matthews 1988).

Reproductive Ecology of Great Plains Fishes

Plains stream fishes exhibit a myriad of subtle specializations that combine to adapt them for surviving their harsh environment. These include barbels bearing chemosensory and tactile sense organs (Moore 1950) that allow reduced dependence on vision, reduced scales embedded in a thickened epidermis that is less sensitive to abrasion (F.B. Cross, personal communication), and tolerance to high temperature and hypoxia (Matthews and Hill 1980, Matthews 1987, Smale and Rabeni 1995b). Thus, most Plains fishes are highly tolerant to their fluctuating environment, and in many cases, simply withstand periods of stress until conditions improve. However, because immobile eggs and larvae are fragile, the harsh environmental conditions make reproduction challenging. Here we address various successful reproductive strategies and specializations that allow Plains stream fishes to persist despite the harsh conditions.

Reproductive Strategies

Reproduction in Plains fishes is poorly understood, due to either lack of study (e.g., *Fundulus sciadicus*) or because species reproduce during high and turbid flows when fish are difficult to observe. However, this seasonality also yields insights into the pattern of reproduction. Erosive shifting sand substrate during high flows would destroy delicate eggs and larvae, rendering ineffective the broadcast spawning of demersal eggs, which sink to the bottom. A mode of reproduction discovered somewhat fortuitously by Moore (1944) for *N. girardi* and subsequently for *M. aestivalis,* and by others for *H. placitus* (Botrell et al. 1964, Sliger 1967, Lehtinen and Layzer 1988; Taylor and Miller 1990), is to fertilize eggs and deposit them in the water column during high flows. The eggs absorb water, quickly enlarge two to three times their size at spawning, become semibuoyant, and are carried downstream with high flows. This strategy has the advantage of developing eggs away from the erosive substrate. We suspect further research may reveal that this reproductive strategy is used by an entire guild of Plains stream species. Another strategy used by widespread and currently abundant Plains cyprinids such as *Cyprinella lutrensis* (Gale 1986; K.R. Bestgen, personal observation) and *Pimephales promelas* (Gale and Buynak 1982) is to spawn during low flow after runoff and attach eggs to solid surfaces above erosive shifting substrate.

Spawning semibuoyant eggs during high flows would be especially adaptive if eggs developed and hatched quickly, thereby subjecting them to rigorous environmental conditions for shorter periods and retaining larvae relatively close to spawning sites. Incubation time for eggs of *N. girardi, M. aestivalis tetranemus,* and probably *H. placitus* are only 24–28 hours at 24–28°C (Moore 1944, Botrell et al. 1964). The newly hatched larvae are capable of vertical swimming and make vigorous attempts to reach the surface after contacting the bottom of laboratory containers. Within 2–4 days, they are capable of horizontal swimming and feeding. Adhesive eggs of *Cyprinella lutrensis* also hatch in only 4 days at 26–28°C (Gale 1986), and larvae are relatively mobile and capable of swimming just after hatching (K.R. Bestgen, unpublished data), suggesting that many Plains stream fishes have evolved reduced incubation times and precocious larvae.

Timing of Reproduction

Because flow regimes of Plains streams can vary substantially even within a few days, spawning must be timed such that habitat conditions are suitable for egg development and survival of larvae. For Plains stream species such as *N. girardi, M. aestivalis tetranemus,* and *H. placitus,* high flows resulting from spring runoff or summer thunderstorms appear to be optimal for spawning, perhaps because semibuoyant eggs are maintained in the water column or because sufficient backwater or floodplain habitat for larvae is assured.

Spawning of semibuoyant eggs during high flows may also enhance recruitment by dispersing young throughout long reaches, as was illustrated by rapid range expansion of *N. girardi* introduced into the Pecos River, New Mexico, in 1978. From a presumed upstream introduction near Ft. Sumner, *N. girardi* dispersed downstream throughout 260 km of river by 1982. Spawning by the introduced population coincided with high late-spring flow releases for irrigation in 1986 (Bestgen et al. 1989) and resulted in a strong year-class of *N. girardi* that made up 9% of all specimens collected in a riverwide survey. Introduction and rapid dispersal of *H. placitus,* beginning about 1964 throughout the same reach (Bestgen and Platania 1991, Bestgen and Propst 1996), was probably also aided by the same mode of reproduction. This species made up more than 80% of samples at some Pecos River sites in 1986 (Bestgen et al. 1989).

Timing of reproduction for other Plains stream fishes may be regulated more by temperature level and fluctuations than by flow regime. For example, female *Cyprinella lutrensis* held in the laboratory under constant light spawned batches of eggs every 1-4 days at constant temperatures less than 30°C, whereas spawning was sometimes daily at temperatures of 30-34°C (Gale 1986). Increases of 1°C above constant 33°C and 34°C reduced clutch size and egg survival but only occasionally reduced spawning frequency, whereas abrupt reductions in temperature below this threshold generally resulted in large clutches of live eggs.

Fractional or Extended Spawning

Another adaptation to highly variable environmental regimes is to spawn more than once during a reproductive season, thereby enhancing chances for survival of at least some offspring. Gale (1986) found that the consummate fractional spawner, *C. lutrensis,* continued to produce clutches nearly continuously for almost 2 years, given suitable water temperatures. For example, two red shiners each produced more than 100,000 eggs and still had seemingly inexhaustible numbers of eggs in their ovaries. Females apparently have indeterminate fecundity, so reproduction is controlled not by egg supply but rather length of spawning season, temperature, and food supply. Matthews (1988) and we have found individuals of *C. lutrensis* and other Plains cyprinids as small as 15 mm TL in winter through early spring, suggesting that fish spawn through late fall.

Because *C. lutrensis* females can spawn at continuous intervals of 1-4 days (Gale 1986), they can rapidly divert resources to produce viable eggs when environmental conditions are favorable. Females can also apparently hold ripe eggs when conditions are poor, because clutches are larger after periods of high temperatures that cause egg mortality. This flexibility is likely adaptive for *C. lutrensis* and other Plains species because even slight temperature shifts may signal egg-destructive droughts (higher temperatures) or spates (lower), and spawning can be reduced accordingly.

Some Plains stream fishes appear to combine fractional spawning with a buoyant egg strategy. For example, Taylor and Miller (1990) observed multiple modes of age-0 fish in length-frequency distributions of *H. placitus* duringMay and August of one summer. Age-0 *N. girardi* from the introduced Pecos River population also exhibited a wide size range in late summer (Bestgen et al. 1989), indicating possible fractional spawning for that species. Spawning by *M. a. tetranemus* was documented from mid-May to late August (Botrell et al. 1964). However, none of these studies indicated whether young produced late in the year were offspring from multiple spawning bouts by individuals or were a result of later spawning by younger or smaller individuals in the population.

Fish Movement as a Reproductive Strategy

Fishes that produce semibuoyant eggs must also have a mechanism for repopulating upstream reaches, because most eggs and larvae are swept downstream long distances by strong flows, and most adults die by age 2. Despite this scenario, upstream reaches are continually populated, which suggests that upstream fish dispersal is an important component of the reproductive life history. The importance of dispersal was first realized by Dr. Frank Cross (Univ. of Kansas, Lawrence, personal communication), who based it on years of study of Plains stream fish distribution patterns. He observed that mainstream and tributary populations of *M. a. tetranemus* and *N. girardi* disappeared simultaneously in the Arkansas River drainage, Kansas, following the radical changes in flows of the mainstem Arkansas River (Cross et al. 1985). For example, because flows and habitat were relatively unchanged in formerly occupied tributaries such as the Ninnescah River, he reasoned that disappearance of *N. girardi* there was due to loss of mainstem populations, which provided a source for recolonization of tributaries. Without these sources, all populations collapsed.

Observations of wild fish suggest that upstream repopulation is by movement of adults. A school of about 45 *H. placitus* was observed moving upstream in the South Canadian River, New Mexico, during low flows in spring 1989 (K.R. Bestgen, unpublished data). The fish traversed several riffle-pool sequences and moved upstream about 250 m in 15 minutes, suggesting that substantial short-term movements by Plains stream fishes are possible. Upstream movements of Plains taxa in the Rio Grande and Pecos rivers, New Mexico, may also explain occasional presence of immense numbers of fish downstream from diversion dams (Koster 1957, Bestgen and Platania 1990, 1991). Indeed, high vagility would seem a requirement for species inhabiting highly variable and harsh habitat conditions.

In summary, a suite of unique reproductive strategies appears to adapt Plains stream fishes for success in their harsh environment. Some species either produce semibuoyant eggs that drift downstream or attach adhesive eggs to firm substrates, strategies that enhance egg survival in streams with erosive shifting-sand substrate. Other adaptive behaviors include timing of spawning to coincide with high flows and optimal temperatures that enhance egg and larval survival. Many species populations spawn for extended periods from spring to late fall, which may be a bet-hedging strategy to ensure that at least some offspring survive the rigors of the variable environment. Rapid egg development and precocious larvae may be adaptations to environmental conditions that can change drastically over short periods.

Factors Influencing Population and Assemblage Variability

Magnitude and variation of stream discharge are thought to be major factors shaping the physical environment of streams and the structure and function of their biota (Poff and Ward 1989, Bayley and Li 1992). Populations of fishes in stochastically fluctuating and often intermittent Plains streams are predicted to have small body size, be highly vagile, and be well adapted to withstand floods and physicochemical extremes during droughts (Larimore et al. 1959, Poff and Ward 1989). In turn, assemblages of Plains stream fishes are predicted to have low species richness, exhibit high variation in relative abundance due to stochastic abiotic factors, and be trophically simple and persistent. Here, we assess the extent to which these predictions match what is known of populations and assemblages of Plains stream fishes.

Our comparison of life history attributes showed that small-bodied fishes predominated throughout all reaches of the Platte River, relative to the more perennial and mesic Illinois River (Fig. 6.3), which supports the prediction of small body size. Many of these fishes are also highly vagile at several life stages. As described previously, downstream reaches are apparently repopulated by some species via dispersal of buoyant eggs and larvae from upstream. In turn, juveniles and adults often move as part of a reproductive strategy, and to avoid habitats that are drying and becoming unsuitable during drought. Rapid recolonization of dewatered stream reaches of Brier Creek, a "prairie-margin" stream in south central Oklahoma, after drought (Ross et al. 1985, Matthews 1987) further highlights the vagility of Plains stream fishes.

Both direct and indirect evidence support the prediction that Plains stream fishes are well adapted to withstand floods and survive harsh conditions during droughts. For example, both Matthews (1987) and Smale and Rabeni (1995a) found that fishes inhabiting prairie streams of the Central Lowlands in Oklahoma, Missouri, and Arkansas were more tolerant of hypoxia than those in nearby mesic upland Ozark streams. Upland fishes also had lower tolerance to thermal maxima than prairie fishes (Matthews 1987), but differences in critical thermal maxima between the groups were relatively small. Smale and Rabeni (1995b) concluded that differences in fish assemblage composition among harsh and benign streams in Missouri could be explained more by tolerance to hypoxia than high temperature. Many investigators (e.g., Matthews et al. 1988, Fausch and Bramblett 1991) reported little change in species composition after persistent drought and massive floods, providing indirect evidence that Plains stream fishes are highly adapted to withstand these harsh conditions.

The low species richness, high variation in abundance, and simple trophic structure of Plains stream fish assemblages also match predictions made by Poff and Ward (1989). Thus, the Platte River has a relatively depauperate fish fauna with few specialized taxa compared with the more mesic Illinois River (78 vs. 129 species), despite the basin being about three times larger. Although samples of Plains stream fish assemblages are often dominated by a few taxa, most other species vary relatively unpredictably through time and among different reaches.

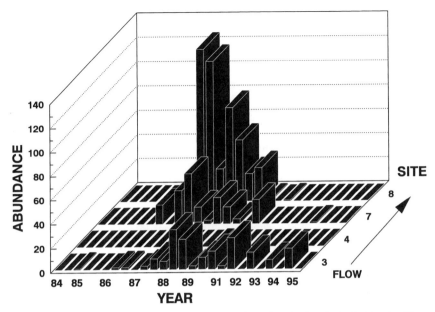

Figure 6.4. Abundance of *Fundulus zebrinus* at four sites in the Cache la Poudre River near Fort Collins, Colorado, over three seasons of 12 years. Collections were made by two electrofishing passes over 150-m reaches in spring, summer, or fall but not all seasons in all years. Independent sampling at two sites with simple versus complex habitat showed that two additional electrofishing passes and three seining passes detected no additional species and that 57–77% of individuals sampled were captured on the first two passes (K.D. Fausch and K.R. Bestgen, unpublished data). Site 3 is upstream, and site 8 is downstream.

For example, abundance of the widespread and common Plains taxon *Fundulus zebrinus* varied among seasons and years across four sites over 12 years in the Cache la Poudre River between Fort Collins and Greeley, Colorado (Fig. 6.4). *F. zebrinus* was most abundant in spring during most years at site 3 but during summer at site 8 and fall at site 7. On an annual basis, *F. zebrinus* was persistent at site 3 from 1987 to 1995 but fluctuated markedly at sites 7 and 8, being persistent and abundant from 1987 to 1989 but nearly absent before and after. Nearly complete absence of the species from site 4 is inexplicable, given the persistent population upstream at site 3 and a large population in the river and a small tributary between the two sites since at least 1990. Riverwide, *F. zebrinus* was rare in 1984 and 1985, increased dramatically at most of the 10 sites sampled during 1986–1990, then declined through 1995 at all but site 3. Low 1984–1985 abundance may have been due to high snowmelt runoff in both 1983 and 1984. In contrast, decreases after 1990 are inexplicable although probably not due to flows, which were benign except in 1995.

Similar conclusions have been drawn for entire fish assemblages. For example, the fish assemblage of Brier Creek, Oklahoma, was persistent from year to year on a streamwide scale (Ross et al. 1985, Matthews et al. 1988). However, stability in ranked relative abundances of species was generally low across sites and among years compared with Piney Creek, Arkansas, a more mesic upland stream. Similarly, species persistence and assemblage similarity over an 11-year period were higher riverwide than for 12 sites along a 70-km reach of the Purgatoire River in southeastern Colorado (Bramblett and Fausch 1991b; S.C. Lohr and K.D. Fausch, unpublished data), because rare species that occurred sporadically at individual sites were consistently present at the larger spatial scale. Persistence and stability were also typically lower among pools in intermittent tributaries than in the main river, probably due to complex interactions among droughts that eliminated pools, variability in flood maxima that provided opportunities for recolonization from the main river and adjacent reaches, and fish recruitment.

In contrast to dynamics at large spatial and temporal scales, assemblage structure in individual pools and riffles over short time scales may be strongly influenced by biotic factors. For example, multilevel trophic interactions were evident in individual pools of Brier Creek during spring through fall, because predatory *Micropterus* spp. controlled distribution and abundance of herbivorous *Campostoma anomalum,* which, in turn, affected distribution and abundance of algae, compared with pools where no *Micropterus* were found (Power et al. 1985). Similarly, *Lepomis cyanellus* were capable of strong predation on *F. zebrinus* during summer in outdoor tanks that mimicked the sloping vegetated littoral zone of pools in intermittent Purgatoire River tributaries (Lohr and Fausch in press). These experimental results also provided a mechanism to explain the complementary distribution of the two species reported from the field (Fausch and Bramblett 1991).

Partitioning of scarce resources by stream fishes along habitat or trophic axes generally implies that biotic interactions are important in structuring those divisions (e.g., Gorman 1988a, 1988b). However, habitat partitioning and trophic specialization are generally weak among Plains stream fishes (Matthews and Hill 1980, Bramblett and Fausch 1991a). Coevolution of species to partition habitats or trophic resources in Plains streams is difficult to envision, because resources often fluctuate widely or are transient. Instead, most Plains stream fishes are classified as generalists that occupy habitats and consume food resources in proportion to what is available (cf. Bramblett and Fausch 1991b), suggesting evolution of broad niches in response to widely fluctuating environmental conditions.

In summary, both abiotic and biotic processes likely influence structure and function of fish assemblages in Great Plains streams, but their relative importance depends on environmental variability and scale. For example, several studies report that biotic interactions such as predation are important in structuring fish assemblages and trophic webs in individual pools of Plains or prairie streams (Power et al. 1985, Lohr and Fausch in press), but floods and droughts reset assemblages over larger scales and longer periods in these same watersheds (Matthews et al. 1988, Bramblett and Fausch 1991a, Fausch and Bramblett 1991; Lohr and Fausch, unpublished data). Overall, abiotic factors may be more influential than biotic factors in controlling assemblage structure and function, species richness, and evolution of niche and life history attributes of Plains stream fishes.

Conservation of Plains Stream Fishes

State, federal, and nongovernmental natural resource agencies are currently mounting efforts to conserve declining populations of Plains stream fishes. This work is challenging for three reasons. First, as discussed previously, Plains streams represent a variety of habitats with different geomorphology and flow regime that have been subjected to different disturbances, so few broad generalizations about human effects, and potential solutions, are possible. Second, because few Plains fishes are sought by anglers, they have generally been neglected by researchers and managers, and relatively little is known about their distribution, ecology, and current status. Therefore, most agencies find that basic surveys are prerequisite to informed management decisions. Finally, most species occur primarily on private land, and much of the water in channels is controlled by prior appropriation under western water law, making management of habitats for these fishes difficult.

Despite these myriad challenges, five main issues relating to the distribution and ecology of Plains stream fishes emerge that have broad implications for effective conservation. First, it is obvious that without water, there can be no fish. Cross and colleagues (Cross et al. 1985, Cross and Moss 1987, see also Ferrington 1993) chronicled the demise of fish in western Kansas due to upstream water use and pumping of groundwater from the Ogallala aquifer, which together have dried up 160 km of mainstem Arkansas River and many tributaries in the western Kansas River Basin.

Second, although many Plains fish species are apparently tolerant of high and fluctuating temperatures and hypoxia (Matthews 1987, Smale and Rabeni 1995a), small populations of unique species that occur in spring-fed, transition zone, or upland streams are more sensitive to habitat perturbations and thus are easily extirpated. For example, several glacial relicts in Colorado transition zone and Plains streams were apparently extirpated early (e.g., *Nocomis biguttatus, Notropis heterolepis;* see Table 6.3), other endemics and glacial relicts have restricted ranges (*C. plumbeus, P. eos, E. cragini*), and others were likely extirpated by early degradation before being detected. Therefore, a high priority should be placed on identifying and protecting such unique environments and the species they harbor.

Third, the highly variable nature of original temperature and flow regimes in mainstem rivers and tributaries of the Great Plains selected for fishes with high vagility that could rapidly recolonize and populate rewatered reaches. However, these life history traits are generally not adaptive under current water management regimes. Thus, although local extirpations due to dewatering or harsh conditions were probably common in undisturbed Plains streams, dispersal from adjacent reaches and basins likely linked populations together in a spatially dynamic metapopulation (Harrison 1991, Fausch and Young 1995). Diversion dams, reservoirs, and persistently dewatered reaches that fragment contiguous riverine reaches impede upstream dispersal of adults and juveniles and interrupt downstream transport of eggs and larvae (Fig. 6.5), exposing them to possible predation and potentially intolerable conditions. Hampering either upstream or downstream dispersal undoubtedly prevents recolonization of extirpated populations.

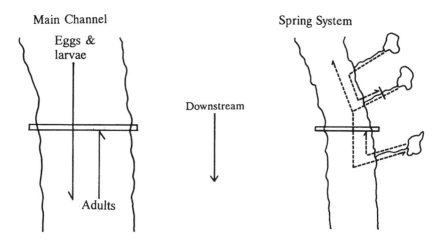

Figure 6.5. Schematic diagram showing how low-head dams for irrigation diversion or other uses can be potential barriers to dispersal of small-bodied Plains stream fishes. Eggs and larvae of Plains stream fishes drift downstream over diversions, but upstream dispersal by the resulting juveniles and adults is blocked. Similarly, recolonization of unique spring habitats by upstream dispersants is prevented.

Examples of extinctions and recolonizations, and circumstantial evidence pointing to the importance of dispersal in Plains streams, are common. Peripheral upstream and downstream populations of a suite of taxa (*M. aestivalis, Notropis jemezanus* [Rio Grande shiner], *N. orca* [phantom shiner], *N. simus* [bluntnose shiner], and *Hybognathus amarus* [Rio Grande silvery minnow]) were eliminated from the mainstem Rio Grande River in New Mexico, probably due to habitat dissection and flow regime alteration by the many diversion dams and reservoirs (Bestgen and Platania 1990). *H. amarus* now survives only in the middle Rio Grande, and the other species were eventually extirpated throughout the basin (Bestgen and Platania 1991). Similarly, Winston et al. (1991) reported extirpation of upstream populations of four taxa (*M. aestivalis, H. placitus, Notropis potteri* [chub shiner], and *N. bairdi,* which is closely related to *N. girardi* and may also spawn semibuoyant eggs) after closure of a dam on a tributary of the Red River in Oklahoma. Among their main hypotheses were that semibuoyant eggs produced by these upstream populations did not survive when transported into the reservoir or that populations throughout the reach were originally maintained by upstream dispersal of adult or juvenile fishes. These studies suggest that mainstem river reaches that provide naturally variable flow regimes and unrestricted dispersal routes may be critical for survival of an entire guild of Plains stream fishes that have evolved buoyant egg spawning.

Dispersal is also important among fishes inhabiting headwater tributaries, where extirpation is common as pools dry up during droughts. Because species have different colonizing abilities, some arrive sooner after flow resumes or penetrate intermittent reaches farther than others (Larimore et al. 1959, Fausch and Bramblett 1991).

Recolonization by vagile species occurs within hours or days after drought ends in streams with unimpeded access (Matthews 1987, see also Peterson and Bayley 1993) but would be prevented in those with barriers to dispersal.

Dispersal may be especially important for unique species inhabiting transition zone and spring-stream habitats. Single individuals of some species are regularly captured in mainstem rivers or Plains tributaries, far from known source populations. These fish are usually considered "waifs," driven downstream by high flows or other unfavorable conditions with little chance for survival (e.g., *E. cragini,* Matthews and McDaniel 1981, Pigg et al. 1985). In contrast, we suspect these individuals are actively dispersing from population sources and are important for recolonizing suitable habitats long distances away. If so, they may serve to link a set of subpopulations together into a metapopulation that functions at a landscape scale (Schlosser 1995a, 1995b, Schlosser and Angermeier 1995). Moreover, because recolonization may require complex dispersal upstream and downstream through inhospitable habitat from sources in distant watersheds, even small barriers such as low diversion dams may isolate potential habitats and prevent refounding of subpopulations (Fig. 6.5).

Irrigation canals may also offer avenues for fish dispersal, especially across drainage divides, and provide potentially important aquatic habitat. For example, Platania (1990) reported that the Cache la Poudre River Basin of north central Colorado alone had more than 500 km of canals. Extensive networks of ditches and canals occur throughout the arid West and can support entire assemblages of small-bodied Plains stream fishes (cf. Bestgen and Platania 1991), probably because some of the habitat is not unlike highly fluctuating and turbid Plains streams. Four canals that Platania (1990) studied supported 24 fish species, of which 15 native taxa made up 98% of all specimens. Those accessible to invading species and having year-round flow supplemented by groundwater had the most diverse fauna. Although few studies have been conducted on fishes in canals, such habitat may become increasingly important as flows in natural streams decline.

The fourth major issue with implications for conservation of Plains stream fishes is that introductions of non-native small-bodied fishes from other Plains watersheds are likely to have more detrimental effects than introductions of large sport fishes from mesic environments of the eastern United States. Although large percids, centrarchids, and moronids stocked for sport fishing flourish in artificial reservoir habitats, and regulated reaches of mainstem rivers in the Central Lowlands to the east, they are unlikely to reproduce and survive in most western Great Plains streams under current water management regimes. In contrast, Plains stream fishes that may have only recently diverged from closely related taxa are susceptible to hybridization when non-native congeners are transplanted from other basins. Moreover, interspecific competition and predation by such non-native species, although not yet experimentally demonstrated, are likely important causes of declines in native species.

Transplants of non-native Plains fishes are thought to have resulted largely from introductions of unused bait fish, originally transported by commercial suppliers from other Plains regions. Continued transfers of bait fish suggest that this industry is not well regulated, perhaps because the consequences of species introductions are not fully appreciated by management agencies. Nevertheless, most fishes in the American Southwest that are listed as endangered, threatened, or of special concern owe their current status largely to degraded habitat and effects of non-native fishes (Williams et al. 1989).

In contrast, current efforts to prevent loss of small-bodied Plains fishes include laws prohibiting seining of bait fish where declining species occur (e.g., in the Pecos River, New Mexico, to protect the threatened *Notropis simus pecosensis*). However, this strategy may be counterproductive. Such regulations likely increase the chances that additional non-native Plains species will be transplanted, because more bait fishes will be imported from sources outside the basin to supply demand (Bestgen and Platania 1991). Even monospecific cultures of widespread native species, such as *Pimephales promelas,* include small numbers of other species. Research to determine sources and composition of imported bait, effects of bait harvesting on stream fish populations in healthy habitat, and angler bait preferences would aid in forming policies to reduce the potentially deleterious effects of transferring non-native species.

The fifth major issue is that effects of human perturbations are not likely to be predictable from our understanding of fish ecology in more mesic streams of the eastern United States. For example, degradation of habitat by siltation from runoff and channelization, introduction of moderate amounts of oxygen-demanding organic wastes, and certain patterns of flow fluctuations are expected to be less important factors in the ultimate demise of fishes adapted for Plains rivers and tributaries (although often a factor for unique species in transition zone and spring streams) because these species are generally tolerant of harsh conditions (Bramblett and Fausch 1991b). In contrast, reduction in temperature and turbidity, degradation of the channel via scour of sand substrates, alteration of the natural pattern of floods and droughts, such as occurs in cool or cold, clear, tailwaters below reservoirs, and accidental introduction of non-native species adapted for such altered habitat would likely reduce or extirpate many Plains stream fishes (cf. Bestgen and Platania 1990, 1991), regardless of the need for dispersal. Clearly, this trend is opposite those in Ohio (Trautman 1981) and elsewhere in the Midwest and East (Karr et al. 1987), where increased turbidity, sewage effluents, and flow variability have favored more tolerant species such as those adapted to Plains streams (Cross and Moss 1987). However, relatively small changes in climate are predicted to have strong effects on Plains fish assemblages, because most species regularly experience conditions near their limits of thermal (Matthews and Zimmerman 1990) and oxygen tolerance (Smale and Rabeni 1995b). Overall, because we know less about mechanisms that degrade stream environments for fishes of the Great Plains than in other regions, further research is a critical prerequisite for sophisticated biomonitoring and habitat restoration (Bramblett and Fausch 1991b).

What Do We Need to Know?

Given our limited understanding of Plains stream fishes and the urgency of management efforts in many areas, we see four fronts on which research should proceed rapidly to allow effective conservation. First, careful sampling to understand distribution and status of species is still needed in many regions, although efforts are underway in others. Sampling must be done with care in all habitat types, using gear and protocols capable of detecting species with high probability if they are present and establishing their absence if not. Although many sites may be dry when visited during low flow, biologists should not necessarily assume that such channels do not provide important spawning or rearing habitat, or dispersal routes, during higher flows.

Second, investigators conducting sampling should collect and submit properly labeled voucher specimens for curation to museums, publish the results of their sampling in retrievable journals, and enter these verified records in electronic catalogs. Just as museum collections and published records have proven critical in reconstructing the historical distribution of many species to this point (and the lack of them has frustrated attempts for the rest), so will they also be critical in the future as humans rapidly degrade environments for these fishes.

Third, in addition to broad surveys of distribution and abundance, careful measurement of fish assemblages in specific watersheds will be needed to detect long-term anthropogenic changes. Because fluctuations in populations of some Plains species are drastic but poorly understood (Bramblett and Fausch 1991a) (Fig. 6.4), predicting responses of fish assemblages to water management regimes and other disturbances is not possible without data spanning several generations of the longest-lived species involved (Schlosser 1990). However, we know of only two published studies of central and southern Great Plains stream fish assemblages that now span a decade or more (Matthews et al. 1988, Bramblett and Fausch 1991a, Fausch and Bramblett 1991), although other studies and additional years of data are yet to be published (K.R. Bestgen and K.D. Fausch, unpublished data; S.C. Lohr and K.D. Fausch, unpublished data; W.J. Matthews, personal communication). It is somewhat fortuitous that in transition zone and small Plains tributaries responses to anthropogenic disturbances may be relatively rapid for most species due to their early maturity and short lifespan (Fig. 6.3).

Finally, we must learn much more about the ecology of many Plains species and their unique adaptations to local environments to be effective in conservation. For example, recent efforts to develop a recovery plan for *E. cragini* in Colorado (Loeffler and Krieger 1994) are hampered because we know little about its life-history, such as where or on what substrate Arkansas darter spawn (Distler 1972), its lifespan and age at maturity, thermal and hypoxia tolerance, habitat requirements, and whether dispersal is important in sustaining populations. Moreover, populations at the western edge of the range may differ markedly in life history from those in other regions. Such basic life-history attributes, especially the reproductive ecology and dispersal of fishes inhabiting turbid mainstem river habitats, are unknown for many species. We believe that this ecological knowledge will prove critical in understanding how fishes of the Great Plains use habitat across landscape scales, information that is clearly needed to allow predicting effects of further changes in land and water use.

Acknowledgments. We thank the many biologists cited herein for their dedication to research on fishes in Plains streams. We are grateful to F.L. Knopf for inviting us to write this chapter, which allowed us to synthesize the historical work. F.B. Cross and W.J. Matthews provided insightful reviews, and D. Rust and D. Crawford drafted maps. This chapter is Contribution no. 87 from the Larval Fish Laboratory, Colorado State University.

References

Bailey, R.M. 1954. Distribution of the American cyprinid fish *Hybognathus hankinsoni* with comments on its original description. Copeia 1954:289-291.

Bayley, P.B., and H.W. Li. 1992. Riverine fishes. Pp. 251-281 *in* P. Calow and G.E. Petts, eds. The rivers handbook. Blackwell Sci. Publ., London.

Baxter, G.T., and J.R. Simon. 1970. Wyoming fishes. Wyoming Game and Fish Dept., Bull. 4, Cheyenne.

Becker, G.C. 1983. Fishes of Wisconsin. Univ. Wisconsin Press, Madison.

Beckman, W.C. 1952. Guide to the fishes of Colorado. Univ. Colorado Mus., Boulder.

Benke, A.C., R.L. Henry III, D.M. Gillespie, and R.J. Hunter. 1985. Importance of snag habitat for animal production in southeastern streams. Fisheries (Bethesda) 10(5):8-13.

Bestgen, K.R. 1989. Distribution and notes on the biology of the northern redbelly dace, *Phoxinus eos,* in Colorado. Southwest. Nat. 34:225-231.

Bestgen, K.R., K.D. Fausch, and S.C. Riley. 1991. Rediscovery of a relict southern population of lake chub, *Couesius plumbeus,* in Colorado. Southwest. Nat. 36:125-127.

Bestgen, K.R., and S.P. Platania. 1990. Extirpation of *Notropis simus simus* (Cope) and *Notropis orca* Woolman (Pisces: Cyprinidae) from the Rio Grande in New Mexico, with notes on their life history. Occas. Pap. Mus. Southwest. Biol., Univ. New Mexico. 6:1-8. Albuquerque.

Bestgen, K.R., and S.P. Platania. 1991. Status and conservation of the Rio Grande silvery minnow, *Hybognathus amarus.* Southwest. Nat. 36:225-232.

Bestgen, K.R., S.P. Platania, J.E. Brooks, and D.L. Propst. 1989. Dispersal and life history traits of Arkansas River shiner, *Notropis girardi* introduced into the Pecos River, New Mexico. Am. Midl. Nat. 122:228-235.

Bestgen, K.R., and D.L. Propst. 1996. Redescription, geographic variation, and taxonomic status of the Rio Grande silvery minnow, *Hybognathus amarus* (Girard 1856). Copeia 1996:41-55.

Botrell, C.E., R.H. Ingersol, and R.W. Jones. 1964. Notes on the embryology, early development, and behavior of *Hybopsis aestivalis tetranemus* (Gilbert). Trans. Am. Microscop. Soc. 83:391-399.

Bramblett, R.G., and K.D. Fausch. 1991a. Fishes, macroinvertebrates, and aquatic habitats of the Purgatoire River in Piñon Canyon, Colorado. Southwest. Nat. 36:281-294.

Bramblett, R.G., and K.D. Fausch. 1991b. Variable fish communities and the index of biotic integrity in a western Great Plains river. Trans. Am. Fish. Soc. 120:752-769.

Brown, A.V., and W. J. Matthews. 1995. Stream ecosystems of the central United States. Pp. 79-110 *in* C.E. Cushing, K.W. Cummins, and G. W. Minshall, eds. Ecosystems of the world. Vol. 22: River and stream Ecosystems. Elsevier, New York.

Brown, C.J.D. 1971. Fishes of Montana. Big Sky Books, Montana State Univ., Bozeman.

Brown, K.L. 1987. Colonization by mosquitofish (*Gambusia affinis*) of a Great Plains river basin. Copeia 1987:336-351.

Burr, B.M., and L.M. Page. 1986. Zoogeography of fishes of the lower Ohio-upper Mississippi basin. Pp. 287-324 *in* C.H. Hocutt and E.O. Wiley, eds. The zoogeography of North American freshwater fishes. John Wiley & Sons, New York.

Cockerell, T.D.A. 1911. A new minnow from Colorado. Science 34:615.

Cook, J.A., K.R. Bestgen, D.L. Propst, and T.L. Yates. 1992. Allozymic divergence and systematics of the Rio Grande silvery minnow, *Hybognathus amarus,* (Teleostei: Cyprinidae). Copeia 1992:36-44.

Cope, E.D., and H.C. Yarrow. 1875. Report upon the collections of fishes made in portions of Nevada, Utah, California, Colorado, New Mexico, and Arizona during the years

1871, 1872, 1873, and 1874. Pp. 635–703 *in* Report upon geographical and geological explorations and surveys, west of the one hundredth meridian, in charge of First Lieutenant George M. Wheeler. Vol. V. Zoology. U.S. Govt. Print. Off., Washington, DC.

Cross, F.B. 1967. Handbook of fishes of Kansas. Univ. Kansas Mus. Nat. Hist. Publ. 45, Lawrence.

Cross, F.B. 1970. Fishes as indicators of Pleistocene and Recent environments in the central plains. Univ. Kansas Dept. Zool. Spec. Publ. 3:241–257.

Cross, F.B., and J.T. Collins. 1995. Fishes in Kansas. Public Educ. Ser. 14. Univ. Kansas Nat. Hist. Mus., Lawrence.

Cross, F.B., O.T. Gorman, and S.G. Haslouer. 1983. The Red River shiner, *Notropis bairdi,* in Kansas with notes on depletion of its Arkansas River cognate, *Notropis girardi.* Trans. Kansas Acad. Sci. 86:93–98.

Cross, F.B., and S.G. Haslouer. 1984. *Pimephales vigilax* (Pisces, Cyprinidae) established in the Missouri River basin. Trans. Kansas Acad. Sci. 87:105–107.

Cross, F.B., R.L. Mayden, and J.D. Stewart. 1986. Fishes in the western Mississippi drainage. Pp. 363–412 *in* C.H. Hocutt and E.O. Wiley, eds. The zoogeography of North American freshwater fishes. John Wiley & Sons, New York.

Cross, F.B., and R.E. Moss. 1987. Historic changes in fish communities and aquatic habitats in plains streams of Kansas. Pp. 155–165 *in* W.J. Matthews and D.C. Heins, eds. Community and evolutionary ecology of North American stream fishes. Univ. Oklahoma Press, Norman.

Cross, F.B., R.E. Moss, and J.T. Collins. 1985. Assessment of dewatering impacts on stream fisheries in the Arkansas and Cimarron rivers. Mus. Nat. Hist., Univ. Kansas, Lawrence.

Distler, D.A. 1972. Observations on the reproductive habits of captive *Etheostoma cragini* Gilbert. Southwest. Nat. 16:439–441.

Echelle, A.A., and P.J. Conner. 1989. Rapid, geographically extensive genetic introgression after secondary contact between two pupfish species (*Cyprinodon,* Cyprinodontidae). Evolution 43:717–727.

Ellis, M.M. 1914. Fishes of Colorado. Univ. Colorado Stud. 11, Boulder.

Eschner, T.R., R.F. Hadley, and K.D. Crowley. 1983. Hydrologic and morphologic changes in channels of the Platte River basin in Colorado, Wyoming, and Nebraska: a historical perspective. Geol. Surv. Prof. Paper. 1277-A, Washington, DC.

Fausch, K.D., and R.G. Bramblett. 1991. Disturbance and fish communities in intermittent tributaries of a western Great Plains river. Copeia 1991:659–674.

Fausch, K.D., J.R. Karr, and P.R. Yant. 1984. Regional application of an index of biotic integrity based on stream fish communities. Trans. Am. Fish. Soc. 113:39–55.

Fausch, K.D., J. Lyons, J.R. Karr, and P.L. Angermeier. 1990. Fish communities as indicators of environmental degradation. Am. Fish. Soc. Symp. 8:123–144.

Fausch, K.D., and M.K. Young. 1995. Evolutionarily significant units and movement of resident stream fishes: a cautionary tale. Am. Fish. Soc. Symp. 17:360–370.

Felley, J.D., and E.G. Cothran. 1981. *Notropis bairdi* (Cyprinidae) in the Cimarron River, Oklahoma. Southwest. Nat. 25:564.

Ferrington, L.C., Jr. 1993. Endangered rivers: a case history of the Arkansas River in the central plains. Aquat. Conserv. Mar. Freshwater Ecosystems 3:305–316.

Gale, W.F. 1986. Indeterminate fecundity and spawning behavior of captive red shiners— fractional, crevice spawners. Trans. Am. Fish. Soc. 115:429–437.

Gale, W.F., and G. Buynak. 1982. Fecundity and spawning frequency of the fathead minnow—a fractional spawner. Trans. Am. Fish. Soc. 111:35–40.

Gilbert, C.R. 1980. *Notropis girardi* Hubbs and Ortenburger, Arkansas River shiner. P. 268 *in* D.S. Lee, C.R. Gilbert, C.H. Hocutt, R.E. Jenkins, D.E. McAllister, and J.R. Stauffer, Jr., eds. Atlas of North American freshwater fishes. North Carolina State Mus. Nat. Hist., Raleigh.

Gorman, O.T. 1988a. An experimental study of habitat use in an assemblage of Ozark minnows. Ecology 69:1239–1250.

Gorman, O.T. 1988b. The dynamics of habitat use in a guild of Ozark minnows. Ecol. Monogr. 58:1-18.

Gregg, K.L. 1952. The road to Santa Fe: the journal and diaries of George Champlin Sibley and others pertaining to the surveying and marking of a road from the Missouri frontier to the settlements of New Mexico, 1825-1827. Univ. New Mexico Press, Albuquerque.

Harrison, S. 1991. Local extinction in a metapopulation context: an empirical evaluation. Biol. J. Linnean Soc. 42:73-88.

Hatch, M.D., W.H. Baltosser, and C.G. Schmitt. 1985. Life history and ecology of the bluntnose shiner (Notropis simus pecosensis) in the Pecos River of New Mexico. Southwest. Nat. 30:555-562.

Hendricks, L.J. 1950. The fishes of Boulder County, Colorado. M.S. thesis. Univ. of Colorado, Boulder.

Hubbs, C. 1994. Springs and spring runs as unique aquatic systems. Copeia 1994:989-991.

Jackson, D., and M.L. Spence, eds. 1970. The expeditions of John Charles Frémont. Vol. 1. Univ. Illinois Press, Urbana.

Johnson, R.E. 1942. The distribution of Nebraska fishes. PhD dissertation. Univ. Michigan, Ann Arbor.

Jordan, D.S. 1891. Report on explorations in Colorado and Utah during the summer 1889, with an account of fishes found in each of the river basins examined. U.S. Fish. Comm. Bull. 9:1-40.

Juday, C. 1904. Fishes of Boulder County. Univ. Colorado Stud. 2:113-114.

Juday, C. 1905. List of fishes collected in Boulder County, Colorado, with description of a new species of Leuciscus. U.S. Bur. Fish. Bull. 24:225-227.

Karr, J.R. 1991. Biological integrity: a long-neglected aspect of water resource management. Ecol. Appl. 1:66-84.

Karr, J.R., P.R. Yant, K.D. Fausch, and I.J. Schlosser. 1987. Spatial and temporal variability of the index of biotic integrity in three midwestern streams. Trans. Am. Fish. Soc. 116:1-11.

Kircher, J.E., and M.R. Karlinger. 1983. Effects of water development on surface-water hydrology, Platte River basin in Colorado, Wyoming, and Nebraska upstream from Duncan, Nebraska. U.S. Geol. Surv. Prof. Pap. 1277-B, Washington, DC.

Koster, W.J. 1957. Guide to the fishes of New Mexico. Univ. New Mexico Press, Albuquerque.

Larimore, R.W., W.F. Childers, and C. Heckrotte. 1959. Destruction and re-establishment of stream fish and invertebrates affected by drought. Trans. Am. Fish. Soc. 88:261-285.

Lee, D.S., and C.R. Gilbert. 1980. Culaea inconstans (Kirtland) Brook stickleback. P. 562 in D.S. Lee, C.R. Gilbert, C.H. Hocutt, R.E. Jenkins, D.E. McAllister, and J.R. Stauffer, Jr., eds. Atlas of North American freshwater fishes. North Carolina State Mus. Nat. Hist., Raleigh.

Lee, D.S., C.R. Gilbert, C.H. Hocutt, R.E. Jenkins, D.E. McAllister, and J.R. Stauffer, Jr., eds. 1980. Atlas of North American freshwater fishes. North Carolina State Mus. Nat. Hist., Raleigh.

Lehtinen, S.F., and J.B. Layzer. 1988. Reproductive cycle of the plains minnow, Hybognathus placitus (Cyprinidae), in the Cimarron River, Oklahoma. Southwest. Nat. 33:27-33.

Li, H.W. 1968. Fishes of the South Platte River basin. M.S. thesis. Colorado State Univ., Ft. Collins.

Loeffler, C.W., and D.A. Krieger. 1994. Arkansas darter, Etheostoma cragini, recovery plan. Colorado Div. Wildl., Denver.

Loeffler, C.W., D. Miller, R. Shuman, D. Winters, and P. Nelson. 1982. Arkansas River threatened fish survey, July 1982. Perf. Rep. SE-8-1. Colorado Div. Wildl., Denver.

Lohr, S.C. and K.D. Fausch. In press. Effects of green sunfish (Lepomis cyanellus) predation on survival and habitat use of plains killifish (Fundulus zebrinus). Southwest. Nat.

Lynch, J.D. 1988. Introduction, establishment, and dispersal of western mosquitofish in Nebraska (Actinopterygii: Poeciliidae). Prairie Nat. 20:203-216.

Matthews, W.J. 1987. Physicochemical tolerance and selectivity of stream fishes as related to their geographic ranges and local distributions. Pp. 111-120 in W.J. Matthews and D.C. Heins, eds. Community and evolutionary ecology of North American stream fishes. Univ. of Oklahoma Press, Norman.

Matthews, W.J. 1988. North American prairie streams as systems for ecological study. J. N. Am. Benthol. Soc. 7:387-409.

Matthews, W.J., R.C. Cashner, and F.P. Gelwick. 1988. Stability and persistence of fish faunas and assemblages in three mid-western streams. Copeia 1988:947–955.

Matthews, W.J., and L.G. Hill. 1980. Habitat partitioning in the fish community of a southwestern river. Southwest. Nat. 25:51–66.

Matthews, W.J., J.J. Hoover, and W.B. Milstead. 1985. Fishes of Oklahoma springs. Southwest. Nat. 30:23–32.

Matthews, W.J., and R. McDaniel. 1981. New locality records for some Kansas fishes, with notes on the habitat of the Arkansas darter (*Etheostoma cragini*). Trans. Kansas Acad. Sci. 84:219–222.

Matthews, W.J., and E.G. Zimmerman. 1990. Potential effects of global warming on native fishes of the southern Great Plains and the southwest. Fisheries (Bethesda) 15(6):26–32.

Mead, J.R. 1896. A dying river. Trans. Kansas Acad. Sci. 14:111–112.

Meffe, G.K. 1985. Predation and species replacement in American Southwestern fishes: a case study. Southwest. Nat. 30:173–187.

Metcalf, A.L. 1966. Fishes of the Kansas River system in relation to the zoogeography of the Great Plains. Univ. Kansas Publ. Mus. Nat. Hist. 17:23–189.

Miller, D.L. 1984. Distribution, abundance, and habitat of the Arkansas darter *Etheostoma cragini* (Percidae) in Colorado. Southwest. Nat. 29:496–499.

Minckley, W.L., and G.K. Meffe. 1987. Differential selection by flooding in stream fish communities of the arid American southwest. Pp. 93–104 *in* W.J. Matthews and D.C. Heins, eds. Community and evolutionary ecology of North American stream fishes. Univ. Oklahoma Press, Norman.

Moore, G.A. 1944. Notes on the early life history of *Notropis girardi*. Copeia 1944:209–214.

Moore, G.A. 1950. The cutaneous sense organs of barbelled minnows adapted to life in the muddy waters of the Great Plains region. Trans. Am. Microscop. Soc. 69:69–95.

Nadler, C.T., and S.A Schumm. 1981. Metamorphosis of South Platte and Arkansas Rivers, eastern Colorado. Phys. Geog. 2:95–115.

Peterson, J.T., and P.B. Bayley. 1993. Colonization rates of fishes in experimentally defaunated warmwater streams. Trans. Am. Fish. Soc. 122:199–207.

Pigg, J., W. Harrison, and R. Gibbs. 1985. Records of the Arkansas darter, *Etheostoma cragini* Gilbert, in Harper and Beaver counties in Oklahoma. Proc. Oklahoma Acad. Sci. 65:61–63.

Platania, S.P. 1990. Ichthyofauna of four irrigation canals in the Fort Collins region of the Cache la Poudre River valley. M.S. thesis. Colorado State Univ., Ft. Collins.

Platania, S.P., T.R. Cummings, and K.J. Kehmeier. 1986. First verified record of the stonecat, *Noturus flavus* (Ictaluridae), in the South Platte River system, Colorado, with notes on an albinistic specimen. Southwest. Nat. 31:553–555.

Poff, N.L., and J.V. Ward. 1989. Implications of streamflow variability and predictability for lotic community structure: a regional analysis of streamflow patterns. Can. J. Fish. Aquat. Sci. 46:1805–1818.

Power, .E., W.E. Matthews, and A.J. Stewart. 1985. Grazing minnows, piscivorous bass a... stream algae: dynamics of a strong interaction. Ecology 66:1448–1456.

Propst, D.L. 1982. Warmwater fishes of the Platte River basin, Colorado; distribution, ecology, and community dynamics. PhD dissertation. Colorado State Univ., Ft. Collins.

Propst, D.L., and C.A. Carlson. 1986. The distribution and status of warmwater fishes in the Platte River drainage, Colorado. Southwest. Nat. 31:149–168.

Rabeni, C.F. 1996. Prairie legacies—fish and aquatic resources. Pp. 111–124 *in* F.B. Samson and F.L. Knopf, eds. Prairie conservation: preserving North America's most endangered ecosystem. Island Press, Covallo, CA.

Rahel, F.J., and W.A. Hubert. 1991. Fish assemblages and habitat gradients in a Rocky Mountain-Great Plains stream: biotic zonation and additive patterns of community change. Trans. Am. Fish. Soc. 120:319–332.

Robins, C.R., R.M. Bailey, C.E. Bond, J.R. Brooker, E.A. Lachner, R.N. Lea, and W.B. Scott. 1991. Common and scientific names of fishes from the United States and Canada. Am. Fish. Soc. Spec. Publ. 20. Bethesda, MD.

Ross, S.T., W.J. Matthews, and A.A. Echelle. 1985. Persistence of stream fish assemblages: effects of environmental change. Am. Nat. 126:24–40.

Sanders, R.M., Jr., D.G. Huggins, and F.B. Cross. 1993. The Kansas River system and its biota. Pp. 295-326 *in* L.W. Hesse, C.B. Stalnaker, N. G. Benson, and J.R. Zuboy, eds. Restoration planning for the rivers of the Mississippi River ecosystem. Biol. Rep. 19. U.S. Dept. of Interior, National Biol. Surv., Washington, DC.

Schlosser, I.J. 1982. Fish community structure and function along two habitat gradients in a headwater stream. Ecol. Monogr. 52:395-414.

Schlosser, I.J. 1987. A conceptual framework for fish communities in a small warmwater stream. Pp. 17-24 *in* W.J. Matthews and D.C. Heins, eds. Community and evolutionary ecology of North American stream fishes. Univ. Oklahoma Press, Norman.

Schlosser, I.J. 1990. Environmental variation, life history attributes, and community structure in stream fishes: implications for environmental management and assessment. Environ. Manage. 14:621-628.

Schlosser, I.J. 1995a. Critical landscape attributes that influence fish population dynamics in headwater streams. Hydrobiologia 303:71-81.

Schlosser, I.J. 1995b. Dispersal, boundary processes, and trophic-level interactions in streams adjacent to beaver ponds. Ecology 76:908-925.

Schlosser, I.J., and P.L. Angermeier. 1995. Spatial variation in demographic processes in lotic fishes: conceptual models, empirical evidence, and implications for conservation. Am. Fish. Soc. Symp. 17:392-401.

Schrader, L.H. 1989. Use of the index of biotic integrity to evaluate fish communities in western Great Plains streams. M.S. thesis. Colorado State Univ., Ft. Collins.

Scott, M.L., J.M. Friedman, and G.T. Auble. 1996. Fluvial process and the establishment of bottomland trees. Geomorphology 14:327-339.

Scott, M.L., J.M. Friedman, G.T. Auble, P. Anderson, and L.S. Ischinger. In press. Historical perspectives on riparian ecosystems in Colorado. Colo. Riparian Assoc. 7.

Scott, W.B., and E.J. Crossman. 1973. Freshwater fishes of Canada. Fish. Res. Board Can. Bull. 184, Ottawa.

Scudder, G.G.E. 1989. The adaptive significance of marginal populations: a general perspective. Can. Spec. Publ. Fish. Aquat. Sci. 105:180-185.

Sliger, A.S. 1967. The embryology, egg structure, micropyle, and egg membranes of the plains minnow, *Hybognathus placitus* Girard. M.S. thesis. Oklahoma State Univ., Stillwater.

Smale, M.A., and C.F. Rabeni. 1995a. Hypoxia and hyperthermia tolerances of headwater stream fishes. Trans. Am. Fish. Soc. 124:698-710.

Smale, M.A., and C.F. Rabeni. 1995b. Influences of hypoxia and hyperthermia on fish species composition in headwater streams. Trans. Am. Fish. Soc. 124:711-725.

Southwood, T.R.E. 1988. Tactics, strategies, and templets. Oikos 52:3-18.

Taylor, C.M., and R.J. Miller. 1990. Reproductive ecology and population structure of the plains minnow, *Hybognathus placitus* (Pisces: Cyprinidae), in central Oklahoma. Am. Mid. Nat. 123:32-39.

Trautman, M.B. 1981. The fishes of Ohio. Ohio State Univ. Press, Columbus.

Wilde, G.R., and A.A. Echelle. 1992. Genetic status of Pecos pupfish populations after establishment of a hybrid swarm involving an introduced congener. Trans. Am. Fish. Soc. 121:277-286.

Williams, J.E., J.E. Johnson, D.A. Hendrickson, S. Contreras-Balderas, J.D. Willliams, M. Navarro-Mendoza, D.E. McAllister, and J.E. Deacon. 1989. Fishes of North America, endangered, threatened, or of special concern: 1989. Fisheries (Bethesda) 14(6):2-20.

Wiltzius, W.J. 1985. Fish culture and stocking in Colorado, 1872-1978. Div. Rep. 12. Colorado Div. Wildl., Denver.

Winston, M.R., C.M. Taylor, and J. Pigg. 1991. Upstream extirpation of four minnow species due to damming of a prairie stream. Trans. Am. Fish. Soc. 120:98-105.

Zale, A.V., D.M. Leslie, Jr., W.L. Fisher, and S.G. Merrifield. 1989. The physicochemistry, flora, and fauna of intermittent prairie streams: a review of the literature. Biol. Rep. 89(5). U.S. Fish and Wildl. Serv., Washington, DC.

7. Avian Community Responses to Fire, Grazing, and Drought in the Tallgrass Prairie

John L. Zimmerman

Introduction

Vegetation typical of the tallgrass prairie occurs east of the Great Plains in bottomland "openings" and as small "glades" or "balds" within the eastern deciduous forest. West of the Mississippi River, however, coverage by tallgrass prairie expands with the increasingly greater aridity under the deepening rainshadow of the western mountains. Along the Kansas-Missouri border on the western fringe of the deciduous forest, prairie is present across 50–80% of the region (Schroeder 1983). Forested area within this prairie-forest mosaic continues to diminish, decreasing to as little as 7% in the Flint Hills Uplands (Knight et al. 1994). The portion of the Great Plains characterized by the tallgrass prairie community exists in a climate that allows the development of forest as a bordering "gallery" along stream courses. These naturally occurring forest fragments, these island remnants of the continental forest to the east, remain an integral aspect of the tallgrass prairie landscape, contributing a disproportionately greater component, considering the small extent of their coverage, to the regional avian species richness (Faanes 1984, Zimmerman 1993).

Like the gallery forest, which maintains an attenuated relationship with eastern North America, the true tallgrass prairie community is not a unique flora (Wells 1970), having been largely derived from ancestral centers of distribution in southeastern North America (Anderson 1990). The few endemic plant species include no grasses. The avifauna of the tallgrass prairie is also composed of immigrants from faunas in grass-dominated habitats in other regions, such as the coastal plain or early seral communities of eastern forest succession. There are no endemic bird species in the tallgrass prairie (Mengel 1970). This situation may be the result of the prairie community's relatively recent origin in the late Tertiary but is more likely a consequence of the absence of habitat isolation across the broad horizons of this sea of tall grasses.

Although warm-season grasses of the tallgrass prairie, especially big bluestem (*Andropogon gerardii*), are dominant (Abrams and Hulbert 1987), cool-season grasses, forbs, and woody plants contribute to a rich floristic diversity. Where seeps and springs arise, often developing along rock outcrops, woody shrubs form continuous thickets within the grass-dominated landscape. Periodic fires determine the proportionate contributions of these plant life-forms, increasing the dominance of warm-season grasses while decreasing coverage by cool-season grasses, many forbs, and woody plants (Collins and Gibson 1990). The avifauna reflects the structural heterogeneity of the vegetative substrate.

Tallgrass Prairie Avifauna: Species Richness and Relative Abundance

Beginning in 1981, I assessed the breeding-bird community of the tallgrass prairie in June by a series of transect counts conducted in the following treatments on the Konza Prairie Research Natural Area in the Flint Hills of northeastern Kansas: annually burned (transects 1.6 km in total length), areas burned every 4 years (transects 1.6 km in total length), and prairie affected by longer unburned intervals (transect length varying from 3.2 to 6.8 km). All these prescribed burns occur in April. Since 1991, transects totaling 4.9 km in length in those burning treatments grazed by bison (*Bison bison*) have been added to the analysis. From 1993 through 1995, populations of breeding birds were additionally estimated by variable-distance point counts (Ralph et al. 1992) spaced at least 250 m apart in annually burned or unburned prairie that was either grazed (84 ha and 94 ha, respectively) or ungrazed (44 ha and 107 ha, respectively) by cattle. Data were collected by counting the number of individuals of all species noted during 10-minute intervals within each treatment area. Counts were made in mid-May, mid-June, mid-July, and mid-August, essentially encompassing the entire breeding season.

The birds of the tallgrass prairie can be divided into those dependent on the herbaceous grasses and forbs for nesting and/or foraging sites and those similarly dependent on woody plants (Table 7.1). The Brown-headed Cowbird, a social parasite that lays its eggs in the nests of species associated with both vegetation types, is habitat independent.

Of the 12 grass/forb-dependent species, 9 (75%) form a core group that is present every year. This annual constancy in species richness is typical of the grassland breeding bird community (Wiens 1973). Furthermore, the low diversity among grass/forb-dependent species is characteristic of grasslands throughout the world (Cody 1966).

Table 7.1. Birds of the Tallgrass Prairie and Their Relative Abundances (Mean Individuals/km ± SE, n = 10) on Annual June Transects in Unburned and Annually Burned Ungrazed Grasslands (Raptors, Greater Prairie-Chicken, and Swallows Excluded)

	Unburned	Burned
Grass/forb-dependent species		
Core species (present 100% of years)		
Ring-necked Pheasant (*Phasianus colchicus*)	1.0 ± 0.16	0.2 ± 0.10[a]
Upland Sandpiper (*Bartramia longicauda*)	2.9 ± 0.46	5.2 ± 0.70
Mourning Dove (*Zenaida macroura*)	4.0 ± 0.39	2.2 ± 0.45
Common Yellowthroat (*Geothlypis trichas*)	3.1 ± 0.39	0.2 ± 0.17[a]
Dickcissel (*Spiza americana*)	13.9 ± 0.90	12.3 ± 1.86
Grasshopper Sparrow (*Ammodramus savannarum*)	8.3 ± 0.67	6.8 ± 0.90
Henslow's Sparrow (*Ammodramus henslowii*)	2.8 ± 0.59	0
Red-winged Blackbird (*Agelaius phoeniceus*)	1.6 ± 0.22	3.8 ± 0.37
Eastern Meadowlark (*Sturnella magna*)	8.0 ± 0.41	6.8 ± 0.80
Erratic or "satellite" species (present 20–90% of years)		
Killdeer (*Charadrius vociferus*)	0.4 ± 0.03	0
Common Nighthawk (*Chordeiles minor*)	0.3 ± 0.10	1.0 ± 0.36
Lark Sparrow (*Chondestes grammacus*)	0.1 ± 0.03	0
Woody-dependent species		
Core species (present 100% of years)		
Northern Bobwhite (*Colinus virginianus*)	3.0 ± 0.47	1.2 ± 0.44[a]
Eastern Kingbird (*Tyrannus tyrannus*)	1.4 ± 0.18	1.5 ± 0.19
House Wren (*Troglodytes aedon*)	1.3 ± 0.29	0
Brown Thrasher (*Toxostoma rufum*)	2.8 ± 0.23	0.7 ± 0.18[a]
Bell's Vireo (*Vireo bellii*)	2.3 ± 0.52	0
Field Sparrow (*Spizella pusilla*)	2.8 ± 0.45	0
Baltimore Oriole (*Icterus galbula*)	0.4 ± 0.09	0.3 ± 0.10[a]
American Goldfinch (*Carduelis tristis*)	2.3 ± 0.55	0
Erratic or "satellite" species (present 20–90% of years)		
Yellow-billed Cuckoo (*Coccyzus americanus*)	0.6 ± 0.13	0
Red-headed Woodpecker (*Melanerpes erythrocephalus*)	0.3 ± 0.06	0
Red-bellied Woodpecker (*Melanerpes carolinus*)	0.1 ± 0.05	0
Northern Flicker (*Colaptes auratus*)	0.7 ± 0.15	0.2 ± 0.10
Eastern Phoebe (*Sayornis phoebe*)	0.1 ± 0.07	0
Great Crested Flycatcher (*Myiarchus crinitus*)	0.2 ± 0.06	0
Blue Jay (*Cyanocitta cristata*)	0.3 ± 0.08	0
Black-capped Chickadee (*Parus atricapillus*)	0.1 ± 0.05	0
Tufted Titmouse (*Parus bicolor*)	0.1 ± 0.04	0
Eastern Bluebird (*Sialia sialis*)	0.4 ± 0.24	0
American Robin (*Turdus migratorius*)	0.2 ± 0.05	0.1 ± 0.08
Gray Catbird (*Dumetella carolinensis*)	0.2 ± 0.08	0
Loggerhead Shrike (*Lanius ludovicianus*)	0.2 ± 0.15	0
European Starling (*Sturnus vulgaris*)	0.3 ± 0.12	0
Warbling Vireo (*Vireo gilvus*)	0.1 ± 0.03	0
Yellow Warbler (*Dendroica petechia*)	0.1 ± 0.03	0
Northern Cardinal (*Cardinalis cardinalis*)	0.3 ± 0.10	0
Blue Grosbeak (*Guiraca caerulea*)	0.1 ± 0.05	0
Eastern Towhee (*Pipilo erythrophthalmus*)	0.6 ± 0.56	0
Common Grackle (*Quiscalus quiscula*)	0.1 ± 0.02	0.8 ± 0.28
Orchard Oriole (*Icterus spurius*)	0.1 ± 0.09	0
Habitat-independent		
Core species (present 100% of years)		
Brown-headed Cowbird (*Molothrus ater*)	8.4 ± 0.51	6.1 ± 1.24

[a] Present less than every year in burned prairie.

Twenty-nine species are categorized as woody dependent, but only 28% of these species comprise a perennially present core group (Table 7.1). This habitat-related difference in the proportion of core species reflects the more variable year-to-year presence and coverage of woody plants, which in turn results from temporally variability in the incidence and intensity of fires.

Effects of Fire

Spring fires, usually in April, are the most frequent experimental pattern on Konza Prairie as well as in the management of Flint Hills pastures used for cattle production. Fire at this time has direct effects by destroying early nests of Mourning Doves and Greater Prairie-chickens (*Tympanuchus cupido*). Although summer fires could have more severe direct effects on breeding birds, the lower combustibility and higher relative humidities at this season usually result in fires of limited extent. Fall and winter fires do not directly affect bird reproduction; and indeed, most species have migrated away from the prairie or moved to other habitats by this time (Zimmerman 1993).

Periodic fires reduce coverage by woody plants. Thus bird species diversity decreases in response to fire by the elimination of some woody-dependent species and through reduction in the abundances of others (Zimmerman 1992) (Table 7.1). Only the Eastern Kingbird and Baltimore Oriole, which nest above the reach of the flames, are unaffected. In time, these species will also disappear, because there is no recruitment as the older trees suffer disease and wind throw. The longer the prairie remains unburned after a fire, the more similar the populations of woody-dependent species become to those in unburned prairie (Zimmerman 1993).

Certain grass/forb-dependent species, such as Henslow's Sparrow, for which litter and standing dead vegetation are integral dimensions of the ecological niche (Zimmerman 1988), are also negatively affected by fire. The Common Yellowthroat may be similarly impacted, but this dependency has not been documented. The more abundant Dickcissel, Grasshopper Sparrow, and Eastern Meadowlark, which together comprise about 70% of the total individuals in the community, suffered no decrease in numbers on June censuses during this period as a result of fire, whether fire occurred annually or at less frequent intervals (Zimmerman 1993) (Table 7.1).

In nondrought years, fire stimulates the growth of warm-season grasses so that the standing crop biomass increases on ungrazed sites (Knapp and Seastedt 1986). This increase in density and height of the dominant vegetation is correlated with an increase in the relative abundance of grass/forb-dependent species. Thus the parity in numbers between treatments during June is modified by a significant increase in the numbers of these species on point counts in July on burned prairie compared with the numbers in unburned sites not affected by grazing (Fig. 7.1). This increase results primarily from the continued immigration of Dickcissels into the breeding community, an annual pattern that occurs in other habitats as well (Zimmerman 1971).

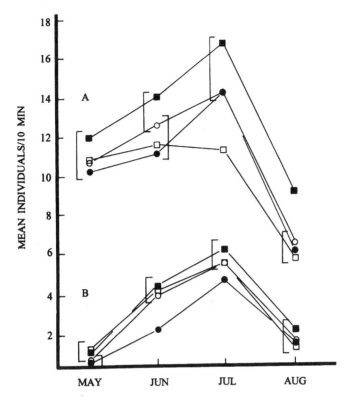

Figure 7.1. Mean relative abundances of all grass/forb-dependent species and for Dickcissels in unburned/ungrazed (open squares, $n = 60$), annually burned/ungrazed (closed squares, $n = 30$), unburned/grazed (open circles, $n = 33$), and annually burned/grazed (closed circles, $n = 27$) prairie. Because there were no consistent significant differences between years, these points reflect pooled data for 1993–1995. Means within the same bracket are statistically similar ($P > .05$). (A) All grass/forb-dependent species; (B) Dickcissel.

Because primary productivity is augmented after fire (Hulbert 1988), I hypothesized that this increase in energy availability for higher trophic levels in burned prairie could be translated into greater productivity of the birds nesting in burned prairie compared with the productivity of these species in unburned prairie. Burned prairie might function as a source population, providing a pool of surplus individuals to supplement the predicted lower productivity in unburned prairie (Thompson and Nolan 1973, Temple 1990). The net effect of this differential production would be the basis for the lack of observed differences in relative abundances of the major grass/forb-dependent species during June transect counts. A comparison of daily nest survival probabilities (Table 7.2) (Mayfield 1961, 1975, Johnson 1979) and fledging weights (Table 7.3) does not support this hypothesis as there is no increase in nest success or potential survivability (as indicated by fledging weights) of the young in burned treatments compared with

Table 7.2. Daily Survival Probabilities for the Combined Egg-Laying, Incubation, and Brooding Periods with 95% Confidence Intervals of Grass/Forb-Dependent Species

Species[a]	Unburned/Ungrazed	Unburned/Grazed	Burned/Ungrazed	Burned/Grazed
DICK	0.971	0.982	0.976	0.960
	0.9617-0.9809	0.9723-0.9906	0.9676-0.9854	0.9397-0.9811
EMDL	0.979	0.977	0.929	—[b]
	0.9588-0.9998	0.9535-0.9997	0.8809-0.9775	
RWBB	0.949	0.953	0.904	0.890
	0.9184-0.9788	0.9230-0.9838	0.8766-0.9306	0.8475-0.9323
MDOV	0.964	0.949	0.970	0.953
	0.9516-0.9768	0.9300-0.9672	0.9477-0.9927	0.9253-0.9841
GRAS	0.954	0.933	—	—
	0.9032-1.0000	0.8807-0.9859		
POOLED[c]	0.967[1]	0.966[1]	0.955[1,2]	0.941[2]
	0.9602-0.9742	0.9576-0.9743	0.9461-0.9647	0.9249-0.9575

[a]DICK, Dickcissel; EMDL, Eastern Meadowlark; RWBB, Red-winged Blackbird; MDOV, Mourning Dove; GRAS, Grasshopper Sparrow.
[b]—, No data.
[c]Probabilities with the same superscript are not significantly different.

unburned treatments for individual species. Nor is there a difference in the mean production of young per attempted nest found between burned and unburned, ungrazed prairies for all species (including cowbirds produced in Dickcissel and Red-winged Blackbird nests) except Eastern Meadowlarks (Table 7.4). Overall, grass/forb-dependent species are equally productive no matter what the annual fire frequency in the habitat might be. This is not too surprising because fire has

Table 7.3. Fledging Weights (g), Mean ± SE

	Unburned: Ungrazed and Grazed	Burned	
		Ungrazed	Grazed
DICK[a]	15.8 ± 0.27	15.9 ± 0.33	15.8 ± 0.40
	n = 55	n = 37	n = 7
BHCB	25.6 ± 0.57	25.6 ± 0.82	23.9 ± 0.73
	n = 63	n = 26	n = 18
EMDL	46.0 ± 1.08	49.0	47.2 ± 3.75
	n = 25	n = 1	n = 2
RWBB	30.4 ± 1.07	29.7 ± 1.18	29.1 ± 1.79
	n = 15	n = 22	n = 8
MDOV	55.5 ± 1.02	57.8 ± 2.75	56.5 ± 3.87
	n = 40	n = 8	n = 8

[a]DICK, Dickcissel; BHCB, Brown-headed Cowbird; EMDL, Eastern Meadowlark; RWBB, Red-winged Blackbird; MDOV, Mourning Dove.

Table 7.4. Mean Productivity/Attempted Nest Found ± SE

	N	Host	Cowbird
Unburned/ungrazed			
Dickcissel	66	0.71 ± 0.15	0.74 ± 0.15
Grasshopper Sparrow	4	0.50 ± 0.50	0.25 ± 0.25
Eastern Meadowlark	10	1.50 ± 0.50	0.40 ± 0.16
Red-winged Blackbird	15	0.53 ± 0.27	0.40 ± 0.21
Mourning Dove	48	0.71 ± 0.14	—[a]
Unburned/grazed			
Dickcissel	43	1.09 ± 0.19	0.84 ± 0.20
Grasshopper Sparrow	7	0.29 ± 0.29	0.29 ± 0.29
Eastern Meadowlark	10	1.20 ± 0.49	0.20 ± 0.13
Red-winged Blackbird	15	0.87 ± 0.32	0.27 ± 0.15
Mourning Dove	35	0.54 ± 0.15	—
Burned/ungrazed			
Dickcissel	59	1.19 ± 0.19	0.58 ± 0.12
Eastern Meadowlark	9	0.11 ± 0.11	—
Red-winged Blackbird	54	0.41 ± 0.16	0.02 ± 0.02
Mourning Dove	9	0.67 ± 0.35	—
Burned/grazed			
Dickcissel	21	0.38 ± 0.13	0.76 ± 0.24
Red-winged Blackbird	27	0.29 ± 0.17	0.07 ± 0.07
Mourning Dove	14	0.57 ± 0.25	—

[a]—, No data.

always been a reality in the environment of the tallgrass prairie. Furthermore, these results on fledging weights and productivity suggest that food availability for the young, even in unburned prairie with lower primary productivity, is sufficient for normal development to fledging in these species. Miller et al. (1994) similarly concluded that reproduction in Savannah Sparrows (*Passerculus sandwichensis*) of northern grasslands was not limited by food.

Effects of Grazing

In grazed prairie, cattle were pastured at a density of 3 ha (7.7 acres) per cow-calf unit, with the animals being free to choose either burned or unburned areas. This grazing intensity results in the removal of 20–25% of the annual aboveground plant growth during the seasonal stocking period, May–October. In the Flint Hills of Kansas, this grazing pressure is considered moderate. Grazing has no effect on the species richness of the grass/forb-dependent species in either unburned or burned prairie in each of the 4 months during the breeding season, except during August in burned prairie (Table 7.5). Neither is there a grazing-induced decrease

Table 7.5. Species Richness of Grass/Forb-Dependent Species in All Burning and Grazing Treatments (Mean ± SE) throughout the Breeding Season[a]

	Unburned/ Ungrazed	Unburned/ Grazed	Burned/ Ungrazed	Burned/ Grazed
May	4.8 ± 0.24	4.4 ± 0.25	4.1 ± 0.23	4.3 ± 0.26
	$n = 60$	$n = 33$	$n = 30$	$n = 27$
June	4.9 ± 0.20	5.1 ± 0.25	5.2 ± 0.19	5.3 ± 0.28
	$n = 60$	$n = 33$	$n = 30$	$n = 27$
July	4.3 ± 0.16	4.5 ± 0.20	4.9 ± 0.13	5.1 ± 0.24
	$n = 60$	$n = 33$	$n = 30$	$n = 27$
August	3.0 ± 0.26	3.0 ± 0.28	3.8 ± 0.28	2.5 ± 0.24[b]
	$n = 60$	$n = 32$	$n = 30$	$n = 27$

[a]Data are pooled for 1993–1995, due to lack of significant differences between years.
[b]Only significant difference related to grazing is in the burned prairie in August (student's $t = 3.49$, df = 55, $P < .01$).

in relative abundances of the grass/forb-dependent species on unburned prairie (Fig. 7.1); however, grazing does depress the numbers of grass/forb-dependent species in burned prairie. Relative abundance of these species is lowest in the burned/grazed treatment throughout the breeding season but only significantly different from burned but ungrazed prairie in June and August (Fig. 7.1).

Combining days of nest exposure and nest losses for all five grass/forb-dependent species with adequate nest records (last two rows in Table 7.2) reveals no effect of grazing on nest survival in unburned prairie. In burned/grazed prairie, however, the probability for nest survival is significantly lower compared with ungrazed prairie. Production of young per attempted nest found in burned/grazed prairie (Table 7.4) is similar to all other treatments for Red-winged Blackbirds and Mourning Doves as well as for cowbirds in all host nests, but production of Dickcissel young in burned/grazed prairie is lower and significantly different ($t = 2.53$, df = 78, $P < .05$) from production in burned but ungrazed prairie. Yet there is no effect of grazing on fledging weights (Table 7.3), which suggests that it must be environmental resources other than food availability that affects the reductions in population size and nest survival probabilities for all grass/forb-dependent species as well as the production of young in Dickcissels in the burned/grazed habitat.

Although nest survival and fledging weight data are not available for the bison-grazed treatments stocked at an intensity equal to that of cattle, a similar response in relative abundance of the grass/forb-dependent birds is apparent in burned prairie (Table 7.6). Bison were introduced to the treatments in 1991, but the 1991 data have been omitted from the analysis of unburned pastures because they were affected by a wildlife that year. Although the mean relative abundance of the grass/forb-dependent species decreased after the introduction of the bison, changes are neither significantly different from the period before introduction nor from prairie that remained ungrazed. In the other pair of pastures, burning was initiated in 1984 and bison introduced in 1991. The relative abundance of birds in

Table 7.6. Relative Abundances (Mean Individuals/km ± SE) of Grass/Forb-Dependent Species on Annual June Transects before and after the Introduction of Bison versus in Ungrazed Prairie[a]

	Before Bison	After Bison
Unburned/ungrazed	1981–1990 Ungrazed 46.0 ± 2.31[1] $n = 10$	1992–1994 Ungrazed 40.8 ± 7.49[1,2] $n = 3$
Unburned/grazed	1981–1990 Ungrazed 41.2 ± 3.80[1,2] $n = 10$	1992–1994 Grazed 30.4 ± 5.43[2] $n = 3$
Burned/ungrazed	1984–1990 Ungrazed 41.2 ± 1.32[1,2] $n = 7$	1991–1993 Ungrazed 32.8 ± 5.28[2] $n = 3$
Burned/grazed	1984–1990 Ungrazed 45.5 ± 4.48[1] $n = 7$	1991–1993 Grazed 25.4 ± 3.08[2] $n = 3$

[a]Means with the same superscript are not significantly different ($P > .05$).

the grazed pasture is significantly lower than the relative abundance before the introduction of bison but not significantly different from abundance in burned and ungrazed prairie.

The primary impact of grazing results from the preference of large herbivores for the more palatable and nutritious grasses in prairie that had been burned at the beginning of the growing season. The greater frequency and higher intensity of grazing on recently burned prairie results in a significant decrease in total peak aboveground standing crop biomass of the herbaceous vegetation under this treatment compared with the total in burned but ungrazed prairie (Fig. 7.2). In unburned prairie, grazing has no effect on the peak standing crop. I suggest that it is this modification of the height and density of aboveground vegetation in burned/grazed prairie that leads to reductions in avian abundance and productivity.

The relative abundance of Dickcissels is lowest in prairie that is both burned and grazed and significantly different from numbers in ungrazed treatments for all months except when populations are low in May and during the waning days of the breeding season in August (Fig. 7.1). The lower production of Dickcissel young per attempted nest found in grazed/burned prairie is significantly different from that in grazed but unburned prairie ($t = 2.46$, df = 62, $P < .05$) (Table 7.4). Dickcissel young production for unburned/ungrazed treatments on Konza Prairie from the late 1970s (Zimmerman 1982) is no different from any of the treatments presented in Table 7.4 except that it is significantly different from the low value for the burned/grazed treatment. Furthermore, the removal of the vegetation in burned/grazed prairie delays the onset of reproductive activity (Fig. 7.3). For the

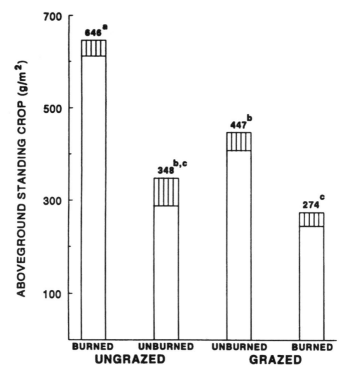

Figure 7.2. Mean peak aboveground standing crop biomass of herbaceous vegetation (open part of each bar is grass, lined portion is forbs) in ungrazed/unburned, ungrazed/ annually burned, grazed/unburned, and grazed/annually burned prairie. Means at the top of each histogram with the same superscript are not significantly different ($P > .05$).

first half of the breeding season, the increase in nesting activity in the grazed/ burned treatment lags 2 or 3 weeks behind the initiation of nesting in the other three treatments. For this neotropical migrant, delayed nesting may adversely affect the ability of both females and young to acquire sufficient productive energy to meet the demands of molt and fall migration because of seasonal limits imposed by changing photoperiod and temperature (Zimmerman 1965).

Factors Affecting Species Richness

The low avian diversity of the prairie bird community has been attributed to interspecific competition in a structurally simple habitat (Cody 1968) or to the paucity of species that have evolved adaptations to environmental bottlenecks, such as climatic extremes (Wiens 1974). Although burned and unburned prairie differ in the trophic base for the avian community by virtue of the significantly higher primary productivity after spring fire, burning treatments have little effect on the abundance of the grass/forb-dependent species or their productivity. Food

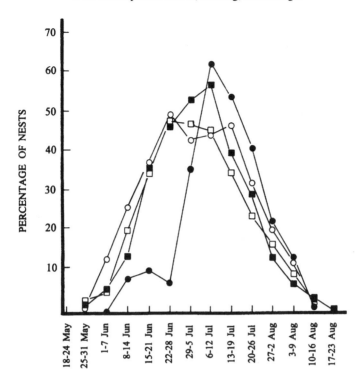

Figure 7.3. Nesting phenology of the Dickcissel. Percentage of total nests in each treatment active during the weeks of the nesting season in unburned/ungrazed (open squares), annually burned/ungrazed (closed squares), unburned/grazed (open circles), and annually burned/grazed (closed circles) prairie. Data pooled for the years 1993–1995.

resources appear to be adequate (Wiens and Dyer 1975), and competition for those resources does not appear to have an effect on bird populations. Drought is a periodic event that affects prairie, and annual relative abundances of the woody-dependent species are correlated with soil moisture availability. Numbers of grass/forb-dependent tallgrass prairie species, however, are unaffected by annual differences in soil moisture on unburned prairie (Zimmerman 1992). That is, the true prairie species have adapted to drought in some manner so that drought does not suppress their numbers. These data support the climatic bottleneck hypothesis for the low diversity of grassland birds. The woody-dependent species are likely more recent invaders into tallgrass prairie and are more affected by the ramifications of changes in soil moisture availability, although the direct proximate factors involved remain unknown.

On burned prairie, however, relative abundances of the grass/forb-dependent species decrease as a function of soil moisture reduction (Zimmerman 1992). This effect is analogous to the combination of fire and grazing in that it is correlated with the decrease in aboveground herbaceous biomass as result of reduced moisture availability (Briggs et al. 1989). Indeed, competition for nest site resources might limit populations under these conditions, affecting both population size and nest survival. The fact that fledging weights are similar between all burning and grazing treatment combinations supports the hypothesis that food availability is not a limiting resource. Although no data are available, it is reasonable to hypothesize that avian productivity on burned prairie during drought-influenced growing seasons is also reduced. I conclude that the few grass/forb-dependent species regularly present in tallgrass prairie are composed of individuals that can withstand periodic drought. Further, the sizes of their populations and the probabilities of nest survival decline when the combined effects of fire and drought or fire and grazing reduce the amount of aboveground vegetation, affecting nonfood-related resources.

Management Implications

Except for a few small remnants sequestered in prairie preserves, most of the tallgrass prairie in the Great Plains has been maintained on private land for pasturing cattle. If the costs of doing business would result in cattle production no longer being tenable, the tallgrass prairie will be converted toward some other economic endeavor despite the fact that the conversion of grass protoplasm into cattle protein is still the best human use of the tallgrass prairie in a hungry world. This is a critical consideration, because populations of grassland birds have suffered significant declines in recent years, primarily from the loss of appropriate habitat (Askins 1993, Knopf 1994).

Fire is the primary management tool for tallgrass prairie because fire controls coverage by woody plants and indirectly increases primary productivity, which translates into better weight gains in cattle. For example, the daily weight gain in steers pastured on burned prairie can be as much as 38% greater in May compared with that of steers grazing unburned prairie; and this difference in growth rate is maintained into summer, although the disparity drops to less than 10% in July (Anderson et al. 1970). Cattle producers must implement periodic prescribed burning to maintain an economically sustainable industry.

Yet, fire adversely affects grass/forb-dependent prairie birds due to intense cropping by domestic and native grazers on recently burned prairie. These herbivores significantly reduce the amount of aboveground herbaceous vegetation, which in turn is correlated with lower abundances of grassland birds, poorer nest survival, and in the case of the Dickcissel, delayed nesting and reduced production of young in successful nests.

There is a dilemma. On the one hand, the habitat for tallgrass birds will only remain available if cattle production continues as a dominant land use in the eastern Great Plains, but this requires periodic burning of pastures. On the other hand, burned/grazed prairie negatively affects populations and productivity of grassland birds. The problem to be solved involves discovering the optimum mix, both in coverage and spatial distribution, of burned prairie (avian sinks), which will allow the land to be maintained in cattle production retaining tallgrass prairie habitat, and unburned prairie, which will permit the maintenance of a sustainable grass/forb-dependent bird community (avian source populations). This is a challenge that can be met through increasingly sophisticated methods and developing theory of landscape ecology. This is a challenge that must be met if we seek to maintain regional biodiversity to enhance our quality of life within the context of our economic endeavors that sustain our life.

Acknowledgments. The data presented in this chapter were collected on the Konza Prairie Research Natural Area, a site owned by The Nature Conservancy and managed by the Division of Biology at Kansas State University for ecological research, with David C. Hartnett as director. I am indebted to Deborah Beutler, John Cavitt, Elmer Finck, Justin Kretzer, Todd Miller, Kim Perez, and Alan Stevens for their assistance in the field. Professor Kimberly G. Smith of the University of Arkansas provided a valuable review of an initial draft of this chapter. Support for this research was provided by USFWS, Region 6, Nongame/ Neotropical Migratory Bird Program and by the National Science Foundation through the continuing Long-term Ecological Research Program funded at Konza Prairie since 1980.

References

Abrams, M.D., and L.C. Hulbert. 1987. Effect of topographic position and fire on species composition in tallgrass prairie in northeast Kansas. Am. Midl. Nat. 117:442–445.

Anderson, K.L., E.F. Smith, and C.E. Owensby. 1970. Burning bluestem range. J. Range Manage. 23:81–92.

Anderson, R.C. 1990. The historic role of fire in North American grassland. Pp. 8–18 *in* S.L. Collins and L.L. Wallace, eds. Fire in North American tallgrass prairie. Univ. Oklahoma Press, Norman.

Askins, R.A. 1993. Population trends in grassland, shrubland, and forest birds in eastern North America. Current Ornithol. 11:1–34.

Briggs, J.M., T.R. Seastedt, and D.J. Gibson. 1989. Comparative analysis of temporal and spatial variability in above-ground production in a deciduous forest and prairie. Holarctic Ecol. 12:130–136.

Cody, M.L. 1966. The consistency of intra- and inter-continental grassland bird species counts. Am. Nat. 100:371–376.

Cody, M.L. 1968. On the methods of resource division in grassland bird communities. Am. Nat. 102:107–147.

Collins, S.L., and D.J. Gibson. 1990. Effects of fire on community structure in tallgrass and mixed-grass prairie. Pp. 81–98 *in* S.L. Collins and L.L. Wallace, eds. Fire in North American tallgrass prairie. Univ. Oklahoma Press, Norman.

Faanes, C.A. 1984. Wooded islands in a sea of prairie. Am. Birds 38:3–6.

Hulbert, L.C. 1988. Causes of fire effects in tallgrass prairie. Ecology 69:46–58.

Johnson, D.H. 1979. Estimating nest success: the Mayfield method and an alternative. Auk 96:651–661.

Knapp, A.K., and T.R. Seastedt. 1986. Detritus accumulation limits productivity of tallgrass prairie. BioScience 36:662–668.

Knight, C.L., J.M. Briggs, and M.D. Nellis. 1994. Expansion of gallery forest on Konza Prairie Research Natural Area, Kansas, USA. Landscape Ecol. 9:117–125.

Knopf, F.L. 1994. Avian assemblages on altered grasslands. Stud. Avian Biol. 15:247–257.

Mayfield, H. 1961. Nesting success calculated from exposure. Wilson Bull. 73:255–261.

Mayfield, H. 1975. Suggestions for calculating nest success. Wilson Bull. 87:456–466.

Mengel, R.M. 1970. The North American central plains as an isolating agent in bird speciation. Pp. 279–340 *in* W. Dort, Jr. and J.K. Jones, Jr., eds. Pleistocene and Recent environments of the central Great Plains. Univ. Kansas Publ. 3. Dept. Geology, Lawrence.

Miller, C.K., T.L. Knight, L.C. McEwen, and T.L. George. 1994. Responses of nesting Savannah Sparrows to fluctuations in grasshopper densities in interior Alaska. Auk 111:962–969.

Ralph, C.J., G.R. Geupel, P. Pyle, T.E. Martin, and D.F. DeSante. 1992. Field methods for monitoring landbirds. USDA For. Serv. Redwood Sci. Lab., Arcata, CA.

Schroeder, W.A. 1983. Presettlement prairie of Missouri. Missouri Dept. Conserv. Nat. Hist. Ser. 2. Jefferson City, MO.

Temple, S.A. 1990. Sources and sinks for regional bird populations. Passenger Pigeon 52:35–37.

Thompson, C.F., and V. Nolan, Jr. 1973. Population biology of the Yellow-breasted Chat (*Icteria virens* L.) in southern Indiana. Ecol. Monogr. 43:145–171.

Wells, P.V. 1970. Historical factors controlling vegetation patterns and floristic distributions in the central plains region of North America. Pp. 211–240 *in* W. Dort, Jr., and J.K. Jones, Jr., eds. Pleistocene and Recent environments in the central Great Plains. Univ. Kansas Publ. 3. Dept. Geology, Lawrence.

Wiens, J.A. 1973. Pattern and process in grassland bird communities. Ecol. Monogr. 43:237–270.

Wiens, J.A. 1974. Climatic instability and the "ecological saturation" of bird communities in North American grasslands. Condor 76:385–400.

Wiens, J.A., and M.I. Dyer. 1975. Rangeland avifaunas: their composition, energetics, and role in the ecosystem. Pp. 146–182 *in* D.R. Smith, tech. coord. Proc. Symp. Manage. of For. and Range Hab. for Nongame Birds. USDA For. Serv. Gen. Tech. Rep. WO-1. Washington, DC.

Zimmerman, J.L. 1965. The bioenergetics of the Dickcissel, *Spiza americana.* Physiol. Zool. 38:370–389.

Zimmerman, J.L. 1971. The territory and its density dependent effect in *Spiza americana.* Auk 88:591–612.

Zimmerman, J.L. 1982. Nesting success of Dickcissels (*Spiza americana*) in preferred and less preferred habitats. Auk 99:292–298.

Zimmerman, J.L. 1988. Breeding season habitat selection by the Henslow's Sparrow (*Ammodramus henslowii*) in Kansas. Wilson Bull. 100:17–24.

Zimmerman, J.L. 1992. Density-independent factors affecting the avian diversity of the tallgrass prairie community. Wilson Bull. 104:85–94.

Zimmerman, J.L. 1993. The birds of Konza: the avian ecology of the tallgrass prairie. Univ. Press Kansas, Lawrence.

8. Effects of Fire on Bird Populations in Mixed-Grass Prairie

Douglas H. Johnson

Introduction

The mixed-grass prairie is one of the largest ecosystems in North America, originally covering about 69 million ha (Bragg and Steuter 1995). Although much of the natural vegetation has been replaced by cropland and other uses (Samson and Knopf 1994, Bragg and Steuter 1995), significant areas have been preserved in national wildlife refuges, waterfowl production areas, state game management areas, and nature preserves. Mixed-grass prairie evolved with fire (Bragg 1995), and fire is frequently used as a management tool for prairie (Berkey et al. 1993).

Much of the mixed-grass prairie that has been protected is managed to enhance the reproductive success of waterfowl and other game birds, but nongame birds now are receiving increasing emphasis. Despite the importance of the area to numerous species of birds and the aggressive management applied to many sites, relatively little is known about the effects of fire on the suitability of mixed-grass prairie for breeding birds. Several studies have examined effects of fire on breeding birds in the tallgrass prairie (e.g., Tester and Marshall 1961, Eddleman 1974, Halvorsen and Anderson 1983, Westenmeier and Buhnerkempe 1983, Zimmerman 1992, Herkert 1994), in western sagebrush grasslands (Peterson and Best 1987), and in shrub-steppe (Bock and Bock 1987).

Studies of fire effects in the mixed-grass prairie are limited. Huber and Steuter (1984) examined the effects on birds during the breeding season after an early-May prescribed burn on a 122-ha site in South Dakota. They contrasted the bird populations on that site to those on a nearby 462-ha unburned site that had been lightly grazed by bison (*Bison bison*). Pylypec (1991) monitored breeding bird populations occurring in fescue prairies of Canada on a single 12.9-ha burned area and on an adjacent 5.6-ha unburned fescue prairie for 3 years after a prescribed burn.

This chapter describes the effects of prescribed fire on common terrestrial birds at a mixed-grass prairie site in east central North Dakota. Birds were censused annually during 1972–1995 on seven plots subjected to various regimes of prescribed fire.

Study Area

The Woodworth Study Area consists of 1,231 ha of mixed-grass prairie pothole habitat in east central North Dakota (Higgins et al. 1992). It is situated on the Missouri Coteau, a morainal belt extending from south central to northwestern North Dakota. The rolling terrain contains 548 wetlands on the study area, totaling 10% of the land area. Most of these are seasonally flooded (classified according to Stewart and Kantrud 1971), but many temporary and semipermanent wetlands also are present.

Before purchase by the U.S. Fish and Wildlife Service in the mid-1960s, land use on the study area was a mixture of grazing by cattle, haying, and crop production (Bayha 1964). Those practices continue on the privately owned portions of the study area. On the Service-owned portions (87% of the area), management of the uplands since acquisition has emphasized restoration of grassland. Some formerly cropped fields were replanted to grasses or grass-legume mixtures. Unplowed grasslands have been managed mostly by prescribed burning.

Study plots were located in relatively homogeneous areas within seven different quarter sections, the units that received various treatments under the management of the study area. I located study plots to avoid large wetlands, in order to concentrate on upland bird communities. Plots were measured and marked by use of compass and pacing. Surveyor's flags were placed at 40-m intervals on a grid throughout each plot to facilitate recording of bird locations. One plot served as a control. The other six plots were subjected to burning under different regimes, intervals between burns averaging 3–5 years. Spring burns were slightly more frequent than fall burns. The seven study plots were denoted by the quarter section in which they were located (e.g., plot 13 was located within quarter section 13).

Plot 13, the control, had been grazed from 1906 to 1961 but has been left idle subsequently. Originally 8.09 ha in size, plot 13 was increased to 10.12 ha in 1973. All or portions of eight seasonal and two temporary wetland basins lie within the plot, totaling about 1.2 ha. Upland vegetation is a mixture of grasses, forbs, and woody plants. Common species are Kentucky bluegrass (*Poa pratensis*), needle-and-thread (*Stipa comata*), stiff sunflower (*Helianthus rigidus*), and Canada golden-rod (*Solidago canadensis*). Wolfberry (*Symphoricarpos occidentalis*), silverberry (*Eleagnus commutata*), and Woods' rose (*Rosa woodsii*) form shrubby patches of various sizes. One thicket of chokecherry (*Prunus virginiana*), surrounded by wolfberry and silverberry, became decadent during the years of the surveys.

Plot 2, 8.68 ha in size, was surveyed during 1977–1982, after which the trail to it became difficult. It had been grazed or hayed from 1906 to 1967, then left idle except for five prescribed burns. Plot 2 contains all or portions of seven seasonal wetlands, totaling about 0.9 ha. Kentucky bluegrass is abundant, and needle-and-thread, yarrow (*Achillea lanulosa*), fringed sage (*Artemisia frigida*), and prairie wild rose (*Rosa arkansana*) are common. Stands of wolfberry and silverberry also occur.

Plot 7, 6.07 ha in extant, was hayed during 1904–1955. It was plowed in 1956 for a year of crop production. Alfalfa (*Medicago sativa*) and possibly some tame grasses were planted in 1958, after which it was grazed or hayed until 1970. Plot 7 has since been subjected to four prescribed burns. Six seasonal and two temporary wetlands, covering 1.1 ha, are included in the plot. Dominant upland plants are needle-and-thread, green needlegrass (*Stipa viridula*), alfalfa, Kentucky bluegrass, rigid goldenrod (*Solidago rigida*), and stiff sunflower. Patches of silverberry and wolfberry have increased in area during the study.

Plot 9, 6.07 ha in size, consists of unbroken sod that had been hayed and probably grazed from 1908 to 1965. There have been five burns since 1965. Two small seasonal and one ephemeral wetland cover 0.2 ha. Dominant plant species are Kentucky bluegrass, needle-and-thread, yellow sweetclover (*Melilotus officinalis*), white prairie aster (*Aster ericoides*), and stiff sunflower. Wolfberry occurs in several patches.

Plot 11, 4.86 ha in size, had been cropped during 1917–1927 and from 1934 to about 1940. It then reverted to grass and was grazed through 1970. Since then, it has been burned four times. Plot 11 contains one ephemeral and portions of two seasonal wetlands, totaling about 0.5 ha. The uplands are dominated by Kentucky bluegrass and, to a lesser extent, smooth brome (*Bromus inermis*).

Plot 16 is 6.07 ha in size. It was grazed from 1906 to 1968, after which it was treated with a total of six prescribed fires. Five seasonal wetlands and small portions of two seasonal to semipermanent wetlands lie within the plot, comprising about 0.6 ha. Common plants are Kentucky bluegrass, quackgrass (*Agropyron repens*), needle-and-thread, and little bluestem (*Schizachyrium scoparium), as well as wolfberry and silverberry.*

Plot 18, also 6.07 ha in size, is unbroken prairie sod that had been grazed from 1906 to 1968. The plot was burned seven times between 1969 and 1990. In addition, it was intensively grazed by sheep during 1973 and 1974. The plot contains four small wetlands—one ephemeral, two temporary, and one seasonal—covering less than 0.1 ha. At the beginning of the surveys, the plot had several thickets of chokecherry and hawthorn (*Crataegus chrysocarpa*). The various fire and grazing treatments, as well as nest-searching with a cable-chain device (Higgins et al. 1969), have reduced the thickets considerably. Other common plants include Kentucky bluegrass, blue grama (*Bouteloua gracilis*), and fringed sage.

I excluded plot 18 in 1974–1975 because it had been crowd-grazed by sheep during the previous growing seasons. I excluded two plots for a year each (plot 9 in 1982, plot 18 in 1993) because portions were burned during the survey period. In 1977, two plots were burned on June 1. Several surveys of both plots had been completed before the burns; other surveys were conducted after the burns. The preburn and postburn results were used separately in the analysis.

Field Methods

Each year during 1972–1995 (1977–1982 for plot 2) the breeding bird community of each plot was estimated by conducting several surveys and mapping territories. Standard survey methods (Hall 1964, Van Velzen 1972) were used, and annual reports were published in *American Birds* or the *Journal of Field Ornithology*. See Johnson (1996a) for a listing of references. About eight visits were made to each plot during late May through mid-June. Surveys were conducted from just before dawn to late morning. Early-morning surveys emphasized concurrent registrations of indicated pairs of the same species, to define multiple territories. Surveys later in the morning, when vocalizations were reduced, focused on reflushing birds to delineate their territories (Wiens 1969).

In most years, one other observer and I conducted independent surveys and compared results. I estimated the number of territories from the locations plotted on field maps. There were two exceptions. In 1972, P.F. Springer conducted all surveys and estimated territories on plot 9 for the *American Birds* report. C.A. Faanes conducted surveys and estimated territories on all plots in 1983. For consistency in the analyses presented here (Best 1975), I reviewed all original field data forms associated with these exceptions and independently estimated the number of territories. To conform with guidelines for publishing results of breeding bird censuses, the number of territories was estimated to the nearest half. Sometimes such rounding either over- or underestimated the true number of territories in a plot, so I modified the counts with a + or –. For analyses here, such modified counts were adjusted by 0.2 territory (e.g., a 1+ was converted to 1.2). Also, partial territories, recorded as + in summary reports, were converted to 0.2 for analysis. Visitor species (those recorded on only one to three of the surveys) were credited with 0.1 territory. For Brown-headed Cowbirds, the average number of females seen during the surveys was used as the number of territories.

The reliability of censuses of small plots is greater for species with small home ranges than for those with large ones. Accordingly, results for sparrows, Red-winged Blackbirds (*Agelaius phoeniceus*), Bobolinks (*Dolichonyx oryzivorus*), and warblers may be more reliable than those for wider-ranging species such as Western Meadowlark (*Sturnella neglecta*) and shorebirds.

Analytic Methods

Analyses were intended to determine effects on bird populations of the length of time since a plot was burned. The response variable was the density of indicated pairs of a species in a plot during a particular year. Analyses also were carried out using the log-transformed values, log(density + 1). For all species, however, models for the untransformed data provided a better fit to the data; results of the analyses of log-transformed data will not be discussed further. Primary interest was in the influence on density of the number of years since the most recent burn. Years since burn were coded in whole years for burns conducted in spring (April–June) and in half-years for burns conducted in fall (August–October).

The first approach used related the density to the number of years since burn. This simple approach ignores effects due to the different plots and to the various years. Effects of these confounding variables can complicate interpretation of the effects of the explanatory variables of interest. The density of birds varies from plot to plot because of intrinsic differences in habitat. Also, densities vary annually due to climatic variation, regional changes in the population size, and other influences. To eliminate the effects of these "nuisance" variables and focus on the treatment effects, I also took a second approach, which involved two steps. First, I modeled the density of a species as a linear function of plot and year:

$$\text{DENSITY}_{it} = \text{PLOT}_i + \text{YEAR}_t + \text{RESIDUAL}_{it}$$

In the second step, residuals from that models, termed *adjusted densities,* were related to the explanatory variable, years since burn. PROC GLM (SAS Institute 1990) was used for the first step. In the second step, the locally weighted regression (loess) procedure of S-PLUS (Statistical Sciences, Inc. 1993) was used. Because there was no variation in the explanatory variable for the control plot

Table 8.1. Total Numbers of Indicated Pairs of Common Terrestrial Bird Species Recorded in Surveys at Woodworth Study Area, North Dakota, 1972–1995

Species	Total Count
Clay-colored Sparrow	275.5
Red-winged Blackbird	199.2
Bobolink	158.2
Brown-headed Cowbird	133.1
Grasshopper Sparrow	115.1
Western Meadowlark	110.7
Common Yellowthroat	77.6
Eastern Kingbird	76.1
Willow Flycatcher	35.6
Savannah Sparrow	34.0
Yellow Warbler	28.9
Upland Sandpiper	25.2
Sedge Wren	19.8
Killdeer	17.6
Marbled Godwit	7.9
Willet	7.9
Baird's Sparrow	6.7

(plot 13), effects were totally confounded with year, so data from that plot were excluded from the analysis of residuals.

The analysis was conducted on 17 of the most common terrestrial species recorded on the plots (Table 8.1). The American Goldfinch (*Carduelis tristis*) was omitted because it nests later in the season, so the surveys were inadequate to assess their breeding densities. The Western Kingbird (*Tyrannus verticalis*), Gray Catbird (*Dumetella carolinensis*), and swallows were omitted because they nest mostly outside the plots and used the plots only for foraging.

Results and Discussion

Killdeers (*Charadrius vociferus*) were most common at Woodworth immediately after a burn (Fig. 8.1). Densities and adjusted densities declined sharply with age of burn. Killdeer prefer open habitats with bare ground or very short vegetation, little residual vegetation, and no woody vegetation (Sample 1989, Best et al. 1995). These habitats include cultivated fields, heavily grazed pastures, bare shorelines of lakes and ponds, and exposed gravel and sand (Stewart 1975, Kantrud 1981). Open habitat with sparse cover is created immediately after a burn, which accounts for Killdeer favoring such plots at Woodworth. Huber and Steuter (1984) reported a noticeable absence of killdeers on unburned areas.

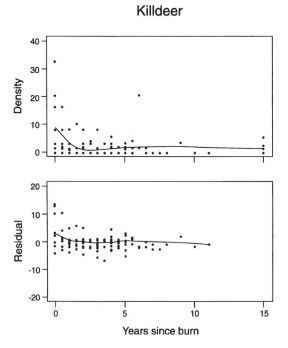

Figure 8.1. (Top) Density (pairs per 100 ha) of Killdeer since plot was burned at Woodworth, North Dakota. (Bottom) Density (pairs per 100 ha) of Killdeer adjusted for year and plot effects since plot was burned.

Figure 8.2. (Top) Density (pairs per 100 ha) of Marbled Godwit since plot was burned at Woodworth, North Dakota. (Bottom) Density (pairs per 100 ha) of Marbled Godwit adjusted for year and plot effects since plot was burned.

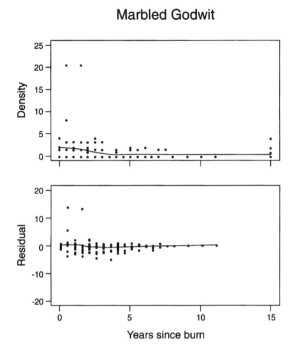

Marbled Godwits (*Limosa fedoa*) were not common at Woodworth, but highest densities were found in plots during the first 2 years after a burn (Fig. 8.2). Godwits often forage in wetlands, but they nest and forage in upland areas, sometimes a considerable distance from water (Stewart 1975). They often feed in upland areas immediately after a burn. Preferred habitats of breeding godwits include idle grasslands and pastures with short vegetative cover (Ryan et al. 1984). They nest in cover of low-to-intermediate density and height (Kantrud and Higgins 1992). At Woodworth, the occasional high densities of Marbled Godwits were in plots within 2 years of a prescribed burn, where the lack of litter and low vegetation profile facilitated feeding. Ryan et al. (1984) recommended periodic disturbance by fire, grazing, or mowing to produce the shorter-grass habitats favored by this species.

Upland Sandpipers (*Bartramia longicauda*) at Woodworth had highest densities and adjusted densities immediately and for about a year after a burn (Fig. 8.3). This appropriately named shorebird spends little time near water, nesting and foraging on dry land. It favors open grassland, with short vegetation and little woody cover (Skinner 1982, Sample 1989), and uses bare-ground habitats such as cropland (Stewart 1975). Although the Upland Sandpiper favors somewhat heavier cover for nest sites (Kantrud and Higgins 1992, Bowen and Kruse 1993), it uses open habitats for foraging (Kantrud 1981). Such selection for feeding areas is consistent with the higher-than-average use of recently burned plots at Woodworth. Huber and Steuter (1984) also reported that Upland Sandpipers used burned fields more often than unburned ones.

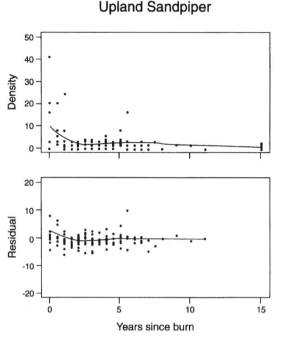

Upland Sandpiper

Figure 8.3. (Top) Density (pairs per 100 ha) of Upland Sandpiper since plot was burned at Woodworth, North Dakota. (Bottom) Density (pairs per 100 ha) of Upland Sandpiper adjusted for year and plot effects since plot was burned.

The Willet (*Catoptrophorus semipalmatus*) was uncommon on the Woodworth study plots; both densities and adjusted densities appeared unrelated to the age of burn (Fig. 8.4). Willets are most common in marshy areas and nearby uplands (Ehrlich et al. 1988). In North Dakota, Stewart (1975) indicated that they use mostly semipermanent and seasonal wetlands. Ryan and Renken (1987), however, found that, relative to the available area, Willets prefer less-permanent wetland types and alkali wetlands. Willets nest in the upland, often a considerable distance from water (Stewart 1975, Ryan and Renken 1987), typically in short, grassy cover (Higgins et al. 1979). Although Willets make little use of uplands except for nesting, they tend to favor short, native vegetation (Ryan and Renken 1987, Kantrud and Higgins 1992). With the Willet's affinity for wetlands and its large territory size, no selection of habitats at Woodworth was evident.

Densities of Eastern Kingbirds (*Tyrannus tyrannus*) at Woodworth were slightly lower than average for the first 3 years after a burn, although there was much variation around the loess line (Fig. 8.5). Adjusted densities were virtually unaffected by the recency of burn. Eastern Kingbirds use woodlands with open canopies and habitats with scattered stands of small trees or shrubs (Stewart 1975). Sample (1989) indicated that the species uses a variety of open habitats, with the nearby presence of tall, woody vegetation. At Woodworth, Eastern Kingbirds require only a small thicket of chokecherry or other woody species, from which they range out to forage into the grassland. Despite repeated fires, Woodworth still retains sufficient thickets for nesting eastern kingbirds. Should the small trees and shrubs be eliminated, use of the area by Eastern Kingbirds will diminish (Arnold and Higgins 1986).

Figure 8.4. (Top) Density (pairs per 100 ha) of Willet since plot was burned at Woodworth, North Dakota. (Bottom) Density (pairs per 100 ha) of Willet adjusted for year and plot effects since plot was burned.

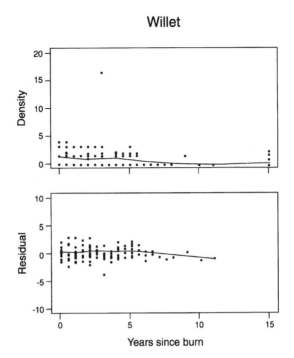

Figure 8.5. (Top) Density (pairs per 100 ha) of Eastern Kingbird since plot was burned at Woodworth, North Dakota. (Bottom) Density (pairs per 100 ha) of Eastern Kingbird adjusted for year and plot effects since plot was burned.

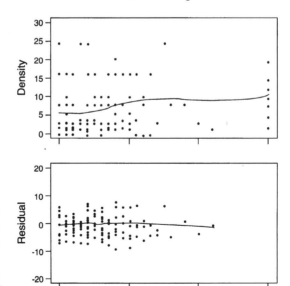

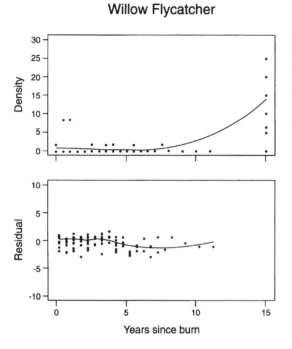

Willow Flycatcher

Figure 8.6. (Top) Density (pairs per 100 ha) of Willow Flycatcher since plot was burned at Woodworth, North Dakota. (Bottom) Density (pairs per 100 ha) of Willow Flycatcher adjusted for year and plot effects since plot was burned.

Willow Flycatchers (*Empidonax traillii*) at Woodworth were seen only in thickets of chokecherry, hawthorn, or similar vegetation. The species occurred regularly only in the control plot, where they used a large thicket in almost every year. Adjusted densities for this species showed no pattern after a burn (Fig. 8.6). The Willow Flycatcher is a species of swamps and thickets, especially of willow (*Salix* spp.) (Ehrlich et al. 1988, Sedgwick and Knopf 1992). Kahl et al. (1985) described the species' breeding habitat as having intermediate-to-tall ground vegetation; a low, open canopy; dense ground vegetation; at least a few woody stems 2.5 cm or greater diameter at breast height; an intermediate-to-high number of smaller woody stems; and a litter layer of intermediate depth and intermediate-to-dense coverage. In North Dakota, Willow Flycatchers use natural prairie thickets, consisting of species such as chokecherry, hawthorn, and wild plum (Stewart 1975). Similar vegetation occurs at Woodworth; these habitats are reduced by repeated fires. Clearly, the species would disappear once the habitat it requires was eliminated (Arnold and Higgins 1986).

Sedge Wrens (*Cistothorus platensis*) were common at Woodworth only in some years. Densities were mostly zero, but none of the positive values occurred within 1 year after a burn (Fig. 8.7). Adjusted values likewise demonstrated little response with time after a burn. In the study plots at Woodworth, the species was rarely found in emergent wetland vegetation during years with average water conditions but was frequent in upland grasses during unusually wet years, such as 1995. The breeding habitat of Sedge Wrens consists of tall and dense vegetation (Skinner 1982, Schramm et al. 1986) found in emergent wetland vegetation

Figure 8.7. (Top) Density (pairs per 100 ha) of Sedge Wren since plot was burned at Woodworth, North Dakota. (Bottom) Density (pairs per 100 ha) of Sedge Wren adjusted for year and plot effects since plot was burned.

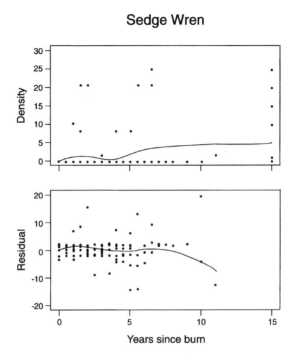

(Stewart 1975), dry marshes or wet meadows (Bent 1968), moist grasslands, old fields, dense cultivated grainfields (Dobkin 1992), or retired cropland (Stewart 1975). Tall and dense vegetation and dense, prostrate residual vegetation appear to be important (Sample 1989). Such habitats are eliminated by a burn, suggesting why the species did not occupy recently burned plots at Woodworth. By contrast, for the tallgrass prairie of Illinois, Schramm et al. (1986) suggested that Sedge Wrens prefer a clear understory and ground area and indicated that the wrens were most attracted to recently burned areas, although they used, and perhaps required, unburned areas nearby to gather material for nest construction. In that study, the recently burned areas had taller and denser vegetation than did the unburned areas.

Yellow Warblers (*Dendroica petechia*) were most common at Woodworth in the control plot. Adjusted densities showed no pattern after a burn (Fig. 8.8). The Yellow Warbler favors thickets of small deciduous trees or tall shrubs (Stewart 1975, Dobkin 1992). Knopf and Sedgwick (1992) defined more precisely the habitat features associated with nest sites, primarily characteristics of the vegetation patch. At Woodworth, Yellow Warblers were most common in the control plot, especially in the large woody thicket. This species likely would disappear once the taller woody vegetation was eliminated (Arnold and Higgins 1986).

The Common Yellowthroat (*Geothlypis trichas*) had highest densities at Woodworth in the control plot. Adjusted densities indicated no trend, except for a modest depression immediately after a burn (Fig. 8.9). The yellowthroat usually nests in tall, dense herbaceous vegetation. Breeding habitat can be either in emergent wetland vegetation or in uplands with lush vegetation, possibly includ-

Yellow Warbler

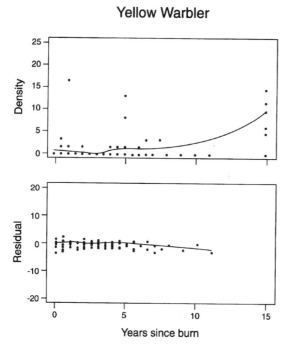

Figure 8.8. (Top) Density (pairs per 100 ha) of Yellow Warbler since plot was burned at Woodworth, North Dakota. (Bottom) Density (pairs per 100 ha) of Yellow Warbler adjusted for year and plot effects since plot was burned.

Common Yellowthroat

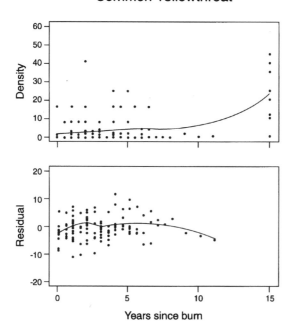

Figure 8.9. (Top) Density (pairs per 100 ha) of Common Yellowthroat since plot was burned at Woodworth, North Dakota. (Bottom) Density (pairs per 100 ha) of Common Yellowthroat adjusted for year and plot effects since plot was burned.

Figure 8.10. (Top) Density (pairs per 100 ha) of Bobolink since plot was burned at Woodworth, North Dakota. (Bottom) Density (pairs per 100 ha) of Bobolink adjusted for year and plot effects since plot was burned.

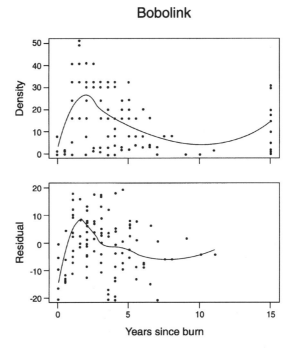

ing shrubs or small trees (Stewart 1975, Kahl et al. 1985). Sample (1989) noted that the species prefers the presence of some standing residual cover and dense, prostrate residual vegetation. At Woodworth, wetland habitats were not usually affected markedly by fires, so yellowthroats used those habitats even immediately after a fire. Upland thickets sometimes were affected by the burns, and the species would avoid those areas until the vegetation had a chance to regrow. Huber and Steuter (1984) found yellowthroats more common, but not significantly so, in unburned than in burned grassland. Repeated fires that eliminate the brushy vegetation will reduce the suitability of the upland habitat for this species (Arnold and Higgins 1986).

Densities of Bobolinks at Woodworth were low immediately after a burn, peaked 1–3 years after a burn, and began to decline about 5 years after a burn (Fig. 8.10). Densities in the control plot remained high, however. Adjusted densities showed the same pattern. Breeding habitats of the Bobolink include mixed-grass and tallgrass prairies, wet-meadow zones of wetlands, domestic haylands, retired croplands, and occasionally active croplands (Stewart 1975). Bobolinks prefer grasslands with a high coverage of fairly lush vegetation of intermediate height, with some residual vegetation (Wiens 1969, Sample 1989). The species favors areas with deep litter and a preponderance of grasses over legumes (Wiens 1969, Bollinger and Gavin 1992). Its preference for lush vegetation and deep litter is consistent with its reduced use of recently burned plots at Woodworth. Huber and Steuter (1984) also found that Bobolinks avoided burned grasslands. By contrast, Johnson and Temple (1986) found Bobolink nests most frequently in

Western Meadowlark

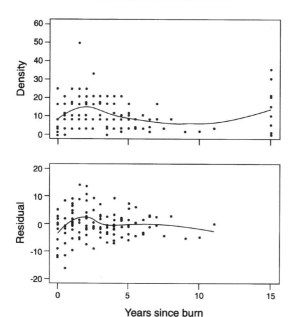

Figure 8.11. (Top) Density (pairs per 100 ha) of Western Meadowlark since plot was burned at Woodworth, North Dakota. (Bottom) Density (pairs per 100 ha) of Western Meadowlark adjusted for year and plot effects since plot was burned.

tallgrass prairie a single growing season after a burn and less frequently in fields burned less recently.

Densities of Western Meadowlarks at Woodworth were low immediately after a burn, highest 2-4 years after a burn, and lower after about 5 years after a burn (Fig. 8.11). The species was frequently common in the control plot, however. The loess line fitted to adjusted densities indicated a depression immediately after a burn and a slight peak about 3 years after a fire. Western Meadowlarks are most common in native grasslands and pastures, but also occur in hayfields, roadsides, retired cropland, and other open areas (Stewart 1975, Lanyon 1994). A preference by meadowlarks for habitats with grass and litter cover was identified by Wiens and Rotenberry (1981) and Sample (1989). Ground cover and litter seem especially important, as nests are often constructed with a dome of interwoven grasses (Lanyon 1994). Fire removes most litter, which suggests why meadowlark densities at Woodworth were somewhat depressed for about a year after a burn. Bock and Bock (1987) noted the adaptability of the species and observed that it was equally common on burned and unburned sites. Pylypec (1991) observed that Western Meadowlarks were less common on a burned area than on an unburned area for 2 years after a fire, but densities in the two areas were comparable in the third year.

Red-winged Blackbird densities at Woodworth varied only slightly with age of burn, although adjusted densities were high about 2-4 years after a burn (Fig. 8.12). Their overall numbers varied dramatically among years, as is demonstrated by the variation in densities for the control plot. Red-winged Blackbirds use a wide variety of habitats for breeding, including marshes, riparian areas,

Figure 8.12. (Top) Density (pairs per 100 ha) of Red-winged Blackbird since plot was burned at Woodworth, North Dakota. (Bottom) Density (pairs per 100 ha) of Red-winged Blackbird adjusted for year and plot effects since plot was burned.

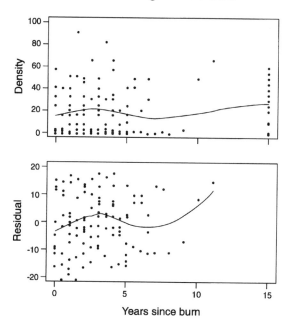

ditches, hayfields, weedy fields, and active and retired croplands (Stewart 1975, Sample 1989). Sample (1989) reported that Red-winged Blackbirds prefer lush habitats with fairly tall, dense vegetation and with standing and prostrate residual vegetation. At Woodworth, Red-winged Blackbirds are most commonly associated with wetlands, and their numbers fluctuated in close relation to water conditions. For that reason, redwing densities did not exhibit any marked response to the fire regime but were widely variable around the loess line. Huber and Steuter (1984) indicated that Red-winged Blackbirds used the burned and unburned treatments similarly during June, but by July use was reduced on the burned treatment. Eddleman (1974) stated that Red-winged Blackbirds increased when long-term protection from burning causes increased shrub and forb cover. Arnold and Higgins (1986) also noted higher densities of Red-winged Blackbirds in grassland areas with more shrubby vegetation. Although Red-winged Blackbirds used brushy areas at Woodworth, they appeared to rely on the wetland habitat.

Densities of Brown-headed Cowbirds (*Molothrus ater*) at Woodworth were depressed for the first year or so after a burn but were constant after that time (Fig. 8.13). Cowbirds are habitat generalists, preferring habitats with low or scattered trees among grassland vegetation (Lowther 1993). The slight reduction in cowbird densities after prescribed burns at Woodworth may reflect the reduction of potential host nests in those plots. Huber and Steuter (1984) found no response by cowbirds to spring burns.

Savannah Sparrows (*Passerculus sandwichensis*) at Woodworth had highest densities 1–5 years after a burn, although they were occasionally common in the control plot (Fig. 8.14). The loess line fitted to adjusted densities indicated that the

Brown-Headed Cowbird

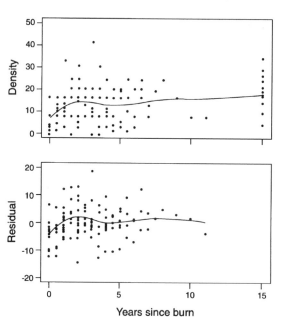

Figure 8.13. (Top) Density (pairs per 100 ha) of Brown-headed Cowbird since plot was burned at Woodworth, North Dakota. (Bottom) Density (pairs per 100 ha) of Brown-headed Cowbird adjusted for year and plot effects since plot was burned.

Savannah Sparrow

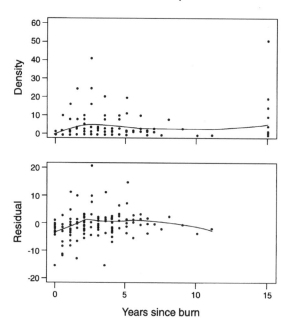

Figure 8.14. (Top) Density (pairs per 100 ha) of Savannah Sparrow since plot was burned at Woodworth, North Dakota. (Bottom) Density (pairs per 100 ha) of Savannah Sparrow adjusted for year and plot effects since plot was burned.

species was less common for about 1 year or so immediately after a burn, with possibly a longer-term depression as well. Favored breeding habitat of the Savannah Sparrow has been described in various ways: dense ground vegetation, especially grasses, and moist microhabitats (Wiens 1969); dry or wet habitats with little woody cover, moderate-to-high cover of fairly short vegetation, and medium litter depth (Sample 1989); and sparse ground cover and moderate aboveground cover (Skinner 1982). Wheelwright and Rising (1993) included grassy meadows, cultivated fields (especially alfalfa), lightly grazed pastures, roadsides, and sedge bogs as Savannah Sparrow habitats. In North Dakota, the species selects tallgrass prairie, lightly grazed mixed-grass prairie, wet-meadow zones bordering wetlands, hayfields, weedy fields, and retired croplands (Stewart 1975). Within plots at Woodworth, Savannah Sparrows were not regularly seen in areas with extensive shrub patches; Arnold and Higgins (1986) reported a similar response. At Woodworth, Savannah Sparrows avoided plots for a year or so after a burn, but no other responses to burns were evident. Halvorsen and Anderson (1983) also reported that Savannah Sparrow densities were reduced on burned fields, a finding they attributed to the reduction of residual cover. Huber and Steuter (1984) likewise noted that the species was absent from a field burned earlier in the spring. Pylypec (1991) indicated that Savannah Sparrows in a fescue prairie were reduced for 3 years after a burn.

Densities of Grasshopper Sparrows (*Ammodramus savannarum*) at Woodworth were depressed for about a year after a burn; they increased after that time and appeared to decline gradually after about 5 years after a burn (Fig. 8.15). Adjusted

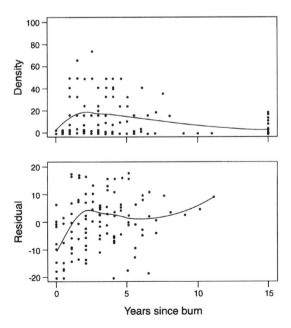

Figure 8.15. (Top) Density (pairs per 100 ha) of Grasshopper Sparrow since plot was burned at Woodworth, North Dakota. (Bottom) Density (pairs per 100 ha) of Grasshopper Sparrow adjusted for year and plot effects since plot was burned.

densities were more marked, with values depressed for 2 years after a burn, and no effect of burning thereafter. Over its broad breeding range, the Grasshopper Sparrow uses a variety of habitats but prefers grassland, old field, and retired cropland (Smith 1963, Stewart 1975, Kahl et al. 1985, Johnson and Schwartz 1993). Dense ground vegetation and shallow-to-moderate litter accumulation were identified as habitat features by Huber and Steuter (1984) and Kahl et al. (1985). Preference for patchy vegetation was noted by Skinner (1982) and Sample (1989). Whitmore (1981) observed that Grasshopper Sparrows favor bunchgrass habitats, with gaps in vegetation that facilitate movement. Smith (1963) indicated that Grasshopper Sparrows abandon fields once they become filled with shrubs. The requirement by the species for some litter accumulation is consistent with the reduced use of Woodworth plots for a year or so after a burn. Although Grasshopper Sparrows used the control plot regularly, they tended to concentrate in areas that lacked heavy shrub cover, similar to the finding of Arnold and Higgins (1986). In Montana, Bock and Bock (1987) observed fewer Grasshopper Sparrows in burned than unburned shrub-steppe habitats. Huber and Steuter (1984) reported Grasshopper Sparrows recolonizing a spring-burned field by mid-July of the same year. In tallgrass prairie in western Minnesota, Johnson and Temple (1986) found Grasshopper Sparrow nests more commonly in fields with four or more growing seasons since a burn. Eddleman (1974) indicated that burning will partially provide the interspersion of cover heights necessary for the species, but a severe lack of litter will result in few nesting sites and a lack of nesting material.

Baird's Sparrows (*Ammodramus bairdii*) occurred infrequently at Woodworth. Densities tended to be highest in plots 2-5 years after a burn, although birds occasionally used the control plot (Fig. 8.16). The adjusted densities indicated a slight depression for the first couple of years after a burn, followed by a modest increase. The low observed densities caution against drawing any firm conclusions, however. The Baird's Sparrow is a mixed-grass prairie species, endemic to the northern Great Plains. In North Dakota, it favors idle or lightly grazed areas, although it also uses more heavily grazed sites and, particularly in more arid locations, lowland areas (Stewart 1975). Baird's Sparrows favor abundant residual vegetation (Salt and Salt 1976) and a dense understory (De Smet and Miller 1989), which is consistent with its occurrence at Woodworth in plots several years after a burn. The species avoids areas with extensive bare ground (Davis and Duncan submitted). It has been reported to favor areas with some but not extensive coverage of shrubs (Arnold and Higgins 1986; Winter 1994; Dale et al. in press). Anstey et al. (1995) detected a positive association between Baird's Sparrow densities and litter depth. Winter (1994) suggested that the species favored intermediate depths of litter. At Woodworth, the species was absent in recently burned plots, which typically have little remaining litter. Pylypec (1991) observed that Baird's Sparrows avoided a recently burned area, returned to the site the second year after burn, and in the third year had densities comparable with an unburned area.

Densities of Clay-colored Sparrows (*Spizella pallida*) at Woodworth increased almost monotonically after a burn (Fig. 8.17). The occasional high densities in

Figure 8.16. (Top) Density (pairs per 100 ha) of Baird's Sparrow since plot was burned at Woodworth, North Dakota. (Bottom) Density (pairs per 100 ha) of Baird's Sparrow adjusted for year and plot effects since plot was burned.

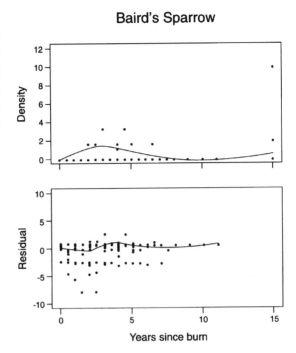

Figure 8.17. (Top) Density (pairs per 100 ha) of Clay-colored Sparrow since plot was burned at Woodworth, North Dakota. (Bottom) Density (pairs per 100 ha) of Clay-colored Sparrow adjusted for year and plot effects since plot was burned.

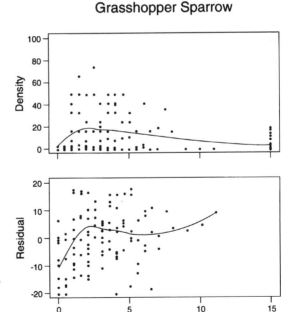

recently burned plots was due to the occupancy by Clay-colored Sparrows of brush thickets that were not markedly affected by the fires. Adjusted densities showed a similar pattern. The enormous variability among those points reflects the high annual variation in abundance of the species. Although Clay-colored Sparrows breed in a variety of habitats, ranging from retired cropland and abandoned fields to forest edges (Stewart 1975, Knapton 1994), in North Dakota the prime habitats are low thickets of wolfberry and other shrubs (Stewart 1975). Those habitats are common at Woodworth, accounting for the abundance of the species. Thickets were especially prominent in the control plot, which had the highest densities of Clay-colored Sparrows. Densities tended to increase with the number of years since the most recent burn. Clay-colored Sparrows were common even in recently burned plots, however, particularly if the burn was incomplete and did little damage to thickets. Huber and Steuter (1984) noted that the species favored the dense grassland that resulted from light grazing and the absence of fire. Halvorsen and Anderson (1983) observed declines exceeding 90% for Clay-colored Sparrows immediately after a burn, a change they attributed to the reduction of residual cover. In Saskatchewan, Pylypec (1991) noted that a burned area supported Clay-colored Sparrows at only one-third the density of an unburned area for the first 3 years after the burn. Long-term idling of grassland habitat will permit the encroachment of shrubby and other woody vegetation, which favors this species (Eddleman 1974, Arnold and Higgins 1986), as was observed in the control plot.

Conclusions

The birds considered in these analyses can be grouped into three major categories, depending on their response to burning and successional changes in vegetation. In the first group are those species that respond positively and immediately to a burned area. Included are three of the common shorebirds at Woodworth: Killdeer, Marbled Godwit, and Upland Sandpiper. All three favor open areas with sparse vegetation, where they forage. The Killdeer and Marbled Godwit likewise nest in these open areas, but the Upland Sandpiper typically nests in heavier vegetation. Other species, not treated here because of limited numbers observed, that likely would favor recently burned mixed-grass prairies include the Horned Lark (*Eremophila alpestris*) and Vesper Sparrow (*Pooecetes gramineus*).

The second category includes those species that use habitats enhanced by long-term protection from fire, specifically the woody vegetation that encroaches in unburned grassland. The most common species at Woodworth in this group are Eastern Kingbird, Willow Flycatcher, Yellow Warbler, Common Yellowthroat, Clay-colored Sparrow, and Brown-headed Cowbird. The Red-winged Blackbird also uses brushy vegetation but at Woodworth relied more on wetland habitats.

In the third category are birds that avoid recently burned areas but favor grassland with little or no woody vegetation. Several of these species are most common 2-5 years after a fire. These might be termed true grassland species.

Included in this category are Bobolink, Western Meadowlark, Grasshopper Sparrow, Baird's Sparrow, and Savannah Sparrow.

Two species analyzed here did not fit into any of the three categories. The Willet, although commonly seen in the uplands, uses mostly wetland habitat except for nesting. No evidence of a response to burning was detected. The Sedge Wren used upland habitats, but usually only when long-term precipitation patterns resulted in luxuriant herbaceous growth. This species showed no response to grassland burning, except for a reduction immediately after a fire.

A Proposed Conservation Strategy

Results presented here suggest a conservation strategy for the northern Great Plains involving prescribed burning. On large areas, such as wildlife refuges, only portions should be burned in any particular year, and these on a rotational basis. The same prescription would apply to smaller areas that can be considered as components in a landscape, such as waterfowl production areas. They should be burned periodically but not all in the same year. That strategy will ensure that in any given year habitats in a variety of successional stages will be available for a variety of breeding bird species.

This prescription will provide habitat for birds in two of the three categories, although those that benefit from long-term protection from fire will suffer. True grassland species should be emphasized in the management of mixed-grass prairies. Birds that favor short and sparse vegetative cover are common and widespread or have positive or neutral population trends, as indicated by Breeding Bird Survey results (Table 8.2). Also, many of these birds use habitats such as cultivated fields and heavily grazed pastures, which are made available from agricultural operations (Best et al. 1995), although the reproductive success in such habitats may be inadequate. Further, a regime of prescribed fires will provide suitable habitat for these species immediately after each fire.

Most of the species that favor woody vegetation also are common and widespread and have had neutral or positive population trends (Table 8.2). They can also rely on habitats provided on private land, including shelterbelts, suburban areas, and wetlands. Moreover, all these species have widespread distributions and are more common elsewhere than the mixed-grass prairies (Price et al. 1995). An exception is the Clay-colored Sparrow. That species has declined during the past 25 years or so, although not significantly, and its center of abundance is in the mixed-grass prairie. A prescribed-burning program that eliminated brushy vegetation would reduce breeding populations of Clay-colored Sparrows. Mitigating this concern is the fact that the species uses brushy habitat, which is common in private pastures, and retired cropland, such as offered by the Conservation Reserve Program (Johnson and Schwartz 1993).

Although true grassland birds suffer short-term habitat losses from a burn, they do require grassland, which in turn requires periodic fire for maintenance. Several of these species have suffered long-term population declines (Table 8.2). More-

Table 8.2. Trends from Breeding Bird Survey for Central Region, 1966–1994 and 1980–1994, and Route Averages for 1966–1994

Species	1966–1994 Route Average	1966–1994 Trend[a]	1980–1994 Trend
Killdeer	8.88	−0.3	−2.0 ↓↓↓
Marbled Godwit	1.36	0.7	NA
Upland Sandpiper	3.92	2.1 ↑↑↑	0.3
Willet	1.03	−1.8	−0.4
Eastern Kingbird	7.78	0.2	0.4
Sedge Wren	1.25	1.3	5.7 ↑↑↑
Yellow Warbler	2.29	−0.2	2.7 ↑↑
Common Yellowthroat	6.87	−0.9 ↓↓↓	−2.1 ↓↓↓
Bobolink	6.46	−2.4 ↓↓↓	−3.0 ↓↓↓
Western Meadowlark	95.69	−0.3	0.3
Red-winged Blackbird	85.09	−0.5 ↓↓	−1.3 ↓↓↓
Brown-headed Cowbird	26.17	−0.5 ↓	−0.2
Savannah Sparrow	6.98	0.5	1.5
Grasshopper Sparrow	8.95	−2.9 ↓↓↓	−1.8 ↓↓
Baird's Sparrow	1.82	−0.9	−0.5
Clay-colored Sparrow	6.67	−1.1	0.9

[a]Average annual change; ↑↑↑, increasing at $P = .01$; ↑↑, increasing at $P < .05$; ↑, increasing at $P < .10$; ↓↓↓, decreasing at $P = .01$; ↓↓, decreasing at $P < .05$; ↓, decreasing at $P < .10$; NA, not available.

over, they typically do not attain high densities or reproduce successfully in habitats other than grassland, as do birds in the other two categories. Furthermore, these species generally have breeding distributions centered in the grasslands of the midcontinent.

Concern about declining bird populations is widespread (Terborgh 1989), but grassland specialists seem particularly threatened. Numbers of many such species have suffered population declines, as judged by results of the Breeding Bird Survey (BBS), more consistently and severely than other groups of species (Peterjohn and Sauer 1993). More troubling, the BBS was begun in the mid-1960s; we lack quantitative evidence of how grassland birds responded to the massive conversion of native prairie to cropland that took place earlier (Johnson 1996b).

Further, uncultivated grasslands are vital for certain species. Although some species such as Horned Lark and Vesper Sparrow have adapted to cropland habitats (Best et al. 1995), many, such as Sprague's Pipit (*Anthus spragueii*) and Burrowing Owl (*Athene cunicularia*), have not. And even those species that use cultivated habitats may not be reproductively successful (Rodenhouse and Best 1983).

The fraction of native mixed-grass prairie that has been protected is small (Samson and Knopf 1994, Noss et al. 1995); most of what remains is privately owned and is used for grazing. Fortunately, much of the publicly owned mixed-grass prairie is included in the National Wildlife Refuge System, either as refuges or as waterfowl production areas. These areas also include many formerly cultivated fields that have been replanted to native or tame grasses and forbs. The areas are managed to benefit wildlife, particularly waterfowl. Also fortunately, a rotational system of prescribed burning, which is necessary to maintain grassland and provide habitat for nesting waterfowl, will provide breeding habitat for many of the terrestrial grassland birds.

Acknowledgments. Establishment and maintenance of the plots were performed by L.M. Kirsch (deceased), H.W. Miller, K.F. Higgins, J.M. Callow, and H.L. Clark, to whom I am most grateful. R.E. Stewart (deceased) and R.L. Kologiski (deceased) surveyed the vegetation in the plots. Other bird observers were P.F. Springer (1972), G.L. Krapu (1973), J.M. Callow (1979-1980), C.A. Faanes (1981-1983, 1987), J.M. Andrew (1981), J.L. Eldridge (1984), J. Hassler (1985), K.A. Luttschwager (1985), T.M. Pabian (1985), J.M. Hicks (1986), and M.D. Schwartz (1987-1989, 1991-1995). I am grateful to B.R. Euliss for help in preparing this report and to L.D. Igl, H.A. Kantrud, and F.L. Knopf for providing comments.

References

Anstey, D.A., S.K. Davis, D.C. Duncan, and M. Skeel. 1995. Distribution and habitat requirements of eight grassland songbird species in southern Saskatchewan. Saskatchewan Wetland Conservation Corp., Regina.

Arnold, T.W., and K.F. Higgins. 1986. Effects of shrub coverages on birds on North Dakota mixed-grass prairies. Can. Field-Nat. 100:10-14.

Bayha, K.D. 1964. Project A-30: history of the Woodworth study area with emphasis on land use. Unpublished report on file at Northern Prairie Science Center, Jamestown, ND.

Bent, A.C. 1968. Life histories of North American cardinals, grosbeaks, buntings, towhees, finches, sparrows, and allies. Dover Publications, New York.

Berkey, G., R. Crawford, S. Galipeau, D. Johnson, D. Lambeth, and R. Kreil. 1993. A review of wildlife management practices in North Dakota: effects on nongame bird populations and habitats. Report submitted to Region 6, U.S. Fish and Wildl. Serv., Denver, CO.

Best, L.B. 1975. Interpretational errors in the "mapping method" as a census technique. Auk 92:452-460.

Best, L.B., K.E. Freemark, J.J. Dinsmore, and M. Camp. 1995. A review and synthesis of habitat use by breeding birds in agricultural landscapes of Iowa. Am. Midl. Nat. 134:1-29.

Bock, C.E., and J.H. Bock. 1987. Avian habitat occupancy following fire in a Montana shrubsteppe. Prairie Nat. 19:153-158.

Bollinger, E.K., and T.A. Gavin. 1992. Eastern bobolink populations: ecology and conservation in an agricultural landscape. Pp. 497-506 *in* J.M. Hagan III and D.W. Johnston, eds. Ecology and conservation of neotropical migrant landbirds. Smithsonian Instit. Press, Washington, DC.

Bowen, B.S., and A.D. Kruse. 1993. Effects of grazing on nesting by Upland Sandpipers in southcentral North Dakota. J. Wildl. Manage. 57:291-301.

Bragg, T.B. 1995. Climate, soils and fire: the physical environment of North American grasslands. Pp. 49-81 *in* K. Keeler and A. Joern, eds. The changing prairie. Oxford Univ. Press, New York.

Bragg, T.B., and A.A. Steuter. 1995. Mixed prairie of the North American Great Plains. Trans. N. Am. Wildl. and Nat. Resour. Conf. 60:335–348.

Dale, B.C., P.A. Martin, and P.S. Taylor. In press. Effects of hay management regimes on grassland songbirds in Saskatchewan.

Davis, S.K., and D.C. Duncan. Submitted. Grassland songbird abundance in native and crested wheatgrass pastures of southern Saskatchewan. J. Field Ornithol.

De Smet, K.D., and W.S. Miller. 1989. Report on the status of the Baird's Sparrow, *Ammodramus bairdii*, in Canada. Unpublished report prepared for the Committee on the Status of Endangered Wildlife in Canada.

Dobkin, D.S. 1992. Neotropical migrant landbirds in the Northern Rockies and Great Plains: a handbook for conservation and management. U.S. Dept. Agri., For. Serv. N. Region. Publ. R1-93-34. Missoula, MT.

Eddleman, W.R. 1974. The effects of burning and grazing on bird populations in native prairie in the Kansas Flint Hills. Unpublished report NSF-URP. Kansas State University, Manhattan.

Ehrlich, P.R., D.S. Dobkin, and D. Wheye. 1988. The birder's handbook: a field guide to the natural history of North American birds. Simon and Schuster, New York.

Hall, G.A. 1964. Breeding-bird censuses—why and how. Audubon Field Notes 18:413–416.

Halvorsen, H.H., and R.K. Anderson. 1983. Evaluation of grassland management for wildlife in central Wisconsin. Proc. N. Am. Prairie Conf. 7:267–279.

Herkert, J.R. 1994. Breeding bird communities of midwestern prairie fragments: the effects of prescribed burning and habitat-area. Nat. Areas J. 14:128–135.

Higgins, K.F., L.M. Kirsch, and I.J. Ball, Jr. 1969. A cable-chain device for locating duck nests. J. Wildl. Manage. 33:1009–1011.

Higgins, K.F., L.M. Kirsch, A.T. Klett, and H.W. Miller. 1992. Waterfowl production on the Woodworth station in south-central North Dakota, 1965–1981. U.S. Fish and Wildl. Serv. Resour. Publ. 180. Washington DC.

Higgins, K.F., L.M. Kirsch, M.R. Ryan, and R.B. Renken. 1979. Some ecological aspects of Marbled Godwits and Willets in North Dakota. Prairie Nat. 11:115–118.

Huber, G.E., and A.A. Steuter. 1984. Vegetation profile and grassland bird response to spring burning. Prairie Nat. 16:55–61.

Johnson, D.H. 1996a. Studies of bird communities at the Woodworth study area. Proc. N. D. Acad. Sci. 50:127–131.

Johnson, D.H. 1996b. Management of northern prairies and wetlands for the conservation of neotropical migratory birds. Pp. 53–67 *in* F. R. Thompson III, ed. Management of midwestern landscapes for the conservation of neotropical migratory birds. U.S. Dept. Agric., For. Serv. Gen. Tech. Rep. NC-187. No. Central For. Exp. Sta., St. Paul, MN.

Johnson, D.H., and M.D. Schwartz. 1993. The Conservation Reserve Program and grassland birds. Conserv. Biol. 7:934–937.

Johnson, R.G., and S.A. Temple. 1986. Assessing habitat quality for birds nesting in fragmented tallgrass prairies. Pp. 245–249 *in* J. Verner, M.L. Morrison, and C.J. Ralph, eds. Wildlife 2000: modeling habitat relationships of terrestrial vertebrates. Univ. Wisconsin Press, Madison.

Kahl, R.B., T.S. Baskett, J.A. Ellis, and J.N. Burroughs. 1985. Characteristics of summer habitats of selected nongame birds in Missouri. Res. Bull. 1056. Univ. Missouri-Columbia.

Kantrud, H.A. 1981. Grazing intensity effects on the breeding avifauna of North Dakota native grasslands. Can. Field-Nat. 95:404–417.

Kantrud, H.A., and K.F. Higgins. 1992. Nest and nest site characteristics of some ground-nesting, non-passerine birds of northern grasslands. Prairie. Nat. 24:67–84.

Knapton, R.W. 1994. Clay-colored Sparrow (*Spizella pallida*). *In* A. Poole and F. Gill, eds. The birds of North America 120. Academy of Natural Sciences, Philadelphia, and American Ornithologists' Union, Washington, DC.

Knopf, F.L. and J.A. Sedgwick. 1992. An experimental study of nest-site selection by Yellow Warblers. Condor 94:734–742.

Lanyon, W.E. 1994. Western Meadowlark (*Sturnella neglecta*). *In* A. Poole and F. Gill, eds. The birds of North America 104. Academy of Natural Sciences, Philadelphia, and American Ornithologists' Union, Washington, DC.

Lowther, P.E. 1993. Brown-headed Cowbird (*Molothrus ater*). *In* A. Poole and F. Gill, eds. The birds of North America 47. Academy of Natural Sciences, Philadelphia, and American Ornithologists' Union, Washington, DC.

Noss, R.F., E.T. LaRoe III, and J.M. Scott. 1995. Endangered ecosystems of the United States: a preliminary assessment of loss and degradation. Biol. Rep. 28. Natl. Biol. Serv., Washington, DC.

Peterjohn, B.G., and J.R. Sauer. 1993. North American breeding bird survey annual summary 1990–1991. Bird Pop. 1:52–67.

Peterson, K.L., and L.B. Best. 1987. Effects of prescribed burning on nongame birds in a sagebrush community. Wildl. Soc. Bull. 15:317–329.

Price, J., S. Droege, and A. Price. 1995. The summer atlas of North American birds. Academic Press, London.

Pylypec, B. 1991. Impacts of fire on bird populations in a fescue prairie. Can. Field-Nat. 105:346–349.

Rodenhouse, N.L., and L.B. Best. 1983. Breeding ecology of Vesper Sparrows in corn and soybean fields. Am. Midl. Nat. 110:265–275.

Ryan, M.R., and R.B. Renken. 1987. Habitat use by breeding Willets in the northern Great Plains. Wilson Bull. 99:175–189.

Ryan, M.R., R.B. Renken, and J.J. Dinsmore. 1984. Marbled Godwit habitat selection in the northern prairie region. J. Wildl. Manage. 48:1206–1218.

Salt, W.R., and J.R. Salt. 1976. The birds of Alberta. Hurtig Pub., Edmonton.

Sample, D.W. 1989. Grassland birds in southern Wisconsin: habitat preference, population trends, and response to land use changes. M.S. thesis. Univ. Wisconsin, Madison.

Samson, F., and Knopf, F. 1994. Prairie conservation in North America. Bioscience 44:418–421.

SAS Institute, Inc. 1990. SAS/STAT user's guide, version 6, fourth ed., vol. 2. SAS Inst. Inc., Cary, NC.

Schramm, P., D.S. Schramm, and S.G. Johnson. 1986. Seasonal phenology and habitat selection of the Sedge Wren *Cistothorus platensis* in a restored tallgrass prairie. Pp. 95–99 *in* G.K. Clambey and R.H. Pemble, eds. Proc. N. Am. Prairie Conf. 9. Tri-College Univ. Center for Environ. Stud., Fargo, ND.

Sedgwick, J.A., and F.L. Knopf. 1992. Describing Willow Flycatcher habitats: scale perspectives and gender differences. Condor 94:720–733.

Skinner, R.M. 1982. Vegetation structure and bird habitat selection on Missouri prairies. PhD dissertation. Univ. Missouri, Columbia.

Smith, R.L. 1963. Some ecological notes on the Grasshopper Sparrow. Wilson Bull. 75:159–165.

Statistical Sciences, Inc. 1993. S-PLUS for Windows reference manual, version 3.1. Statistical Sciences, Inc., Seattle, WA.

Stewart, R.E. 1975. Breeding birds of North Dakota. Tri-College Center for Environ. Stud., Fargo, ND.

Stewart, R.E., and H.A. Kantrud. 1971. Classification of natural ponds and lakes in the glaciated prairie region. Resour. Publ. 92. U.S. Fish and Wildl. Serv., Bur. Sport Fish. and Wildl., Washington, DC.

Terborgh, J. 1989. Where have all the birds gone? Princeton Univ. Press, Princeton, NJ.

Tester, J.R., and W.M. Marshall. 1961. A study of certain plant and animal interrelations on a native prairie in northwestern Minnesota. Minn. Mus. Nat. Hist. Occas. Pap. 8. Minneapolis.

Van Velzen, W.T. 1972. Breeding bird census instructions. Am. Birds 26:1007–1010.

Westenmeier, R.L., and J.E. Buhnerkempe. 1983. Responses of nesting wildlife to prairie grass management in prairie chicken sanctuaries in Illinois. Pp. 36–46 *in* R. Brewer, ed. Proc. Eighth N. Am. Prairie Conf. Western Michigan Univ., Kalamazoo.

Wheelwright, N.T., and J.D. Rising. 1993. Savannah Sparrow (*Passerculus sandwichensis*). *In* A. Poole and F. Gill, eds. The birds of North America 45. Academy of Natural Sciences, Philadelphia, and American Ornithologists' Union, Washington, DC.

Whitmore, R.C. 1981. Structural characteristics of Grasshopper Sparrow habitat. J. Wildl. Manage. 45:811–814.

Wiens, J.A. 1969. An approach to the study of ecological relationships among grassland birds. Ornithol. Monogr. 8. Allen Press, Lawrence, KS.

Wiens, J.A., and J.T. Rotenberry. 1981. Habitat associations and community structure in shrubsteppe environments. Ecol. Monogr. 51:21–41.

Winter, M. 1994. Habitat selection of Baird's Sparrows in the northern mixed-grass prairie. Diplomarbeit, Universitätübingen, Germany.

Zimmerman, J.L. 1992. Density-independent factors affecting the avian diversity of the tallgrass prairie community. Wilson Bull. 104:85–94.

9. Ecology of Small Mammals in Prairie Landscapes

Glennis A. Kaufman and Donald W. Kaufman

Introduction

When one crosses the prairie landscapes of central North America, one becomes aware of the impacts that humans have had on the region. Recent anthropogenic modifications of the presettlement prairie not only have changed vegetation but also altered distributional ranges, spatial use within ranges, and total numbers of many species of animals. Although human activities often reduce ranges and abundances of animals and these reductions usually are the foci of issues of conservation, human impacts do not always lead to such reductions. Anthropogenic changes can and do lead to increases in numbers, distributional ranges, or both for some species. Some of these increases result from altered landscapes that provide conditions more suitable for some species than the conditions available in native environments. Other increases are due to intentional introductions of both domestic and wild species and to accidental introductions of wild species.

Market hunting, ranching, farming, intentional and accidental introductions of non-native species, control of pest species, and urban development had an effect on many native mammals within the prairie region of central North America. However, some mammals increased in abundance, range, or both in response, directly or indirectly, to these factors. Although human activities had discernible impacts on most mammals, size of distributional ranges and abundance of some prairie mammals may have changed little after European settlement. However, any changes in overall abundance are difficult to assess given the paucity of information regarding abundance of most mammalian species in prairies of North America during the 1800s.

This chapter focuses on ecological studies of rodents (Rodentia) and shrews (Soricidae), collectively called small mammals, in the central prairie region of North America. Two of our primary goals are to assess how populations and assemblages of small mammals are affected by natural (nonanthropogenic) features and events that occur in the prairie region and activities and events imposed by humans. Natural features and events that we examined include climate, fire, topography, and vegetative conditions. Human activities included cropping, ranching, and land development. Our third goal is to examine how small mammals (primarily rodents) have an effect on the prairie by acting as grazers of foliage, predators of seeds, dispersers of seeds and spores, and agents of disturbance to the soil.

Before considering these issues, however, we present a general description of the mammals of the prairie region of central North America before European settlement. We also describe postsettlement changes in the abundance and distribution of intermediate- to large-sized terrestrial mammals and provide a list of mammals introduced to the region. Our intent is to provide a context for the assessment of past and present influences of natural and anthropogenic factors and events on small mammals in grasslands, woodlands, and croplands and other altered habitats in the prairie region of central North America.

Mammals of the Prairie Region

The central North American prairie region (Risser et al. 1981) contains a large but varied grassland that encompasses the area from southern Alberta, Saskatchewan, and Manitoba in the north to eastern Coahuila and northern Nuevo Leon and Tamaulipas in the south. The region is bounded mostly on the west by the Rocky Mountains and on the east by the eastern deciduous forest. This central grassland also includes the eastern extension of tallgrass (prairie peninsula) into Wisconsin, Illinois, and Indiana and the coastal prairie region of Texas and Louisiana. In addition to grassy habitats with associated mammals, this region of prairies includes savannahs, forests, and riparian woodlands.

Within this region, types of grasslands historically ranged from shortgrass prairie in the west to tallgrass prairie in the east and from relatively hot grasslands in the south to cold grasslands in the north. Types and ranges of prairies are related primarily to climatic conditions, particularly temperature, precipitation, and timing of precipitation. In addition to climate, fire was a primary ecological driver for maintaining grasslands. The third major determinant of the structure and composition of vegetation across the region was grazing by mammalian herbivores, especially bison and black-tailed prairie dogs, and also pronghorn, wapiti, and mule deer (scientific names for mammals given in Table 9.1).

The Great Plains was home to a variety of types and sizes of mammals before European settlement (common and scientific names for 131 species given in

Table 9.1. Scientific and Common Names of Native Mammals of the Central North American Prairie Region

Order: Didelphimorphia	
Family: Didelphidae	
Didelphis virginiana	Virginia opossum
Order: Insectivora	
Family: Soricidae	
Sorex arcticus	Arctic shrew
Sorex haydeni	Hayden's shrew
Sorex hoyi	Pygmy shrew
Sorex merriami	Merriam's shrew
Sorex nanus	Dwarf shrew
Sorex vagrans	Vagrant shrew
Blarina brevicauda	Northern short-tailed shrew
Blarina hylophaga	Elliot's short-tailed shrew
Cryptotis parva	Least shrew
Notiosorex crawfordii	Desert shrew
Family: Talpidae	
Scalopus aquaticus	Eastern mole
Order: Chiroptera	
Family: Vespertilionidae	
Myotis ciliolabrum	Western small-footed myotis
Myotis evotis	Long-eared myotis
Myotis lucifugus	Little brown myotis
Myotis septentrionalis	Northern myotis
Myotis thysanodes	Fringed myotis
Myotis velifer	Cave myotis
Myotis volans	Long-legged myotis
Lasionycteris noctivagans	Silver-haired bat
Pipistrellus hesperus	Western pipistrelle
Pipistrellus subflavus	Eastern pipistrelle
Eptesicus fuscus	Big brown bat
Lasiurus borealis	Eastern red bat
Lasiurus cinereus	Hoary bat
Nycticeius humeralis	Evening bat
Plecotus townsendii	Townsend's big-eared bat
Antrozous pallidus	Pallid bat
Family: Molossidae	
Tadarida brasiliensis	Brazilian free-tailed bat
Nyctinomops macrotis	Big free-tailed bat
Order: Primates	
Family: Hominidae	
Homo sapiens	Man
Order: Xenarthra	
Family: Dasypodidae	
Dasypus noveminctus	Nine-banded armadillo
Order: Lagomorpha	
Family: Leporidae	
Sylvilagus audubonii	Desert cottontail
Sylvilagus floridanus	Eastern cottontail
Sylvilagus nuttallii	Nuttall's cottontail

(*continued*)

Table 9.1. (*continued*)

Order: Lagomorpha (*continued*)	
Family: Leporidae (*continued*)	
Lepus americanus	Snowshoe hare
Lepus californicus	Black-tailed jackrabbit
Lepus townsendii	White-tailed jackrabbit
Order: Rodentia	
Family: Sciuridae	
Tamias minimus	Least chipmunk
Tamias striatus	Eastern chipmunk
Marmota flaviventris	Yellow-bellied marmot
Marmota monax	Woodchuck
Spermophilus elegans	Wyoming ground squirrel
Spermophilus franklini	Franklin's ground squirrel
Spermophilus mexicanus	Mexican ground squirrel
Spermophilus richardsonii	Richardson's ground squirrel
Spermophilus spilosoma	Spotted ground squirrel
Spermophilus tridecemlineatus	Thirteen-lined ground squirrel
Spermophilus variegatus	Rock squirrel
Cynomys ludovicianus	Black-tailed prairie dog
Sciurus carolinensis	Eastern gray squirrel
Sciurus niger	Fox squirrel
Tamiasciurus hudsonicus	Red squirrel
Glaucomys volans	Southern flying squirrel
Family: Geomyidae	
Thomomys bottae	Botta's pocket gopher
Thomomys talpoides	Northern pocket gopher
Geomys attwateri	Attwater pocket gopher
Geomys breviceps	Baird's pocket gopher
Geomys bursarius	Plains pocket gopher
Geomys knoxjonesi	Jones' pocket gopher
Geomys personatus	Texas pocket gopher
Geomys texensis	Llano pocket gopher
Cratogeomys castanops	Yellow-faced pocket gopher
Family: Heteromyidae	
Perognathus fasciatus	Olive-backed pocket mouse
Perognathus flavescens	Plains pocket mouse
Perognathus flavus	Silky pocket mouse
Perognathus merriami	Merriam's pocket mouse
Chaetodipus hispidus	Hispid pocket mouse
Dipodomys elator	Texas kangaroo rat
Dipodomys ordii	Ord's kangaroo rat
Family: Castoridae	
Castor canadensis	American beaver
Family: Muridae	
Reithrodontomys fulvescens	Fulvous harvest mouse
Reithrodontomys megalotis	Western harvest mouse
Reithrodontomys montanus	Plains harvest mouse
Peromyscus attwateri	Texas mouse
Peromyscus boylii	Brush mouse
Peromyscus leucopus	White-footed mouse

Table 9.1. (*continued*)

Order: Rodentia (*continued*)	
Family: Muridae (*continued*)	
Peromyscus maniculatus	Deer mouse
Peromyscus pectoralis	White-ankled mouse
Peromyscus truei	Pinyon mouse
Baiomys taylori	Northern pygmy mouse
Onychomys leucogaster	Northern grasshopper mouse
Sigmodon hispidus	Hispid cotton rat
Neotoma albigula	White-throated woodrat
Neotoma cinerea	Bushy-tailed woodrat
Neotoma floridana	Eastern woodrat
Neotoma micropus	Southern plains woodrat
Clethrionomys gapperi	Southern red-backed vole
Microtus longicaudus	Long-tailed vole
Microtus ochrogaster	Prairie vole
Microtus pennsylvanicus	Meadow vole
Microtus pinetorum	Woodland vole
Lemmiscus curtatus	Sagebrush vole
Ondatra zibethicus	Muskrat
Synaptomys cooperi	Southern bog lemming
Family: Dipodidae	
Zapus hudsonius	Meadow jumping mouse
Zapus princeps	Western jumping mouse
Family: Erethizontidae	
Erethizon dorsatum	Common porcupine
Order: Carnivora	
Family: Canidae	
Canis latrans	Coyote
Canis lupus	Gray wolf
Canis rufus	Red wolf
Vulpes velox	Swift fox
Vulpes vulpes	Red fox
Urocyon cinereoargenteus	Common gray fox
Family: Ursidae	
Ursus americanus	Black bear
Ursus arctos	Grizzly or brown bear
Family: Procyonidae	
Bassariscus astutus	Ringtail
Procyon lotor	Common raccoon
Family: Mustelidae	
Mustela erminea	Ermine
Mustela frenata	Long-tailed weasel
Mustela nigripes	Black-footed ferret
Mustela nivalis	Least weasel
Mustela vison	Mink
Taxidea taxus	Badger
Spilogale gracilis	Western spotted skunk
Spilogale putorius	Eastern spotted skunk
Mephitis mephitis	Striped skunk
Conepatus mesoleucus	Common hog-nosed skunk

(*continued*)

Table 9.1. (*continued*)

Order: Carnivora (*continued*)	
Family: Mustelidae (*continued*)	
Lutra canadensis	Northern river otter
Family: Felidae	
Felis concolor	Mountain lion
Felis pardalis	Ocelot
Lynx lynx	Lynx
Lynx rufus	Bobcat
Panthera onca	Jaguar
Order: Artiodactyla	
Family: Dicotylidae	
Tayassu tajacu	Collared peccary
Family: Cervidae	
Cervus elaphus	Wapiti
Odocoileus hemionus	Mule deer
Odocoileus virginianus	White-tailed deer
Family: Antilocapridae	
Antilocapra americana	Pronghorn
Family: Bovidae	
Bison bison	Bison
Ovis canadensis	Bighorn sheep

Table 9.1). Inclusion in the list was based primarily on distributions given in Hall (1981), with consideration for changes in accepted species and the ranges of these species (see Jones et al. 1982, 1986, 1992 and references therein). Our estimate of the number of species in the Great Plains is conservative and does not include species primarily associated with other biomes whose estimated ranges only marginally overlapped the border of the prairie (compare with Risser et al. 1981 for tallgrass prairie). Further, we did not include species associated with grassland that only occurred in coastal prairie or only in extreme southern Texas and northern Mexico. Listing of a species in Table 9.1 also does not indicate that it occurs in grassland sites but rather that it occurs within the region where it may be associated closely with forest or wooded riparian vegetation.

Bison, being diurnal, large, and abundant, were the most visible mammals on the grasslands, ranging over most of the central and western region of the prairie. Other diurnal mammals typical of this grassland region included pronghorn, elk, and mule deer as well as the black-tailed prairie dog, which was restricted to the western portion of the prairie. Typical carnivores of the grassland included the plains gray wolf, coyote, raccoon, badger, striped skunk, and black-footed ferret. In contrast to these relatively visible species, most prairie species were and are small, secretive, and nocturnal (bats, shrews, and most rodents).

Only seven species of ungulates, all in the order Artiodactyla, occurred in the central prairie (Table 9.1). Of these, the collared peccary and mountain sheep were limited to a small region of the Great Plains, whereas white-tailed deer were widespread but typically associated with woody vegetation. More carnivores (*n* = 26)

Table 9.2. Wild and Domestic (*) Mammals Introduced to the Central North American Prairie Region

Order: Rodentia	
Family: Muridae	
Rattus norvegicus	Norway rat
Mus musculus	House mouse
Order: Carnivora	
Family: Canidae	
Canis familiaris*	Dog
Family: Felidae	
Felis catus*	Cat
Order: Perissodactyla	
Family: Equidae	
Equus asinus*	Donkey
Equus caballus*	Horse
Order: Artiodactyla	
Family: Suidae	
Sus scrofa*	Pig
Family: Camelidae	
Lama glama*	Llama
Family: Cervidae	
Cervus axis	Axis deer
Cervus dama	Fallow deer
Family: Bovidae	
Bos taurus*	Cattle
Boselaphus tragocamelus	Nilgai
Antilope cervicapra	Blackbuck
Capra hircus*	Goat
Oreamnos americanus	Mountain goat
Ovis aries*	Sheep
Ammotragus lervia	Barbary sheep

than ungulate species occurred within the central prairie. Of these, wolves, coyotes, badgers, and mountain lions occurred over broad areas of grassland within the Great Plains, whereas other species were present within relatively small regions and not associated strongly with grassland. More than 70% of the species recorded (Table 9.1) represent only four orders, Rodentia (60 species), Chiroptera (18 species), Insectivora (11 species), and Lagomorpha (6 species). Most of these species are relatively small in size (exceptions are the beaver, porcupine, snowshoe hare, and white-tailed and black-tailed jackrabbits).

During the past 400 years, human activities, either directly (hunting, trapping, and poisoning) or indirectly (destruction and fragmentation of native prairies), led to the reduction of the area occupied by many large- and intermediate-sized mammals. For herbivorous species, bison and bighorn sheep were extirpated from the prairie, whereas other mammals such as wapiti, pronghorn, mule deer, white-tailed deer, and black-tailed prairie dog were reduced in number and extirpated from large portions of their native ranges (Risser et al. 1981, Jones et al. 1983, 1985). Carnivores such as the grizzly, black bear, mountain lion, gray wolf (extinction of plains subspecies), and black-footed ferret were once widespread within the Great Plains but they were extirpated from all or nearly all the region. Where large- and intermediate-sized herbivorous mammals were extirpated, impacts due to altered grazing patterns surely changed composition of plant species and vegetative structure (Hartnett et al., this volume) and, therefore, the abundance and local distribution of many small mammals. Extirpation of carnivores probably had little direct influence on the distribution and general abundance of small mammals, although rodents are common prey of many carnivores.

Instead of re-establishing native grazers in fragmented prairie across the central region, humans introduced domestic livestock. Cattle were the primary substitute for bison (Hartnett et al., this volume), but sheep, horses, and goats also were introduced into North American prairie (Table 9.2). Uncommon domestic grazers in native rangelands are the donkey, llama, and pig. Domestic grazers do create a mosaic of grazed and ungrazed microhabitats that may substitute reasonably well for conditions created by native herbivores in the presettlement prairie. Additionally, several species of exotic ungulates were introduced into small- to moderate-sized areas, particularly in the southern part of the prairie (Table 9.2) (Decker 1980, Bolen and Robinson 1995), whereas the mountain goat was introduced into South Dakota (Jones et al. 1983). Introduced carnivores are the domestic cat and dog, which as free-ranging or feral individuals hunt in native and altered habitats across the region. Finally, two commensal rodents were introduced into the region from Europe (Hall 1981).

Natural Influences on Small Mammals

Many factors affect the distribution of small mammals in grasslands in central North America. French et al. (1976) suggested that grazing herbivorous microtines and microtine-like small mammals such as *Sigmodon* dominated grasslands of high plant productivity (tallgrass prairie), a mixture of omnivorous and herbivorous rodents occurred in grasslands of lower productivity (mixed-grass prairie), and omnivorous and some granivorous small mammals occurred in grasslands of even lower productivity (shortgrass prairie). (Additional differences in small mammals among prairie types in the Great Plains are described in Grant and Birney [1979].) Productivity of grasslands, in turn, is affected by physical factors such as climate, topography, and fire. Therefore, it is important to understand how these factors directly or indirectly (through vegetation) affect the abundance or distribution of small mammals in different types of grasslands.

Climate

Precipitation is the most important climatic variable in the grasslands, decreasing by about 0.7 mm/year/km from east to west in the central region (Kansas, Nebraska, and parts of Colorado) (Burke et al. 1991). Overall, grasslands differ in the annual amount of precipitation received with a range from less than 40 cm in the shortgrass prairies to more than 100 cm in the tallgrass prairies (Burke et al. 1991) but are similar in that all prairies are characterized by periodic droughts (Anderson 1990). In contrast to the large gradient in precipitation, temperature varies only from 8 to 14°C in the central region of the prairie of North America; an increase in temperature occurs from the northwest to southeast part of the region (Burke et al. 1991). Few studies in the grasslands have examined the effect of temperature and precipitation on abundance of small mammals or the prolonged effect of drought on abundance or on demographic factors that affect abundance of small mammals in grasslands. Further, few studies have assessed the indirect effect of drought on food resource needs of small mammals. However, French et al. (1976) suggested that the effects of drought on small mammals should occur in the season after the year of drought.

Long-term data were collected on the abundance of small mammals during spring and autumn sampling from autumn 1981 to spring 1991 on the Konza Prairie, a tallgrass prairie in northeastern Kansas. No correlative patterns emerged from data for winter temperature or precipitation and spring abundances of eight common species of small mammals (Table 9.3). By contrast, we found significant patterns between autumn abundances and climatic variables for three species of small mammals. First, abundances of the southern bog lemming (a species whose origin is traced to the deciduous forest) (Jones et al. 1983) and Elliot's short-tailed shrew (a species whose origin is traced to mesic grasslands) (J.R. Choate, *in litt.*) were negatively correlated with summer temperature (i.e., the hotter the summer, the fewer bog lemmings and shrews were captured in live traps in autumn [October]). Second, abundance of Elliot's short-tailed shrew also was positively correlated with precipitation during the previous January–August. Further, the number of white-footed mice, a woodland species, caught in the prairie was positively correlated with summer precipitation. Finally, our observations suggested a negative relationship between abundance of thirteen-lined ground squirrels and summer precipitation, a pattern that might result from earlier hibernation in wet years.

The monthly abundance of deer mice in mixed-grass prairie in north central Kansas varied between 1985 and 1989. High relative densities were observed in wet years and low relative densities in dry years (wet and dry years defined by the deviation from the amount of precipitation recorded for a 30-year norm, 1941–1970) (Kaufman 1990). Both monthly relative densities of deer mice and the deviation of these values from the yearly average of abundance were correlated directly with monthly precipitation but not with the average monthly maximum temperature. Despite the positive influence of precipitation on abundance, numbers of deer mice decreased after the few catastrophic rain events that occurred (>2.5 cm in 24 hours) (Kaufman 1990), which suggests that deaths and/or movements of deer mice from saturated areas occur.

Droughts may have catastrophic effects on small mammals either through direct impact (a lack of free water in the environment) or through indirect effects (vegetation and lack of food resources from plants). The lack of water and seeds may affect demographic processes, and these, in turn, affect the abundance of small mammals. Prairie voles were negatively affected by drought (Wooster 1939,

Table 9.3. Correlation Coefficients for Relationships between Abundances of Spring (1982–1991) and Autumn (1981–1990) Populations of Small Mammals and Climate and Plant Characteristics on the Konza Prairie[a]

	Omnivores				Herbivores			Insecti-vores:
	DM	HM	WM	GS	PV	BL	CR	SS
Spring abundance								
Plant biomass	-0.38	0.10	0.65*	-0.52	0.66*	0.35	0.68*	0.23
Grass stems	0.33	0.53	0.97**	0.26	0.63*	0.55*	0.97**	0.11
Winter temperature	0.24	0.09	-0.41	0.23	-0.11	-0.06	-0.47	-0.46
Winter precipitation	-0.14	-0.35	0.28	0.02	-0.24	0.01	0.25	0.25
Autumn abundance								
Plant biomass	-0.25	0.35	0.17	0.06	0.16	-0.05	0.45	0.26
Grass stems	0.17	0.47	0.44	-0.21	0.12	-0.05	0.59*	0.04
Summer temperature	-0.12	0.02	-0.13	0.17	-0.35	-0.62*	0.05	-0.58*
Summer precipitation	0.45	0.26	0.75*	-0.66*	0.46	0.34	0.53	-0.29
Current-year precipitation	0.18	-0.18	0.30	-0.27	0.16	0.25	-0.01	0.63*

[a] Small mammals were the deer mouse (DM), western harvest mouse (HM), white-footed mouse (WM), thirteen-lined ground squirrel (GS), prairie vole (PV), southern bog lemming (BL), hispid cotton rat (CR), and short-tailed shrew (SS). Biomass and stems were indices to plant production and grass stems per unit area at the end of growing seasons (see text). Winter and summer temperatures were average temperature during December–February and June–August. Winter, summer, and current-year precipitation were total precipitation during December–February, June–August, and January–August.

*$P < .05$ for one-tailed tests, with summer temperature and winter precipitation expected to have negative effects and all other factors positive effects.

**$P < .01$ for one-tailed tests.

Martin 1960, Tomanek and Hulett 1970), and these voles were thought to be on the verge of extinction in western Kansas during the drought of 1935–1939 (Black 1937). During a more recent drought in 1988–1989 in north central Kansas, numbers of deer mice decreased to fewer than 1.4 individuals/ha (Kaufman 1990). This "crash" in numbers of deer mice was due primarily to a lack of recruitment of young individuals and, therefore, a change in the age structure of the population on the study site (i.e., a smaller proportion of the population consisted of juveniles and neonates during the drought than in normal and wet years). This pattern may result from different tactics used by female deer mice of different ages. Young females may delay breeding during a dry period, whereas old females may continue to breed under the stressful environmental conditions.

Fire

Axelrod (1985) suggested that fire is second only to climate in determining the spread and maintenance of grasslands. Periodic fire is prevalent in grasslands, although it has been studied most extensively in tallgrass prairie (Collins and Wallace 1990). Fire increases the diversity of both plants and animals through its

Table 9.4. Average Numbers of Small Mammals Caught on Two Sampling Traplines within a Season in Each Burned and Unburned Treatment on the Konza Prairie during 1981–1991[a]

	Burn		Unburn
Deer mouse	17.0	>	6.51
Elliot's short-tailed shrew	1.83	<	3.63
Western harvest mouse	2.42	<	4.81
White-footed mouse	1.95	=	1.70
Prairie vole	0.84	=	1.20
Thirteen-lined ground squirrel	0.82	=	0.59
Southern bog lemming	0.05	<	0.29
Hispid cotton rat	1.20	>	0.16
Hispid pocket mouse	0.10	=	0.07
Plains harvest mouse	0.05	=	0.04
House mouse	0.06	=	0.04
Eastern woodrat	0.04	=	0.06
Least shrew	0.02	=	0.02
Meadow jumping mouse	0.01	=	0.00
All mammals	26.4	>	19.1

[a] Statistically significant differences ($P < .05$) are indicated by the symbols of > or <, whereas nonsignificant differences ($P > .05$) are indicated by =.

creation of a mosaic of patches; at a local scale, the mosaic is caused by differences in fuel, soil moisture, and wind direction, whereas at the landscape level, the mosaic is due to an interspersion of burned and unburned areas (Risser 1990). Fire may affect small mammals in prairies directly or indirectly through changes in environmental conditions (Kaufman et al. 1990b). Most responses of small mammals occur because of the indirect effects of fire on vegetation in tallgrass prairie. We are unaware of any short-term or long-term studies of response of small mammals to fire in mixed-grass or shortgrass prairies.

Common small mammals of the tallgrass prairie have been placed in three categories regarding their response to fire (Kaufman et al. 1990b). Fire-positive species increase in abundance and fire-negative species decrease in abundance on burned areas as compared with the same species in unburned areas. Abundance of fire-neutral species remains similar on burned and unburned areas. Fire-positive species include deer mice, white-footed mice, meadow jumping mice, and prairie voles. Fire-negative species include meadow voles, southern bog lemmings, northern short-tailed shrews, Elliot's short-tailed shrews, and western harvest mice. A recent study (Clark et al. 1990) on the Konza Prairie suggests that thirteen-lined ground squirrels also are a fire-positive species.

Our data from 1981–1991 on the Konza Prairie generally agree with the published studies; however, some differences are notable. Average abundances of all small mammals combined were greater in burned areas than in unburned areas (Table 9.4). Only two species of small mammals, deer mice and cotton rats, had higher abundances in burned areas than in unburned areas, whereas three species,

Elliot's short-tailed shrews, western harvest mice, and southern bog lemmings, had higher abundances in unburned areas than in burned areas. Three of the other eight common species, white-footed mice, prairie voles, and thirteen-lined ground squirrels, had statistically indistinguishable abundances in burned and unburned areas. None of the six rare species on the Konza Prairie showed significantly different abundances on burned and unburned areas (Table 9.4) but responses of these species are difficult to quantify.

Most differences in abundances between burned and unburned areas result from differential movement and recruitment of small mammals (Kaufman et al. 1990b). Analyses of data for small mammals from a 13-ha study site with approximately equal areas of burned and unburned prairie on the Konza Prairie indicated that western harvest mice and prairie voles moved from the burned side to the unburned side after fire (Clark and Kaufman 1990). Observations for southern bog lemmings suggested the same pattern but sample size was too limited for a statistical test. Additionally, most western harvest mice, Elliot's short-tailed shrews, prairie voles, and southern bog lemmings that immigrated into the study site after the fire were first captured and stayed on the unburned side, whereas new deer mice immigrated into the burned site. Few adult rodents and shrews likely are killed by fire directly except under unusual conditions (Erwin and Stasiak 1979, Geluso et al. 1986, Clark and Kaufman 1990, Harty et al. 1991).

Fire-negative species may use dense vegetation for cover and feed on foliage, or they may use the layer of litter in unburned prairie for food resources and nesting materials, whereas fire-positive species generally are associated with an open soil surface and feed on seeds or insects. Species exhibiting these latter behaviors likely would be hindered by dense vegetation or litter (Kaufman et al. 1990b). Abundance of deer mice on the Konza Prairie was correlated negatively with depth of litter, whereas abundances of western harvest mice and Elliot's short-tailed shrews were correlated positively with litter depth (Kaufman et al. 1989).

Laboratory experiments have shown the effect of the litter layer on the choice of foraging and nesting sites of several prairie rodents. Under controlled lighting conditions that simulated moonless nights, deer mice, a fire-positive species, and western harvest mice, a fire-negative species, preferred to forage for seeds under sparse litter rather than intermediate or heavy litter; however, western harvest mice when compared with deer mice consumed proportionately more seeds from patches covered with intermediate and heavy litter (Kaufman and Kaufman 1990a). Western harvest mice also were more likely than deer mice to eat seeds from under the densest litter.

In trials in which only one depth of litter covered a single seed tray (for deer mice: 12 litter depths from 15 to 420 g spread over an area of 27.5×42.7 cm; for western harvest mice: 6 depths from 30 to 420 g spread over the same area), the amount of seeds used by a single mouse was not correlated to the depth of litter, an unexpected result (Kaufman and Kaufman 1990a). This result may have occurred because trials were performed under optimum conditions (i.e., an entire night was available to find and consume seeds and the light conditions were dark so individuals likely perceived a low-risk environment with regard to predation).

Subsequently, deer mice, western harvest mice, white-footed mice, and prairie voles each were tested under both light and dark conditions in more complex trials in which large patches of sparse and moderate litter covered most of the trial arenas (Clark and Kaufman 1991). In experiment 1, deer mice consumed more seeds under sparse litter than under moderate litter during dark nights but switched and consumed more seeds from under moderate litter than sparse litter on bright nights. Western harvest mice tended to use moderate litter over sparse litter on both dark and bright nights, but the pattern was statistically significant only on bright nights. White-footed mice, a fire-positive species, consumed more seeds from under sparse litter than under moderate litter on both bright and dark nights but this pattern was statistically different only on dark nights. Prairie voles, a fire-negative species, selectively foraged in moderate litter on both dark and bright nights.

In experiment 2, deer mice, white-footed mice, and western harvest mice each were tested with four levels of litter (sparse, moderate, heavy, and compacted) available in each arena (Clark and Kaufman 1991). Patterns of use were similar for sparse and moderate litter for the three species, and all three species avoided foraging in heavy and compacted litter. Both deer mice and white-footed mice nested either in litter or in a nest box in the area of sparse litter on both dark and bright nights, whereas western harvest mice nested either in litter or in a nest box in the area of moderate litter. All three species rarely nested in heavy or compacted litter but use of these habitats was slightly greater on bright (17 of 108 nights) than dark nights (8 of 108 nights).

Topography

A unifying characteristic of grasslands is a level to rolling landscape that favors fire spreading across extensive areas (Anderson 1990). Despite the relatively minor relief, often less than 100 m within local areas, variation in conditions ranges from relatively xeric uplands to mesic lowlands connected by moderately to steeply sloped prairie. Despite the topographically rugged landscape of most remaining prairies, little research has been directed toward understanding local differences in use of habitats by small mammals (Kaufman et al. 1995b). Distribution of small mammals within prairies may be affected significantly by land form with its local influences on air and ground temperatures, soil moisture, drainage characteristics, and vegetative composition (Swanson et al. 1988).

One goal of our long-term research on the Konza Prairie is to understand how the topography of the Flint Hills affects the distribution of rodents and shrews in tallgrass prairie. Three major types of topographic habitats are present in the Flint Hills: upland, limestone breaks, and lowland (for detailed description of topography, see Kaufman et al. 1988). From data collected along permanent live-trap sampling transects (for detailed description, see Kaufman et al. 1988) from 1981 to 1991, the maximum number of species of small mammals that occurred at a live-trap station was five in breaks and lowland prairie as compared with three in upland prairie (maximum number possible within a trapping period was eight)

(Kaufman et al. 1995b). Five of the eight common species on the Konza Prairie showed repeatable patterns of preference for a single type of topography at the microscale (Peterson et al. 1985, Brillhart et al. 1995) and landscape scale (Kaufman et al. 1988, Kaufman et al. 1995b). Lowland prairie was selected preferentially by western harvest mice, white-footed mice, and hispid cotton rats, whereas limestone breaks was chosen preferentially by deer mice and thirteen-lined ground squirrels. Currently, our analyses of 10 years of data (Kaufman et al. 1995b) show no significant landscape level of topographic selection of habitat by Elliot's short-tailed shrews, prairie voles, or southern bog lemmings. In contrast to this lack of pattern, other short-term studies on the Konza Prairie have shown that Elliot's short-tailed shrews selected lowland prairie and avoided upland prairie (Clark et al. 1995) and prairie voles selected upland prairie (Bixler and Kaufman 1995).

With additional records of rarer species on the Konza Prairie, we expect that hispid pocket mice and eastern woodrats will select limestone breaks prairie over upland and lowland prairie and that least shrews will select uplands over breaks and lowland prairie (Kaufman et al. 1995b). Other studies have shown the selection of upland prairie habitats by least shrews in tallgrass (Clark et al. 1995) and mixed-grass prairie (Choate and Fleharty 1975).

Vegetation

Vegetative factors such as species composition, plant productivity, vertical structure, and plant litter may affect the distribution and abundance of small mammals. The composition of plant species and plant productivity may affect small mammal use by influencing the quality and availability of plant parts (leaves, roots, seeds) used as food. The vertical structure of vegetation also may influence the selection of habitat by small mammals, as this structure may provide cover for small mammals vulnerable to mammalian and avian predators. Presence of litter or the lack thereof may affect travel through a habitat, or dead plant material may be used as a resource for constructing runways (*Microtus,* Choate and Fleharty 1975), an aboveground nest (*Reithrodontomys,* Shump 1974), or as a lining in an underground nest of a small mammal (see Fire section for effects of litter).

The distribution of small mammals in grasslands appears independent of the composition of plant species. Kaufman and Fleharty (1974) found that vegetative structure was more important than the presence or absence of a particular species of plants on the distribution of rodents in mixed-grass prairie in north central Kansas. Likewise, we found no relationships between distribution of individual species of small mammals and the presence of individual plant species in tallgrass prairie on the Konza Prairie (Kaufman and Kaufman, unpublished data).

Spring abundances of white-footed mice, prairie voles, southern bog lemmings, and cotton rats on the Konza Prairie from 1982 to 1991 were correlated positively with stem density of flowering stalks (seed productivity) of big bluestem (*Andropogon gerardii*), little bluestem (*Schizachyrium scoparium*), and indiangrass (*Sorghastrum nutans*) for the previous growing season (Table 9.3). Except for bog lemmings, spring abundances of these species also were correlated positively with standing crop biomass at the end of the growing season, an index to plant productivity. However, spring abundances of deer mice, western harvest mice, thirteen-lined ground squirrels, and shrews were not correlated to either measure of vegetation. For autumn populations, in contrast to spring abundances, the only significant relationship between small mammals and characteristics of vegetation was a positive correlation between cotton rats and stem density of dominant grasses (Table 9.3).

Bison were a major influence on the vertical structure of grasslands before European settlement of the central North American prairies. Recent studies on the Konza Prairie show that the patches where bison graze are affected by fire history, composition of C_3 and C_4 grasses, ratio of forb to grass cover, and characteristics of C_4 grasses (Vinton et al. 1993, Pfeiffer and Hartnett 1995). In turn, bison affect their grazing patches by altering growth and performance of some grasses, enhancing water availability to and productivity of forbs, and changing species composition (Vinton and Hartnett 1992, Fahnestock and Knapp 1994, Pfeiffer and Steuter 1994, Pfeiffer and Hartnett 1995). Due to the characteristics of grazing patches and the effect of bison grazing on these patches, it is expected that the abundance and distribution of small mammals in grasslands grazed by bison will be affected. We are unaware of any published studies that examine the indirect effect of bison on abundances of rodents or shrews. We are collecting data on abundance and distribution of small mammals on experimental areas subjected to fire and to grazing by bison on the Konza Prairie; too few data exist to make meaningful statements about the interaction of fire and grazing on small mammals, however.

The importance of vegetative cover to small mammals also has been examined experimentally by exclosing large herbivores from grazed areas and by mowing and adding hay to experimental plots. For example, populations of *Microtus* may be absent from prairies that are grazed, but numbers of voles increase in experimental exclosures that keep out large grazing mammals; exclusion of grazers allows plant biomass to accumulate providing the vegetative cover required for breeding populations of voles (Birney et al. 1976). In ungrazed grassland, abundance of prairie voles decreased in response to reduction of cover by mowing, whereas abundance eventually increased on plots where hay was used to augment cover (Kotler et al. 1988). By contrast, numbers of deer mice and western harvest mice declined on the grid that was supplemented with hay but showed no change in response to mowing.

The effect of vegetative cover also was assessed using experiments in which water and nitrogen were added to plots of native prairie. In Colorado, additions of water and nitrogen to shortgrass prairie resulted in variable patterns of abundance in populations of small mammals (Grant et al. 1977). For example, peak numbers of prairie voles increased from about 60/ha in 1971 (spring abundance was <10 individuals/ha) to about 100/ha in 1973 and 1974 (spring abundances were <30 individuals/ha) on experimental plots that received supplemental water and nitrogen. By contrast, a population of prairie voles did not become established until

1973 and 1974 in two experimental plots where only water was added; peak abundance was never more than 30 individuals/ha. In control plots where neither water nor nitrogen was added, breeding populations of voles were never observed, although an occasional vole was captured on one plot.

In contrast to prairie voles, numbers of northern grasshopper mice were fairly stable (annual peak abundances of 5–10 individuals/ha) on control plots and plots that received nitrogen but grasshopper mice were nearly absent from experimental plots that received either water or water and nitrogen (Grant et al. 1977). Evidence of breeding was observed in control and nitrogen-only plots. Fewer repeatable patterns across treatments and years were observed for deer mice and thirteen-lined ground squirrels. Generally, deer mice followed the pattern observed for voles, whereas ground squirrels followed patterns observed for grasshopper mice.

Visual observations suggest that plots with both supplemental water and nitrogen were islands of "tallgrass" prairie within a "sea" of shortgrass prairie (Grant et al. 1977). *Microtus* were not present on the study site before the manipulations occurred, and cover is important in *Microtus* habitats (Birney et al. 1976). Plots that received both water and nitrogen had vegetative cover and biomass similar to a tallgrass prairie site in Oklahoma but were twice those observed on the other experimental plots and controls in the study site in Colorado (Grant et al. 1977). Additionally, the communities of small mammals did not differ between water and nitrogen plots and tallgrass prairie. We are not aware of other experiments in which water and nitrogen have been added to mixed-grass or tallgrass prairies with simultaneous monitoring of populations of small mammals. However, hispid cotton rats occurred on our study area in mixed-grass prairie in north central Kansas in autumn of 1993 after an unusually wet spring and summer (Kaufman and Kaufman, unpublished data). Also in that year, flowering stocks of big bluestem were more than 2 m in height. Cotton rats commonly occur in fencerows and road ditches in this area (Kaufman and Kaufman 1989) but had never been recorded on this 10.8-ha study site (records kept since 1983) despite light-to-moderate grazing by cattle during autumn.

Interactions among Natural Factors

Through the collection of long-term data on the Konza Prairie, we hope to understand how climate, fire, topography, and grazing by bison affect the distribution and abundance of small mammals. Given the number of variables and the amount of time needed to repeat different climatic patterns, we are only beginning to understand the interactive effects of these major determinants of prairie ecosystems. However, after 10 years, we are able to see differential patterns that occur between topography and frequency of fire.

Generally, upland sites in burned tallgrass prairie were underused (more live-trap stations with no species present than expected by random chance) and lowland sites were overused (more stations with two or more species present than expected by chance) by species of small mammals (Kaufman et al. 1995b). The same pattern was found for treatment sites that had been burned within 2–4 years of the observation date. However, this pattern was not true for treatments 8–20 years after a fire. In those treatment plots, few species of small mammals used the lowland prairie, and limestone breaks prairie was chosen by one or two species (overrepresented by stations with one or two species present).

Deer mice preferentially selected limestone breaks over lowland and upland prairie regardless of fire history. However, deer mice preferred lowland over upland prairie in the months after an annual burn but switched their preference to uplands over lowlands in unburned prairie (Kaufman et al. 1995b). This pattern likely reflects the preference of deer mice to use open soil surfaces (conditions created immediately after fire) and to avoid habitats with a litter layer (conditions created by plant production during the first year after fire) (Kaufman et al. 1988).

Ecological Role of Small Mammals in Prairies

Recently, ecology was defined as "the scientific study of the processes influencing the distribution and abundance of organisms, the interaction among organisms, and the interactions between organisms and the transformation and flux of energy and matter" (Likens 1992 as quoted in Lawton 1994). Although some of the factors that influence the distribution and abundance of small mammals in prairies are known, little is known about the interactions among small mammals and other organisms in the prairie or the role of small mammals in the transformation and flux of energy and matter in prairies. The primary role of organisms in structuring communities and ecosystems may be as "engineers" (Lawton 1994). As such, small mammals may directly or indirectly modulate the availability of resources (other than themselves) to other species. They may alter physical states of biotic or abiotic materials and, in so doing, may modify, maintain, and/or create habitat conditions and features. One potential role of small mammals is to regulate the rate of movement of nutrients and energy in prairies (Grant 1974). Other possible roles include grazing foliage of grasses and forbs, eating seeds of grasses and forbs, dispersing seeds from prairie plants, and disturbing the soil surface through burrowing and dust bathing. These latter disturbances can serve as open sites for germination of seeds or may aerate or enrich soils. Additionally, deposition of urine and/or feces in latrines may increase the nutrients in soil in and surrounding the latrines.

Grazing

French et al. (1976) have suggested that small mammals consume only a small portion (4%) of available herbage in tallgrass prairie. However, rodents at a high biomass per area, such as black-tailed prairie dogs, which were distributed widely across the mixed-grass and shortgrass prairies, obviously removed greater than 4% of the aboveground biomass within their colonies (called towns). Short-term removal of prairie dogs on a mixed-grass prairie in South Dakota showed that graminoid biomass and the ratio of grasses to forbs increased within 2 years (Cid et al. 1991). In addition to removal of biomass, grazing and clipping of vegetation by prairie dogs alter other processes and rates within the ecosystem in which they live (Whicker and Detling 1988). For example, intense grazing by prairie dogs may change the phenotypic or genotypic frequency of perennial plants, the species composition of plants from mid-height and tall grasses to shortgrasses and annual forbs (species diversity is maximized by this intermediate level of disturbance), and aspects of the microhabitat of the patch through reduction in canopy cover (Whicker and Detling 1988, 1993, Painter et al. 1993).

Further, the reduction in plant biomass and a concomitant increase in bare soil alter the soil–water balance by changes in rates of transpiration and evaporation of water, higher leaf and air temperatures that affect evaporation of water from leaves, mineralization rate by increases in soil temperature, and nutrient cycling by changing competition between plants and microorganisms for available nutrients (Whicker and Detling 1988, Day and Detling 1994). Intense grazing by prairie dogs also reduces the biomass of roots but increases the density of nematodes (Polley and Detling 1988, 1990, Whicker and Detling 1988).

Because grazing patches of prairie dogs differ from surrounding prairie, it is not surprising that populations of animals (arthropods, birds, small mammals, nematodes, and large mammals) respond in complex patterns to the mosaic of conditions available within a prairie dog town (Whicker and Detling 1988). Grazed plants in prairie dog towns have higher nitrogen than plants in uncolonized grassland, and bison, elk, and pronghorn spend more time grazing plants within colonies than would be expected given their availability within the overall grassland. Such nonrandom patterns of grazing suggest that prairie dogs facilitate the grazing of large ungulates. Experimental removal of prairie dogs resulted in a lower concentration of nitrogen in graminoid shoots, supporting the idea that prairie dogs enhance nitrogen content (Cid et al. 1991). These effects of prairie dogs on ecosystem processes and populations of other organisms within grassland ecosystems illustrate the role of prairie dogs as a keystone species or ecosystem "engineer" in the shortgrass and mixed-grass prairie regions.

Smaller rodents, such as prairie voles, ground squirrels, and pocket gophers, likely affect attributes of the communities of plants, although the spatial scale and patterns probably differ from prairie dogs. Prairie voles reduced the abundance and biomass of a C_3 perennial grass (Kentucky bluegrass, *Poa pratensis*) and reduced the abundance of a C_4 perennial grass (purple lovegrass, *Eragrotis spectabilis*) and a perennial forb (heath aster, *Aster ericoides*) but increased the abundance of another perennial forb (Missouri goldenrod, *Solidago missouriensis*) in an enclosure experiment conducted in the tallgrass prairie (Gibson et al. 1990). Using the same experimental data but a different analytical approach, Kaufman and Bixler (1995) found the negative effect on bluegrass but also indicated that voles reduced the biomass of indiangrass and apparently increased the biomass of Scribner dicanthelium (*Dicanthelium oligosanthes*) and little bluestem. Direct observations of cuttings of grasses by accumulations of haypiles showed that prairie voles cut flowering stalks of big bluestem, indiangrass, little bluestem, tall dropseed (*Sporobolus asper*), and round-headed lespedeza (*Lespedeza capitata*) (Kaufman and Bixler 1995).

Columbian ground squirrels (*Spermophilus columbianus*) also significantly reduced the biomass of some forbs and grasses on plots that received simulated urine, which increased the concentration of nitrogen, phosphorus, and potassium in vegetation on experimental plots as compared with the control plots (Boag and Wiggett 1994). Over a 4-year period, the species composition in fertilized plots diverged from that in the control plots. Some species were eliminated from the fertilized plots, or their coverage was reduced (<1%), but these changes did not occur in control plots.

Plains pocket gophers, a relatively large fossorial rodent that feeds mostly on plant parts below ground, also have an effect on prairie vegetation and processes. These gophers prefer to forage on forbs in areas of soil with high nitrogen and productivity; this foraging subsequently leads to decreases in the aboveground plant biomass (Reichman and Smith 1985, Inouye et al. 1987, Huntly and Inouye 1988). Reduction of biomass indirectly affects short-term availability of nutrients to other plants because it leaves higher levels of nitrogen in the soil as well as allows more light to pass through the canopy of plants (Huntly and Inouye 1988). Belowground grazing by pocket gophers may affect grasses with diffuse systems of roots less than forbs with dendritic roots in brome (*Bromus inermis*) fields (Reichman and Smith 1985). Plains pocket gophers also slow secondary succession of old fields in Minnesota by grazing the roots of young oaks (*Quercus* spp.) and pines (*Pinus* spp.) that are invading prairie sites (Huntly and Inouye 1988). Pocket gophers and fire may be required to maintain the perennial forb *Penstemon grandiflorus* found in sandy prairies of the Great Plains. Although fire and herbivory and deposition of mounds of soil by pocket gophers may be detrimental to some plants of *P. grandiflorus,* the gaps created by these agents within the oak woodland in Minnesota allow this herb to persist in the landscape (Davis et al. 1991a, 1991b).

Seed Predation

Small mammals in native grasslands may use a greater proportion of available seeds than the proportion of herbage used (French et al. 1976) but few studies have been conducted regarding the impacts of rodents on seeds in prairies. Although anecdotal information is available concerning the use of seeds of grasses and forbs by many prairie rodents, the magnitude of seed predation and its effect on sexual reproduction by prairie plants are generally unknown. Additionally, the use of seeds by birds, invertebrates, and rodents within grassland ecosystems has not been evaluated. Thus, little is known about the relative impacts of these different seed predators in controlling the composition and spatial patterning of plant species. The lack of study of seed predation by rodents within the North American prairies is puzzling because a keystone guild of rodents controls the zone of transition between grassland and desert shrub habitat by seed predation and soil disturbance in the Chihuahuan Desert (Brown and Heske 1990). Long-term removal of three species of kangaroo rats (*Dipodomys* spp.) increased germination of annual and perennial grasses, with a subsequent increase in herbaceous cover of grasses and forbs. The shift to grassland conditions that followed the removal of the keystone rodents resulted in the invasion of grassland rodents onto treatment plots where the kangaroo rats were absent.

We used experimental seed trays to evaluate the use of seeds by vertebrates and invertebrates in tallgrass prairie on the Konza Prairie (A.W. Reed, G.A. Kaufman, and D.W. Kaufman, unpublished data). The study used two different types of seeds and two types of seed trays: one allowed invertebrates but not vertebrates to feed and the other allowed vertebrates but prevented invertebrates from foraging. We monitored use of seeds by both diurnal and nocturnal seed predators. Both diurnal and nocturnal vertebrates removed more seeds than invertebrates; vertebrates had a preference for the larger seed. Further, rodents removed more seeds than birds. Obviously, this type of experiment must be repeated across seasons, years, and grassland types before rodents can be shown to be the most important consumers of seeds in the native grasslands.

Remnants of flowers, seed heads, fruits, and seeds left under trap shelters and near artificial burrows in the mixed-grass prairie of north central Kansas point to the importance of small mammals, especially deer mice, as predators on numerous forb and grass seeds (Kaufman and Kaufman, unpublished data). As a result, we examined the use of individually marked infructescences (hereafter referred to as seed heads) to assess both predation and dispersal (see Seed Dispersal section) of achenes (hereafter referred to as seeds). For Fremont's clematis (*Clematis fremontii*), a herbaceous forb with a narrow range of distribution in northern Kansas and southern Nebraska, rodents harvested seed heads from 95% of the plants and also harvested 95% of all marked seed heads (G.A. Kaufman, D.M. Kaufman, and D.W. Kaufman, unpublished data). Plants whose seed heads were not harvested tended to be those that had only one seed head as compared with harvested plants with multiple seed heads (range, 2–16 heads per plant). When rodents ate seeds from heads that still were attached to the plants, 92% of the seeds were eaten. A lower, but not statistically lower, proportion of seeds (87%) was eaten when heads were removed from the plant and seeds consumed on the ground. In total, we estimated that rodents consumed about 30,000 seeds of Fremont's clematis on 0.9 ha of prairie between July 10 and October 3.

We also examined use of purple coneflower (*Echinacea angustifolia*), which had only one seed head per plant, and compared its use with Fremont's clematis. Rodents, primarily deer mice, used seeds of this forb during autumn and winter (G.A. Kaufman, D.M. Kaufman, and D.W. Kaufman, unpublished data). We estimated that rodents harvested 18–21% of the seed heads and about 10,000 seeds were eaten on the 1.6-ha site. However, general usage of this plant may be underestimated because the study occurred during a drought and the number of deer mice on our study site was at a record low of 1.3 individuals/ha. Further, casual observations made in other years suggest that use of this plant was greater when deer mice were abundant. To understand the impact of rodents on individual species of plants, especially forbs, long-term studies should incorporate use of seeds under various climatic conditions and should include research on other species of grassland plants.

Seed Dispersal

Many animals are dispersers of plant seeds but few studies of mammals as dispersal agents of seeds have been conducted in North America (Willson 1993). Willson (1993) noted that ursids, procyonids, and canids often consumed fruits and, therefore, acted as dispersal agents for at least part of the seeds consumed with the fruits; artiodactyls and reintroduced equids also may be important dispersal agents for some fleshy fruits. Perhaps dispersal by these mammals is overestimated, as deer mice frequently removed and ate seeds contained within fecal pellets of black-tailed jackrabbits and dung of cattle during periods of cold weather (Riegel 1941, Brown 1946). Experimental studies in wooded and grassland habitats in Illinois showed that removal of natural seeds from depots by rodents is dependent on density and species of seeds, habitat, and season (Willson and Whelan 1990). We also have observed that deer mice in winter collected coyote scats and placed them near or in burrows and then mined this material for at least some kinds of seeds; we have found plum (*Prunus* spp.) pits from such scats that were gnawed into and the endosperm removed (Kaufman and Kaufman, unpublished data). Also in winter, deer mice mined cattle dung to the extent that material was shredded and seeds likely were consumed (Kaufman and Kaufman, unpublished data). By contrast, Willson (1993) found no studies that quantified the amount of seeds that might escape when rodents harvest fruits.

In our studies of clematis and coneflower, rodents, primarily deer mice, harvested seed heads produced by these two prairie forbs. Even from harvested heads, however, not all seeds were eaten, and therefore, some of seeds were dispersed as a by-product of the foraging activity of the rodents.

When studying seed dispersal, it is important to assess both the quantity and quality of that dispersal (Schupp 1993). First, the quantity of seed dispersal consists of the number of visits and the number of seeds dispersed per visit. Because both of the seed heads from clematis and coneflower are large, we assume that only one seed head was handled per visit. This assumption is valid for purple coneflower because only one seed head occurred per plant for more than 99% of our marked plants, and no remaining receptacles were found on, under, or within 10 cm of the plant. Because many Fremont's clematis have multiple fruits per plant, more than one seed head may have been examined on each plant during each foraging trip or period. Many remaining receptacles were on or under the same plant, so seeds may have been collected from multiple seed heads in one visit. The number of seeds dispersed by rodents is a function of the number of seeds missed while handling seed heads and consuming seeds.

We found that 8% of seeds from Fremont's clematis remained uneaten when rodents ate seeds from a head that was still attached to the plant, whereas 13% remained uneaten when the rodents ate seeds from a head removed from the plant.

Of about 50,000 clematis seeds produced on the 0.9-ha site, approximately 5,000 were uneaten and, therefore, were potentially dispersed by rodents. Similarly, an average of 3% of the purple coneflower seeds per head remained uneaten when deer mice harvested the heads and carried the heads away from the plant to feed on the seeds. By contrast, rodents were more likely to leave seed heads intact if they left the seed heads less than 1 m from the plant (53%) than if they carried them more than 1 m (3%). Of about 63,000 coneflower seeds produced on the 1.6-ha study site, nearly 10,000 were eaten, more than 3,000 were dispersed, and more than 50,000 seeds were not eaten or dispersed by rodents.

The second aspect of seed dispersal, quality, is dependent on both the treatment and deposition of the seeds (Schupp 1993). The quality of the treatment by rodents should be high because uneaten seeds are left intact and not damaged by ingestion. Even when complete seeds are ingested, however, rodents may not alter the rate of germination. In fact, the germination of 17 kinds of seeds that were fed to deer mice and passed through the digestive tract undamaged was not enhanced (Krefting and Roe 1949).

The quality of deposition consists of movement patterns and deposition rates (Schupp 1993). The quality of deposition by rodents may be high for some prairie forbs such as Fremont's clematis and purple coneflower. For both forbs, the greatest recorded distance that seed heads were carried away from the parent plant was greater than 25 m (G.A. Kaufman, D.M. Kaufman, and D.W. Kaufman, unpublished data). Because no seed heads of purple coneflower were eaten on or under the parent plant or within 10 cm of the plant, the quality of deposition of coneflower seeds likely is greater than that for Fremont's clematis. For clematis, a large number of seed heads was eaten on the parent plant (29%) and under or within 20 cm of the canopy of the parent plant or another close clematis (41%); however, 30% were carried more than 20 cm from the plant. Quality of seed dispersal by mammals such as carnivores is thought to be high because of the distance that seeds can be carried away from the parent plant (Willson 1993); however, seeds may be deposited in inappropriate habitats where germination is unlikely. By contrast, seeds generally deposited by rodents within 0.2–25.0 m of the parent plant are likely to be deposited in an appropriate germination site.

Soil Disturbance

Small mammals may affect geomorphic processes through the mechanical transport of organic and inorganic materials (Swanson et al. 1988). Although most small mammals disturb the soil surface when they excavate burrows or take dust baths, soil mounds and earthcores made by pocket gophers likely are the most obvious soil disturbances made by small mammals (along with the mounds of prairie dogs), and they are the most studied disturbance in the central North American prairies. Although we highlight the information available for pocket gophers, one also should consider the potential for these processes occurring over and over again at small spatial scales due to the actions of other prairie rodents that build and maintain burrow systems.

Pocket gophers move large amounts of soil when they excavate a tunnel in new habitat. Andersen (1987) estimated that plains pocket gophers can move up to 40 liters of soil per day with 40–85% of this soil backfilled into abandoned burrows. Although a large proportion of excavated soil is used as backfill, substantial amounts of soil are deposited on the surface as mounds. These mounds can affect vegetation dependent on location and season of deposition, abundance of the mounds, and physical and chemical composition of soils in the mounds (Collins and Glenn 1988, Huntly and Reichman 1994). For example, annual forbs are more abundant and perennial grasses less abundant on mounds than surrounding areas in old fields, diversity of plant species is greater on and near the mounds than in surrounding areas of old fields, and higher primary productivity occurs adjacent to the mounds (within 10 cm) and decreases above and away (20–40 cm) from mounds in tallgrass prairie (Huntly and Inouye 1988, Reichman et al. 1993). The persistence of mounds over time is variable; mounds disappear quickly in sandy coastal prairie in Texas but last a considerable time in sandy old fields in Minnesota and in loamy soils in Kansas (Texas: Williams et al. 1986; Minnesota: Huntly and Inouye 1988; Kansas: Kaufman and Kaufman, personal observations). These differences likely reflect differences in climatic regimes and soil characteristics.

Mounds of pocket gophers affect both the chemical and physical properties of the soil. Soil on mounds differs in texture, water-holding capacity, and nutrients, whereas backfill within tunnels differs in nutrient contents and bulk density from undisturbed surrounding soil (Andersen 1987, Huntly and Inouye 1988). Pocket gophers also increase the amount of light that reaches the soil surface and its patchiness through the removal of plant biomass (Huntly and Inouye 1988).

The open area within a pocket gopher burrow can affect vegetation directly above the burrow. Reduction of plant biomass above tunnels ranged from 30 to greater than 80% in old fields in Minnesota and tallgrass prairie in Kansas (Reichman and Smith 1985, Reichman et al. 1993). Additionally, concentrated excrement within dens of pocket gophers and uneaten caches of food can enhance the nutrients available to plants in patches of soil around burrow systems. The heterogeneous mosaic of physical disturbances and grazing that pocket gophers create indirectly affects other animals. Grasshopper (*Melanoplus* spp.) abundances are correlated with the number of mounds of pocket gophers in old fields (Huntly and Inouye 1988). This correlation likely is due to the amount of bare soil created that can serve as sites for ovipositing of eggs. Additionally, survival of eggs and nymphs is higher on bare soil, which would result in higher recruitment of grasshoppers to the area. Gophers also may affect abundance of vertebrates that use their abandoned burrows but this has not been studied in detail.

The burrow system of the prairie vole also may affect vegetation in the grassland. Species richness of plants was higher on soil above prairie vole mounds than in adjacent undisturbed tallgrass prairie (Gibson 1989), although perennial grasses (big bluestem [58%] and indiangrass [9%]) remained important on this disturbed area. Our field observations indicate that vegetation often is larger and greener on prairie vole mounds in tallgrass prairie on the Konza Prairie. The causes of this plant response in tallgrass prairie have not been investigated; however, a study of prairie vole mounds in a fescue (*Festuca elatior*) field in Kentucky (Kalisz and Davis 1992) demonstrated that biomass of grasses and concentration of nutrients (nitrogen, phosphorus, and potassium) in the foliage were higher on active and inactive colony sites than on control areas. The concentration of silicon showed the opposite pattern. Generally, soil characteristics (pH and concentration of various nutrients) did not differ among active and inactive colony sites and reference areas; however, mineralizable nitrogen followed a pattern of active colony sites > inactive colony sites > reference areas in soil over depths of 0–30 cm. Additionally, plant litter above burrows of prairie voles has higher concentrations of nutrients than in litter on reference areas. Kalisz and Davis (1992) suggest that urine and feces from prairie voles and belowground nesting materials may be an important source of available nitrogen for vegetation on a colony. Although these patterns may go unnoticed and, therefore, unstudied in native prairie, local enrichments of communities of plants likely occur in most prairie ecosystems and are not restricted to actions of only herbivorous (folivorous) rodents.

Human Influences on Small Mammals

What is the relationship between ecosystem function and species richness in natural ecosystems (Lawton 1994)? Is there a minimal diversity necessary for proper functioning of an ecosystem, and beyond this level, are most species redundant in their roles (Walker 1992, Lawton and Brown 1993)? Does each species make a contribution to the performance of an ecosystem (Ehrlich and Ehrlich 1981), or does the function of the ecosystem change unpredictably when the diversity of species changes because roles of individual species are complex and varied (Lawton 1994)? Is the functioning of an ecosystem insensitive to species deletions or additions? These questions require answers for all ecosystems but are impossible to answer without taking into consideration how humans have already modified natural ecosystems such as the grasslands. Three major human influences within the grassland ecosystem are cropping, ranching, and development of rural and urban areas for housing and industrial use.

Cropping

The historic area of the tallgrass, mixed-grass, and shortgrass prairies of central North America was reduced by the production of food for humans (Samson and Knopf 1994). Because of the change from a perennial grassland to annual production, generally in monoculture, of domestic grasses and forbs, populations and communities of small mammals likely show strong patterns of either negative or positive responses to agroecosystems.

Life histories and habitat requirements of small mammals likely determine how specific agricultural practices have an effect on the abundance and spatial distribution of individual species of small mammals. Deer mice inhabited all the agricultural habitats sampled (native prairie, planted grass, old fields, fencerows, and wheat stubble) and generally were the most abundant small mammal in north central Kansas (Kaufman et al. 1990a). By contrast, the plains harvest mouse (a species endemic to grasslands) (Risser et al. 1981) was restricted to upland mixed-grass prairie and likely would be displaced if that habitat was developed for cropland. Generally, most small mammals (northern grasshopper mice, western harvest mice, prairie voles, thirteen-lined ground squirrels, and cotton rats) have intermediate values of habitat breadth and occur in several habitats in north central Kansas.

Cultivation may have an effect on communities of small mammals in some grassland areas (Getz and Brighty 1986, Kaufman and Kaufman 1990b), while having little impact on other grassland communities (Fleharty and Navo 1983). For example, over an annual cycle, communities of small mammals in wheat fields and fallow wheat fields were different from communities in native mixed-grass prairie in north central Kansas (Kaufman and Kaufman 1990b). Abundances of deer mice were greater in both wheat fields and fallow wheat fields when compared with limestone breaks prairie, their preferred native habitat. Therefore, conversion of native prairie to croplands has promoted the numerical dominance of deer mice in communities of small mammals in habitats for wheat production over that found in mixed-grass prairie. Deer mice also were common in the high-intensity corn-soybean agroecosystem in Illinois (Getz and Brighty 1986). By contrast, abundances of other small mammals were reduced in wheat fields and corn-soybean fields as compared with native prairie, which suggests cropping had a large impact on communities of native small mammals (Navo and Fleharty 1983, Getz and Brighty 1986, Kaufman and Kaufman 1990b). In western Kansas, differences were not found between croplands and prairie as assemblages of small mammals were similar between irrigated corn fields and sandsage prairie (Fleharty and Navo 1983).

Refugia such as fencerows, roadside rights-of-way or ditches, railroad rights-of-way, and waterways allow some species of small mammals that are negatively affected by annual cropping systems to exist within landscapes otherwise dominated by crop fields. Total abundance of small mammals and species richness were higher in fencerows that occur between crop fields and adjacent native prairie than in either the native prairie or the crop fields (Kaufman and Kaufman 1989). Fencerows likely are important to prairie voles and western harvest mice, species that respond favorably to dense herbaceous vegetation, because these species occur in lowland areas of mixed-grass prairie and, in many areas, most lowland prairie has been converted to croplands. Roadsides also are important refugia for small mammals in intensively farmed areas of Illinois (Getz and Brighty 1986). Railroad rights-of-way appear to provide important habitat for Franklin's ground squirrel, a species whose origin is traced to the prairie (Risser et al. 1981) but whose range was and continues to be affected by cultivation (Johnson and Choromanski-Norris 1992). Our general observations of captures of small mammals in planted waterways and in small, uncultivated patches of prairie also suggest that these habitats provide conditions for species not typically caught in crop fields.

What effects do restoration or reclamation of grasslands (such as the Conservation Reserve Program [CRP]) and abandonment of cultivation (old fields) have on the communities of native small mammals? The answer likely varies as to the type of vegetation that is planted. In CRP land on the southern High Plains of Texas where the dominant grass is weeping lovegrass (*Eragrostis curvula,* an exotic from Africa) rather than the native short grasses such as blue grama (*Bouteloua gracilis*) and buffalograss (*Buchloe dactyloides*), the abundances of small mammals were depressed significantly as compared with abundances in shortgrass prairie (Hall and Willig 1994). The community of small mammals observed on CRP land that was in its third year of the program was more similar to the native community of small mammals than communities found the first or second year after planting the CRP land. Values of species diversity remained similar among the CRP lands (1–3 years old) and native prairie; however, the species of small mammals present in the CRP fields changed as plant succession occurred in the fields. For example, deer mice, white-footed mice, and northern grasshopper mice were at their highest abundances in the first-year fields, harvest mice and cotton rats were at their highest abundances in the second-year fields, and Ord's kangaroo rats were found only in the third-year fields and native shortgrass prairie.

A similar transition of species of small mammals was observed over a 12-year period in a hayfield in Iowa that was reconstructed by plowing and then planting five species of tall grasses (Schwartz and Whitson 1987). Meadow voles and house mice were abundant in the weedy stage, white-footed mice in the grassland stage, and meadow voles, short-tailed shrews, and masked shrews in the early-prairie stage. In eastern Kansas, changes in small mammals associated with plant succession in an abandoned old field, a hayfield, and a pasture were monitored for 15 years (Swihart and Slade 1990). Patterns of use of these habitats were related to the types of vegetation present and season of the year. Generally, southern bog lemmings, when present, used grassy habitat, deer mice and harvest mice used areas dominated by forbs, white-footed mice used wooded habitat, prairie voles used areas dominated both by grasses and by forbs, and cotton rats selectively used all three habitats within at least one season of the 15 years.

Have agricultural practices allowed the invasion and expansion of non-native species of small mammals into native grasslands? Few studies have been directed at this question, probably because no major invasions by small mammals that are not native to North America have occurred. Recently, Kaufman and Kaufman (1990c) investigated whether the house mouse, a human commensal rodent from Europe, is a common member of assemblages of small mammals in native prairies in eastern and north central Kansas. House mice were rare (<0.5%) in assemblages of small mammals in grasslands, roadside ditches, mowed hayfields, and woodlands but slightly more common (2.6–4.1%) in crop fields and crop field fencerows. This pattern of rarity in native grasslands and greater abundance in crop fields or disturbed habitats has been found in a variety of grasslands (Illinois: Getz and Brighty 1986; Iowa: Schwartz and Whitson 1987; Kansas: Fleharty and Navo 1983; Texas: Hall and Willig 1994). House mice, although occurring across the prairie region, have not invaded native prairie or woodland habitats at levels that would prove detrimental to native rodents.

Ranching

The eradication of large towns of prairie dogs by poisoning likely was the largest impact that ranching had on rodents of the mixed-grass and shortgrass prairie of central North America. Estimates of the coverage of prairie dog towns range from 40 million to 100 million ha in these prairies about 1900, but this coverage was reduced to 600,000 ha by 1960 and continues to be reduced as eradication programs continue (Miller et al. 1994). Although it was once estimated that prairie dogs reduced productivity of rangelands by 50–75% (Merriam 1902), modern research has shown that 300 prairie dogs consume about the same amount of forage as one cow with a calf (Uresk and Paulson 1988), and market weights of steers are not affected by the presence of prairie dogs (Hansen and Gold 1977, O'Meilia et al. 1982). Additionally, large native herbivores (bison, elk, and pronghorn) preferred to graze on prairie dog towns where grasses are more nutritious (Whicker and Detling 1988). Secondarily, poisoning programs and the spread of plague in prairie dog colonies (Barnes 1993, Cully 1993) led to the near extinction of the black-footed ferret; however, what remains unknown is what effect the removal of prairie dogs has on the distribution and abundance of other small mammals.

Annual burning of rangelands, a common ranching practice in the Flint Hills, may be the factor that most affects small mammals of the tallgrass prairies of central North America. However, this postulate remains to be tested. Our results on ungrazed tallgrass prairie on the Konza Prairie suggest that large-scale burning at an annual frequency will have a negative impact on many if not all small mammals. Data collected from 1981 to 1991 on small mammals on treatment sites that were burned either annually or periodically in spring show that the average number of species present is lower on annual burns (3.3 species) than on periodic burns (4.0). Species diversity (H) and evenness (J) also are significantly lower on annual burns ($H = 0.66$, $J = 0.52$) than on periodic burns ($H = 0.88$, $J = 0.66$), although the proportion of deer mice is similar in both types of communities (annual = 72%, periodic = 0.65%). However, this proportional abundance of deer mice belies the fact that absolute density is almost always lower on annual burns compared with prairie with less-frequent fire.

Clark et al. (1989) examined the effects of fire and grazing by cattle on habitat selection by four common small mammals in tallgrass prairie in the Flint Hills. Deer mice used both grazed and ungrazed, burned prairie but avoided ungrazed, unburned prairie (grazed, unburned prairie was not available). White-footed mice selected grazed, burned areas where shrubs were common but avoided ungrazed, burned areas. Elliot's short-tailed shrews and prairie voles selected ungrazed, unburned habitats and avoided grazed, burned habitats. Although individual species showed selection of or avoidance for particular habitats, total abundance of small mammals did not differ among the three habitats, and species diversity and richness were similar among the three habitats. Total biomass of small mammals was decreased (24%) on grazed versus ungrazed tallgrass prairie in Oklahoma, whereas the pattern was not observed in grazed versus ungrazed shortgrass prairie in Colorado (Grant et al. 1982). One question that still remains unanswered within prairie ecosystems is whether cattle are the ecological equivalent of bison which evolved in the North American prairie. That is, do bison and cattle have the same effect on plant productivity, species composition, and other factors that may affect abundance and distribution of small mammals?

Another common ranching practice is the haying of prairie plants across much of the central North American region. How does haying affect the abundance and distribution of small mammals? Mowing of tallgrass prairie in Nebraska caused the habitat to be unsuitable for two species of *Microtus* (meadow and prairie voles) because of lack of cover (Lemen and Clausen 1984). A short-term study in eastern Kansas suggests that abundance of small mammals is lower in hayfields (54 individuals) than in contiguous native prairie (96), with more than a twofold decrease in numbers of three species (cotton rats, prairie voles, and Elliot's short-tailed shrews) present in both habitats but with similar numbers of deer mice in both habitats (Sietman et al. 1994). Five species were present in both habitats but western harvest mice were absent from the hayfields and hispid pocket mice were absent from native prairie. Our own observations in annually hayed mixed-grass prairie of north central Kansas suggest that most small mammals are affected negatively. Annual haying prevents litter buildup and also may prevent sufficient production of seeds to support seed-eating rodents.

Development

Within the Great Plains, fragmentation of the prairie by cultivation, ranching, and rural and urban development likely is the greatest challenge for some species of small mammals because of their limited mobility. The effect of fragmentation is dependent on the size, shape, density, dispersion, and isolation of fragments as well as the types of vegetative cover that isolate the fragments from each other and from the original intact landscape (Lord and Norton 1990). Fragmentation of natural environments by man-made landscapes that cannot be used by some species of small mammals would result in a lack of a source for immigrants, an obstruction of a route for recolonization of a habitat, and an increase in the probability of extinction when demographic units become too small or disconnected to remain viable. For example, the fragmentation of colonies of prairie dogs by cultivation and eradication programs makes the remaining prairie dogs in small and isolated colonies more susceptible to disease and other mortality factors (Miller et al. 1994). In turn, negative effects of habitat loss and fragmentation on one species can lead to the threatened or endangered status of a second species, such as black-footed ferrets, because of the dependence on the original impacted species.

The effect of experimental fragmentation of habitat on three species of small mammals was studied in eastern Kansas (Diffendorfer et al. 1995). Prairie voles and deer mice moved longer distances as fragmentation increased. The proportion of interpatch movements of cotton rats, prairie voles, and deer mice decreased as fragmentation increased from continuous large patches to fragmented small patches; however, spacing of traps between blocks was not equal. The pattern for deer mice did not hold when corrected for trap spacing using simulated blocks. Expectations of directional movements of individuals from sources to sinks were not supported for any of the three species. However, differences in the likelihood of each species to move from patch to patch and in the scale of the landscape suggest that more studies are needed to understand how the scale of fragmentation affects individual species.

Boundary habitats such as those that develop along fencerows or in road ditches may be important within an otherwise homogeneous matrix. For example, fencerows within croplands may not only act as refugia for some species of small mammals (Kaufman and Kaufman 1989) but also serve as movement corridors between native habitats. Road ditches also may serve as corridors for some small mammals (jumping mice [Choate et al. 1991] and meadow voles and southern bog lemmings [Welker and Choate 1994]). By contrast, some linear habitats (roads) may act as barriers (Oxley et al. 1974). In western Kansas, no individuals of seven species of small mammals crossed an interstate or a state highway, unless displaced, as compared with five thirteen-lined ground squirrels (14%) and one deer mouse (2%) that crossed gravel or limestone roads when native grasses occurred on both sides of all roads (Kozel and Fleharty 1979). Hispid pocket mice, western harvest mice, cotton rats, and prairie voles never crossed any of these types of roads unless they were displaced to the opposite side of the road, and western harvest mice did not return to the original side even after being displaced from their home range (Kozel and Fleharty 1979).

Additions of habitat can influence fragmentation of native habitats. Trees in riparian communities of the Great Plains likely were reduced immediately after European settlement because of their use for fuel and construction (Fleharty 1995). Subsequently, more trees were planted on the Great Plains as settlers planted trees around homesteads and farms and on timber claims (Timber Claim Act of 1862). Later, programs to control soil erosion encouraged farmers to plant trees (Samson and Knopf 1994) as well as windbreaks as habitat for wildlife, primarily birds (Capel 1988). Because of plantings of trees, suppression of fires in grasslands, and subsequent invasion of trees on the prairie, the number of trees present on the Great Plains likely is greater than that found before European settlement (Fleharty 1995). Recently, ten species of small mammals were shown to make some use of shelterbelts in the Great Plains (Johnson and Beck 1988). This increase in trees within the Great Plains likely has promoted the movement of mammals from east to west (Knopf and Scott 1990). This pattern likely has occurred in some small mammals (least shrew) (Armstrong 1972) that readily use wooded habitats that have developed along streams, from tree plantings, and from the invasion of prairie and croplands by native or introduced trees and shrubs.

The white-footed mouse in Kansas is illustrative of how one small mammal has expanded its distribution by using habitats created by humans. Although they still are most abundant in wooded riparian habitats and associated wooded ravines (Fleharty and Stadel 1968, Choate and Fleharty 1975, Van Deusen and Kaufman 1977, Clark et al. 1987, Kaufman et al. 1993, 1995a, McMillan and Kaufman 1994), white-footed mice now occur across Kansas in areas where no trees occurred before European settlement. For example, white-footed mice occupy shelterbelts (Fleharty and Stadel 1968, Van Deusen and Kaufman 1977) and woody plantings for wildlife (plum-cedar plantings and plum thickets) (D.W. Kaufman, G.A. Kaufman, and B.K. Clark, unpublished data) as well as roadside ditches with invading trees and shrubs and planted hedgerows (G.A. Kaufman and D.W. Kaufman, unpublished data). As a result, white-footed mice have been captured in almost all patches of trees sampled in north central Kansas. White-footed mice can reach high densities where they spill over into grasslands (planted grass and roadside ditches), and croplands (yellow clover [*Meliotus officinalis*] and fallow fields) (D.W. Kaufman, G.A. Kaufman, and B.K. Clark, unpublished data). Although less studied than white-footed mice, eastern woodrats also are found in sites with planted or invading trees where trees would not have occurred before European settlement (G.A. Kaufman and D.W. Kaufman, unpublished data).

Conclusions

Climate (both temperature and precipitation), fire, topography, and the structure and associated litter layer of vegetation affect small mammal species differently. In addition, small mammals play important ecological roles in prairies as grazers, seed predators and dispersers, and disturbance agents. Therefore, conservation of the central North American prairie demands holistic management practices. Just as prairies cannot be managed for a single plant or group of plants without consider-ing the animals, neither can it be managed for a single animal or group of animals without considering native plants. This same mindset also must be used in the restoration of prairies (i.e., the restoration of prairies involves more than restoring only the grasses and forbs).

Landscape fragmentation, destruction of diverse vegetative cover, and planting of annual crops in monocultures have had major impacts on populations and communities of small mammals. As a result, certain of the small mammals whose origins trace to the grassland (Risser et al. 1981) are uncommon or rare today. For example, in Kansas, the plains pocket mouse, Franklin's ground squirrel, and silky pocket mouse tend to be rare. Only those species that have adapted to human changes to the environment (thirteen-lined ground squirrels [road ditches and grazed prairie], plains pocket gophers [old fields and alfalfa fields], and prairie voles [road ditches, waterways, and fencerows]) are common in the prairie region. Some other grassland rodents (black-tailed prairie dogs and hispid pocket mice) are only locally common.

The mindset that must be developed for preservation and conservation of relatively intact prairie sites, management of parks, preserves, and grazing lands, and restoration of prairies destroyed by various human activities is that prairies were not and are not homogeneous environments. Rather prairies are a dynamic mosaic of populations of organisms, both plants and animals, and ecological processes. Prairies provide mammals and other animals habitats and microhabitats created by spatial and temporal variation in climatic patterns, fires, and spatial variation in soils and geomorphologic conditions that locally alter the composition, vertical structure and architecture, and dynamics of vegetation. The dynamic mosaic of prairie is further modified by activities of mammals and other animals whose impacts vary broadly in space, time, and intensity. Some species have impacts so large in space and intensity that we consider them ecological "engineers" or keystone species, whereas others have only minor impacts. However, impacts of these minor species may be important to understand both how prairies function and why the animals must be included in any equation concerning the conservation of native grasslands.

Finally, activities associated with a growing human population eventually will lead to greater destruction of the native prairie of central North America. Destruction and modification of the Great Plains landscape may be pushed to levels where the impacts are catastrophic and cannot be predicted from linear-curvilinear response curves between the total abundances and distributional ranges of small mammals. Fragmentation may be one reason why the response of some species of small mammals, as well as other animals, may not remain predictable from past observations of the effects of humans. Global warming is also an anthropogenic factor whose influence on the small mammals of the prairie region cannot be predicted. The magnitude of changes in timing and amount of precipitation and in averages and extremes of summer and winter temperatures that may occur in the near future are unknown. The patterns of influence of various natural and anthropogenic influences examined herein suggest that the abundance and size and spatial distribution of ranges of small mammals likely will change with global warming. Further, changes in abundance and distribution will vary among species of rodents and shrews in the Great Plains but should push average ranges of each species farther north. Populations of some species will be affected negatively by the environmental changes caused by global warming, and some species may decrease to the point of local or regional extirpation. Regardless of specific outcomes, the potential of global climate change and the continued destruction and modification of native grasslands by humans makes it imperative that ecologists and conservationists continue to voice concern about the future of small mammals in prairie landscapes in the Great Plains.

Acknowledgments. We thank J.R. Choate, E.D. Fleharty, and B.R. McMillan for critical review of an earlier draft of this chapter. Research on the Konza Prairie was and continues to be supported by the NSF Konza Prairie Long Term Ecological Research Program (DEB-8012166, BSR-8514327, DEB-9011662) and was supported by NSF grant BSR-8307571 and the Kansas Agricultural Experiment Station. Partial support for research in north central Kansas on the mixed-grass prairie was provided by the Kansas Agricultural Experiment Station (R-231). The Konza Prairie is owned by The Nature Conservancy and is managed for ecological research by the Division of Biology, Kansas State University. This is contribution 96-390-B, Kansas Agricultural Experiment Station, Kansas State University, Manhattan.

References

Andersen, D.C. 1987. *Geomys bursarius* burrowing patterns: influence of season and food patch structure. Ecology 68:1306-1318.

Anderson, R.C. 1990. The historical role of fire in the North American grassland. Pp. 8-18 *in* S.L. Collins and L.L. Wallace, eds. Fire in North American tallgrass prairies. Univ. Oklahoma Press, Norman.

Armstrong, D.M. 1972. Distribution of mammals in Colorado. Kansas Mus. Nat. Hist. Monogr. 3, Lawrence.

Axelrod, D.I. 1985. Rise of the grassland biome, central North America. Bot. Rev. 51:163-202.

Barnes, A.M. 1993. A review of plague and its relevance to prairie dog populations and the black-footed ferret. Pp. 28-37 *in* J.L. Oldemeyer, D.E. Biggins, B.J. Miller, and R. Crete, eds. Management of prairie dog complexes for the reintroduction of the black-footed ferret. Biol. Rep. 13. U.S. Dept. Interior, Washington, DC.

Birney, E.C., W.E. Grant, and D.D. Baird. 1976. Importance of vegetative cover to cycles of *Microtus* populations. Ecology 57:1043-1051.

Bixler, S.H., and D.W. Kaufman. 1995. Local distribution of prairie voles (*Microtus ochrogaster*) on Konza Prairie: effect of topographic position. Trans. Kansas Acad. Sci. 98:61-67.

Black, J.D. 1937. Mammals of Kansas. Kansas State Board Agric., 30th Biennial Rep. 35:116-217.

Boag, D.A., and D.R. Wiggett. 1994. The impact of foraging by Columbian ground squirrels, *Spermophilus columbianus,* on vegetation growing on patches fertilized with urine. Can. Field-Nat. 108:282-287.

Bolen, E.G., and W.L. Robinson. 1995. Wildlife ecology and management. 3rd ed. Prentice Hall, Englewood Cliffs, NJ.

Brillhart, D.E., G.A. Kaufman, and D.W. Kaufman. 1995. Small-mammal use of experimental patches of tallgrass prairie: influence of topographic position and fire history. Pp. 59-65 *in* D.C. Hartnett, ed. Prairie biodiversity. Proc. 14th N. Am. Prairie Conf. Kansas State Univ., Manhattan.

Brown, H.L. 1946. Rodent activity in a mixed prairie near Hays, Kansas. Trans. Kansas Acad. Sci. 48:448-456.

Brown, J.H., and E.J. Heske. 1990. Control of a desert-grassland transition by a keystone rodent guild. Science 250:1705-1707.

Burke, I.C., T.G. F. Kittel, W.K. Lauenroth, P. Snook, C.M. Yonker, and W.J. Parton. 1991. Regional analysis of the central Great Plains. BioScience 41:685-692.

Capel, S.W. 1988. Design of windbreaks for wildlife in the Great Plains of North America. Agric. Ecosys. Environ. 22/23:337-347.

Choate, J.R., and E.D. Fleharty. 1975. Synopsis of native, recent mammals of Ellis County, Kansas. Occas. Pap. 37. Texas Tech Univ. Museum, Lubbock.

Choate, J.R., D.W. Moore, and J.K. Frey. 1991. Dispersal of the meadow jumping mouse in northern Kansas. Prairie Nat. 23:127-130.

Cid, M.S., J.K. Detling, A.D. Whicker, and M.A. Brizuela. 1991. Vegetational responses of a mixed-grass prairie site following exclusion of prairie dogs and bison. J. Range Manage. 44:100-105.

Clark, B.K., and D.W. Kaufman. 1990. Short-term responses of small mammals to experimental fire in tallgrass prairie. Can. J. Zool. 68:2450-2454.

Clark, B.K., and D.W. Kaufman. 1991. Effects of plant litter on foraging and nesting behavior of prairie rodents. J. Mammal. 72:502-512.

Clark, B.K., D.W. Kaufman, E.J. Finck, and G.A. Kaufman. 1989. Small mammals in tall-grass prairie: patterns associated with grazing and burning. Prairie Nat. 21:177-184.

Clark, B.K., D.W. Kaufman, G.A. Kaufman, and E.J. Finck. 1987. Use of tallgrass prairie by *Peromyscus leucopus*. J. Mammal. 68:158-160.

Clark, B.K., D.W. Kaufman, G.A. Kaufman, and S.K. Gurtz. 1995. Population ecology of Elliot's short-tailed shrew and least shrew in ungrazed tallgrass prairie manipulated by experimental fire. Pp. 87-92 *in* D.C. Hartnett, ed. Prairie biodiversity. Proc. 14th N. Am. Prairie Conf. Kansas State Univ., Manhattan.

Clark, B.K., D.W. Kaufman, G.A. Kaufman, S.K. Gurtz, and S.H. Hand. 1990. Population ecology of thirteen-lined ground squirrels in ungrazed tallgrass prairie manipulated by fire. Pp. 51-54 *in* D.D. Smith and C.A. Jacobs, eds. Recapturing a vanishing heritage. Proc. 12th N. Am. Prairie Conf. Univ. N. Iowa, Cedar Falls.

Collins, S.L., and S.M. Glenn. 1988. Disturbance and community structure in North American prairies. Pp. 131-143 *in* H.J. During, M.J.A. Werger, and J.H. Willems, eds. Diversity and pattern in plant communities. Academic Publ., The Hague.

Collins, S.L., and L.L. Wallace, eds. 1990. Fire in North American tallgrass prairie. Univ. Oklahoma Press, Norman.

Cully, J.F., Jr. 1993. Plague, prairie dogs, and black-footed ferrets. Pp. 38-49 *in* J.L. Oldemeyer, D.E. Biggins, B.J. Miller, and R. Crete, eds. Management of prairie dog complexes for the reintroduction of the black-footed ferret. Biol. Rep. 13. U.S. Dept. Interior, Washington, DC.

Davis, M.A., J. Villinski, K. Banks, J. Buckman-Fifield, J. Dicus, and S. Hofmann. 1991a. Combined effects of fire, mound-building by pocket gophers, root loss and plant size on growth and reproduction in Penstemon grandiflorus. Am. Midl. Nat. 125:150-161.

Davis, M.A., J. Villinski, S. McAndrew, H. Scholtz, and E. Young. 1991b. Survivorship of Penstemon grandiflorus in an oak woodland: combined effects of fire and pocket gophers. Oecologia 86:113-118.

Day, T.A., and J.K. Detling. 1994. Water relations of Agropyron smithii and Bouteloua gracilis and community evapotranspiration following long-term grazing by prairie dogs. Am. Midl. Nat. 132:381-392.

Decker, E. 1980. Exotics. Pp. 249-256 *in* J.L. Schmidt and D.L. Gilbert, eds. Big game of North America, ecology and management. Stackpole Books, Harrisburg, PA.

Diffendorfer, J.E., M.S. Gaines, and R.D. Holt. 1995. Habitat fragmentation and movement of three small mammals (*Sigmodon, Microtus,* and *Peromyscus*). Ecology 76:827-839.

Ehrlich, P.R., and A.H. Ehrlich. 1981. Extinction: the causes and consequences of the disappearance of species. Random House, New York.

Erwin, W.J., and R.H. Stasiak. 1979. Vertebrate mortality during the burning of a reestablished prairie in Nebraska. Am. Midl. Nat. 101:247-249.

Fahnestock, J.T., and A.K. Knapp. 1994. Plant responses to selective grazing by bison: interactions between light, herbivory and water stress. Vegetatio 115:123-131.

Fleharty, E.D. 1995. Wild animals and settlers on the Great Plains. Univ. Oklahoma Press, Norman.

Fleharty, E.D., and K.W. Navo. 1983. Irrigated cornfields as habitat for small mammals in the sandsage prairie region of western Kansas. J. Mammal. 64:367-379.

Fleharty, E.D., and D.L. Stadel. 1968. Distribution of *Peromyscus leucopus* (woods mouse) in western Kansas. Trans. Kansas Acad. Sci. 71:231-233.

French, N.R., W.E. Grant, W. Grodzinski, and D.M. Swift. 1976. Small mammal energetics in grassland ecosystems. Ecol. Monogr. 46:201-220.

Geluso, K.N., G.D. Schroder, and T.B. Bragg. 1986. Fire avoidance behavior of meadow voles (Microtus pennsylvanicus). Am. Midl. Nat. 116:202-205.

Getz, L.L., and E. Brighty. 1986. Potential effects of small mammals in high-intensity agricultural systems in east-central Illinois, U.S.A. Agric. Ecosys. Environ. 15:39-50.

Gibson, D.J. 1989. Effects of animal disturbance on tallgrass prairie vegetation. Am. Midl. Nat. 121:144–154.

Gibson, D.J., C.C. Freeman, and L.C. Hulbert. 1990. Effects of small mammal and invertebrate herbivory on plant species richness and abundance in tallgrass prairie. Oecologia 84:169–175.

Grant, W.E. 1974. The functional role of small mammals in grassland ecosystems. PhD dissertation. Colorado State Univ., Fort Collins.

Grant, W.E., and E.C. Birney. 1979. Small mammal community structure in North American grasslands. J. Mammal. 60:23–36.

Grant, W.E., E.C. Birney, N.R. French, and D.M. Swift. 1982. Structure and productivity of grassland small mammal communities related to grazing-induced changes in vegetative cover. J. Mammal. 63:248–260.

Grant, W.E., N.R. French, and D.M. Swift. 1977. Response of a small mammal community to water and nitrogen treatments in a shortgrass prairie ecosystem. J. Mammal. 58:637–652.

Hall, D.L., and M.R. Willig. 1994. Mammalian species composition, diversity, and succession in Conservation Reserve Program grasslands. Southwest. Nat. 39:1–10.

Hall, E.R. 1981. The mammals of North America. 2nd ed. John Wiley & Sons, New York.

Hansen, R.M., and I.K. Gold. 1977. Black-tailed prairie dogs, desert cottontails and cattle trophic relations on shortgrass range. J. Range Manage. 30:210–214.

Harty, F.M., J.M. Ver Steeg, R.R. Heidorn, and L. Harty. 1991. Direct mortality and reappearance of small mammals in an Illinois grassland after a prescribed burn. Nat. Areas J. 11:114–118.

Huntly, N., and R. Inouye. 1988. Pocket gophers in ecosystems: patterns and mechanisms. BioScience 38:786–793.

Huntly, N., and O.J. Reichman. 1994. Effects of subterranean mammalian herbivores on vegetation. J. Mammal. 75:852–859.

Inouye, R.S., N.J. Huntly, D. Tilman, and J.R. Tester. 1987. Gophers (Geomys bursarius), vegetation, and soil nitrogen along a successional sere in east central Minnesota. Oecologia 72:178–184.

Johnson, R.J., and M.M. Beck. 1988. Influences of shelterbelts on wildlife management and biology. Agric. Ecosys. Environ. 22/23:301–335.

Johnson, S.A., and J. Choromanski-Norris. 1992. Reduction in the eastern limit of the range of the Franklin's ground squirrel (Spermophilus franklinii). Am. Midl. Nat. 128:325–331.

Jones, J.K., Jr., D.M. Armstrong, and J.R. Choate. 1985. Guide to mammals of the Plains states. Univ. Nebraska Press, Lincoln.

Jones, J.K., Jr., D.M. Armstrong, R.S. Hoffmann, and C. Jones. 1983. Mammals of the northern Great Plains. Univ. Nebraska Press, Lincoln.

Jones, J.K., Jr., D.C. Carter, H.H. Genoways, R.S. Hoffmann, and D.W. Rice. 1982. Revised checklist of North American mammals north of Mexico, 1982. Occas. Pap. 80. Texas Tech Univ. Museum, Lubbock.

Jones, J.K., Jr., D.C. Carter, H.H. Genoways, R.S. Hoffmann, D.W. Rice, and C. Jones. 1986. Revised checklist of North American mammals north of Mexico, 1986. Occas. Pap. 107. Texas Tech Univ. Museum, Lubbock.

Jones, J.K., Jr., R.S. Hoffmann, D.W. Rice, C. Jones, R.J. Baker, and M.D. Engstrom. 1992. Revised checklist of North American mammals north of Mexico, 1991. Occas. Pap. 146. Texas Tech Univ. Museum, Lubbock.

Kalisz, P.J., and W.H. Davis. 1992. Effect of prairie voles on vegetation and soils in central Kentucky. Am. Midl. Nat. 127:392–399.

Kaufman, D.W., and S.H. Bixler. 1995. Prairie voles impact plants in tallgrass prairie. Pp. 117–121 in D.C. Hartnett, ed. Prairie biodiversity. Proc. 14th N. Am. Prairie Conf. Kansas State Univ., Manhattan.

Kaufman, D.W., B.K. Clark, and G.A. Kaufman. 1990a. Habitat breadth of nongame rodents in the mixed-grass prairie region of north central Kansas. Prairie Nat. 22:19-26.

Kaufman, D.W., E.J. Finck, and G.A. Kaufman. 1990b. Small mammals and grassland fires. Pp. 46-80 in S.L. Collins and L.L. Wallace, eds. Fire in North American tallgrass prairies. Univ. Oklahoma Press, Norman.

Kaufman, D.W., and E.D. Fleharty. 1974. Habitat selection by nine species of rodents in north-central Kansas. Southwest. Nat. 18:443-452.

Kaufman, D.W., and G.A. Kaufman. 1989. Nongame wildlife management in central Kansas: implications of small mammal use of fencerows, fields, and prairie. Trans. Kansas Acad. Sci. 92:185-205.

Kaufman, D.W., and G.A. Kaufman. 1990a. Influence of plant litter on patch use by foraging Peromyscus maniculatus and Reithrodontomys megalotis. Am. Midl. Nat. 124:195-198.

Kaufman, D.W., and G.A. Kaufman. 1990b. Small mammals of wheat fields and fallow wheat fields in north-central Kansas. Trans. Kansas Acad. Sci. 93:28-37.

Kaufman, D.W., and G.A. Kaufman. 1990c. House mice (Mus musculus) in natural and disturbed habitats in Kansas. J. Mammal. 71:428-432.

Kaufman, D.W., G.A. Kaufman, and E.J. Finck. 1989. Rodents and shrews in ungrazed tallgrass prairie manipulated by fire. Pp. 173-177 in T.B. Bragg and J. Stubbendieck, eds. Prairie pioneers: ecology, history, and culture. Proc. 11th N. Am. Prairie Conf. Univ. Nebraska Press, Lincoln.

Kaufman, D.W., G.A. Kaufman, and E.J. Finck. 1993. Small mammals of wooded habitats of the Konza Prairie Research Natural Area, Kansas. Prairie Nat. 25:27-32.

Kaufman, D.W., G.A. Kaufman, and E.J. Finck. 1995a. Temporal variation in abundance of Peromyscus leucopus in wooded habitats of eastern Kansas. Am. Midl. Nat. 133:7-17.

Kaufman, G.A. 1990. Population ecology, social organization, and mating systems in the deer mouse (Peromyscus maniculatus bairdii) in mixed-grass prairie in Kansas. PhD dissertation. Kansas State Univ., Manhattan.

Kaufman, G.A., D.W. Kaufman, D.E. Brillhart, and E.J. Finck. 1995b. Effect of topography on the distribution of small mammals on the Konza Prairie Research Natural Area, Kansas. Pp. 97-102 in D.C. Hartnett, ed. Prairie biodiversity. Proc. 14th N. Am. Prairie Conf. Kansas State Univ., Manhattan.

Kaufman, G.A., D.W. Kaufman, and E.J. Finck. 1988. Influence of fire and topography on habitat selection by Peromyscus maniculatus and Reithrodontomys megalotis in ungrazed tallgrass prairie. J. Mammal. 69:342-352.

Knopf, F.L., and M.L. Scott. 1990. Altered flows and created landscapes in the Platte River Headwaters, 1840-1990. Pp. 47-70 in J.M. Sweeney, ed. Management of dynamic ecosystems, North Cent. Sect. The Wildl. Soc., West Lafayette, IN.

Kotler, B.P., M.S. Gaines, and B.J. Danielson. 1988. The effects of vegetative cover on the community structure of prairie rodents. Acta Theriol. 33:379-391.

Kozel, R.M., and E.D. Fleharty. 1979. Movements of rodents across roads. Southwest. Nat. 24:239-248.

Krefting, L.W., and E.I. Roe. 1949. The role of some birds and mammals in seed germination. Ecol. Monogr. 19:271-286.

Lawton, J.H. 1994. What do species do in ecosystems? Oikos 71:367-374.

Lawton, J.H., and V.K. Brown. 1993. Redundancy in ecosystems. Pp. 255-270 in E.D. Schulze and H.A. Mooney, eds. Biodiversity and ecosystem function. Springer-Verlag, New York.

Lemen, C.A., and M.K. Clausen. 1984. The effects of mowing on the rodent community of a native tall grass prairie in eastern Nebraska. Prairie Nat. 16:5-10.

Lord, J.M., and D.A. Norton. 1990. Scale and the spatial concept of fragmentation. Conserv. Biol. 4:197-202.

Martin, E.P. 1960. Distribution of native mammals among the communities of the mixed prairie. Fort Hays Studies Sci. Ser. 1:1-26.

McMillan, B.R., and D.W. Kaufman. 1994. Differences in use of interspersed woodland and grassland by small mammals in northeastern Kansas. Prairie Nat. 26:107-116.

Merriam, C.H. 1902. The prairie dog of the Great Plains. Pp. 257-270 *in* Yearbook of the United States Department of Agriculture 1901. U.S. Gov. Print. Off., Washington, DC.

Miller, B., G. Ceballos, and R. Reading. 1994. The prairie dog and biotic diversity. Conserv. Biol. 8:677-681.

Navo, K.W., and E.D. Fleharty. 1983. Small mammals of winter wheat and grain sorghum croplands in west-central Kansas. Prairie Nat. 15:159-172.

O'Meilia, M.E., F.L. Knopf, and J.C. Lewis. 1982. Some consequences of competition between prairie dogs and beef cattle. J. Range Manage. 35:580-585.

Oxley, D.J., M.B. Fenton, and G.R. Carmody. 1974. The effects of roads on populations of small mammals. J. Appl. Ecol. 11:51-59.

Painter, E.L., J.K. Detling, and D.A. Steingraeber. 1993. Plant morphology and grazing history: relationships between native grasses and herbivores. Vegetatio 106:37-62.

Peterson, S.K., G.A. Kaufman, and D.W. Kaufman. 1985. Habitat selection by small mammals of the tall-grass prairie: experimental patch choice. Prairie Nat. 17:65-70.

Pfeiffer, K.E., and D.C. Hartnett. 1995. Bison selectivity and grazing response of little bluestem in tallgrass prairie. J. Range Manage. 48:26-31.

Pfeiffer, K.E., and A.A. Steuter. 1994. Preliminary response of Sandhills prairie to fire and bison grazing. J. Range Manage. 47:395-397.

Polley, H.W., and J.K. Detling. 1988. Herbivory tolerance of Agropyron smithii populations with different grazing histories. Oecologia 77:261-267.

Polley, H.W., and J.K. Detling. 1990. Grazing-mediated differentiation in Agropyron smithii: evidence from populations with different grazing histories. Oikos 57:326-332.

Reichman, O.J., J.H. Benedix, Jr., and T. Seastedt. 1993. Distinct animal-generated edge effects in a tallgrass prairie. Ecology 74:1281-1285.

Reichman, O.J., and S.C. Smith. 1985. Impact of pocket gopher burrows on overlying vegetation. J. Mammal. 66:720-725.

Riegel, A. 1941. Some coactions of rabbits and rodents with cactus. Trans. Kansas Acad. Sci. 44:96-103.

Risser, P.G. 1990. Landscape processes and the vegetation of the North American grassland. Pp. 133-146 *in* S.L. Collins and L.L. Wallace, eds. Fire in North American tallgrass prairies. Univ. Oklahoma Press, Norman.

Risser, P.G., E.C. Birney, H.D. Blocker, S.W. May, W.J. Parton, and J.A. Wiens. 1981. The true prairie ecosystem. Hutchinson Ross Publ. Co., Stroudsburg, PA.

Samson, F.B., and F.L. Knopf. 1994. Prairie conservation in North America. BioScience 44:418-421.

Schupp, E.W. 1993. Quantity, quality and the effectiveness of seed dispersal by animals. Vegetatio 107/108:15-29.

Schwartz, O.A., and P.D. Whitson. 1987. A 12-year study of vegetation and mammal succession on a reconstructed tallgrass prairie in Iowa. Am. Midl. Nat. 117:240-249.

Shump, K.A., Jr. 1974. Nest construction by the western harvest mouse. Trans. Kansas Acad. Sci. 77:87-92.

Sietman, B.E., W.B. Fothergill, and E.J. Finck. 1994. Effects of haying and old-field succession on small mammals in tallgrass prairie. Am. Midl. Nat. 131:1-8.

Swanson, F.J., T.K. Kratz, N. Caine, and R.G. Woodmansee. 1988. Landform effects on ecosystem patterns and processes. BioScience 38:92-98.

Swihart, R.K., and N.A. Slade. 1990. Long-term dynamics of an early successional small mammal community. Am. Midl. Nat. 123:372-382.

Tomanek, G.W., and G.K. Hulett. 1970. Effects of historical droughts on grassland vegetation in central Great Plains. Pp. 203-210 *in* W. Dort, Jr. and J.K. Jones, Jr., eds. Pleistocene and Recent environments of the central Great Plains. Univ. Press Kansas, Lawrence.

Uresk, D.W., and D.B. Paulson. 1988. Estimated carrying capacity for cattle competing with prairie dogs and forage utilization in western South Dakota. Pp. 387–390 *in* Symposium on management of amphibians, reptiles, and small mammals in North America. USDA, For. Serv., Washington, DC.

Van Deusen, M., and D.W. Kaufman. 1977. Habitat distribution of *Peromyscus leucopus* within prairie woods. Trans. Kansas Acad. Sci. 80:151–154.

Vinton, M.A., and D.C. Hartnett. 1992. Effects of bison grazing on *Andropogon gerardii* and *Panicum virgatum* in burned and unburned tallgrass prairie. Oecologia 90:374–382.

Vinton, M.A., D.C. Hartnett, E.J. Finck, and J.M. Briggs. 1993. Interactive effects of fire, bison (Bison bison) grazing and plant community composition in tallgrass prairie. Am. Midl. Nat. 129:10–18.

Walker, B.H. 1992. Biodiversity and ecological redundancy. Biol. Conserv. 6:18–23.

Welker, T.L., and J.R. Choate. 1994. Ecogeography of southern bog lemming and meadow vole in north central Kansas. Prairie Nat. 26:283–286.

Whicker, A.D., and J.K. Detling. 1988. Ecological consequences of prairie dog disturbances. BioScience 38:778–785.

Whicker, A.D., and J.K. Detling. 1993. Control of grassland ecosystem processes by prairie dogs. Pp. 18–27 *in* J.L. Oldemeyer, D.E. Biggins, B.J. Miller, and R. Crete, eds. Management of prairie dog complexes for the reintroduction of the black-footed ferret. Biol. Rep. 13. U.S. Dept. Interior, Washington, DC.

Williams, L.R., G.N. Cameron, S.R. Spencer, B.D. Eshelman, and M.J. Gregory. 1986. Experimental analysis of the effects of pocket gopher mounds on Texas coastal prairie. J. Mammal. 67:672–679.

Willson, M.F. 1993. Mammals as seed-dispersal mutualists in North America. Oikos 67: 159–176.

Willson, M.F., and C.J. Whelan. 1990. Variation in postdispersal survival of vertebrate-dispersed seeds: effects of density, habitat, location, season, and species. Oikos 57:191–198.

Wooster, L.D. 1939. The effects of drouth on rodent populations. Turtox News 17:1–2.

10. Stopover Ecology of Transitory Populations: The Case of Migrant Shorebirds

Susan K. Skagen

Introduction

The overviews of prairie vertebrates thus far in this book focus primarily on resident or breeding species of several taxa. A myriad of avian species, including warblers seeking the boreal forests and waterfowl flying to northern wetlands, temporarily inhabit the central Plains as they rest and refuel for their long journeys. This chapter spotlights transitory shorebirds that migrate between arctic and subarctic breeding grounds and Central and South American wintering areas with stopover sites in wetlands in the Great Plains (Myers et al. 1987, Skagen and Knopf 1993). Because these migrants stop in the prairie region only to rest and forage to replenish fat reserves before resuming their journeys, they provide a unique perspective of prairie ecology.

Thirty-seven species of shorebirds, including 7 species of plovers (Family Charadriidae), 2 species of stilts and avocets (Family Recurvirostridae), and 28 species of sandpipers (Family Scolopacidae), commonly cross the interior Plains during spring and fall migrations (Appendix 10.1). Fourteen of these species breed in the Great Plains. There is considerable diversity in morphology and habitats: shorebirds range in size from 130 to 650 mm in body length, from 13 to 219 mm in bill length, and from 20 to 700 g in body mass and use a broad range of habitats, including unvegetated mud substrates, sandy beaches, grassy uplands, flooded agricultural fields, and open water. Migration behavior also is highly variable, as illustrated by the distances traveled by shorebird species that migrate across the

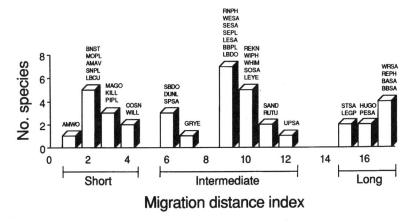

Figure 10.1. Classification of shorebirds by migration distance (short, intermediate, and long) based on a migration distance index, the weighted average of D_s, D_m, and D_e, where D_s is the shortest distance between breeding and wintering areas, D_m is the distance between estimated midpoints, and D_e is the distance between extremes of breeding and wintering areas. All measurements were based on maps in Hayman et al. (1986) and the National Geographic Society (1981). See Appendix 10.1 for species identification. (From Skagen and Knopf 1993.)

Great Plains (Fig. 10.1). Some species, such as the Lesser Golden Plover (*Pluvialis dominica*) and the White-rumped Sandpiper (*Calidris fuscicollis*), travel as many as 14,000–16,000 km between South American wintering grounds and arctic breeding areas. Many of the long-distance migrants move northward directly across the Plains (Fig. 10.2), following the most direct route between summering and wintering grounds.

Collectively, shorebirds inhabit the prairie region from mid-March through mid-October or later. The first waves of northbound migrants, Baird's Sandpipers (*C. bairdii*), appear in the central interior as early as mid-March; the last transient White-rumped Sandpipers fly north from the Dakotas by early June. Arctic breeding species again appear in the northern Plains in early to mid-July, when the first adult Baird's Sandpipers and Stilt Sandpipers (*Calidris himantopus*) begin southward migration (Jehl 1979; M. DeLeon, personal communication). Least Sandpipers (*C. minutilla*), American Avocets (*Recurvirostra americana*), and Long-billed Dowitchers (*Limnodromus scolopaceus*) remain in the southern Plains (northern Oklahoma) until mid-October (Skagen, unpublished data), gradually moving to their wintering grounds in the southern United States, along the Gulf of Mexico coast, and in Central and South America.

Shorebirds are adapted to a highly mobile way of life. Huge energy demands are associated with flights of several thousand kilometers (Morrison 1984, Myers et al. 1987), and migrants must periodically replenish fat reserves while en route. In addition, shorebirds must also build up protein reserves to assist survival and reproduction on breeding grounds (Davidson and Evans 1986). Energy adapta-

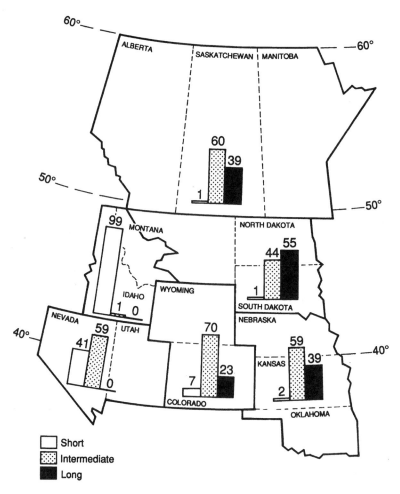

Figure 10.2. Geographic distribution of shorebirds by migration distance during spring migration. Bar graphs and numbers depict percentage of short, intermediate, and long-distance migrants within six regions designated by solid lines. Data from *American Birds* spring migration reports 1980–1990. See Appendix 10.1 for classification of shorebird species by migration distance. (From Skagen and Knopf 1993.)

tions that allow for this highly transitory life cycle include values of basal metabolic rate (BMR), existence metabolism (EM), and daily energy expenditure (DEE) that are considerably higher than predicted based on body weights alone (Kersten and Piersma 1987). Drent and Daan (1980) hypothesized a "maximum sustained working capacity" equal to about 4 BMR, a level of energy expenditure above which there would be detrimental effects on the animal. A higher-than-expected BMR associated with a higher maximum sustained working capacity could be an evolutionary response to the high rates of energy expenditure required during long migratory flights (Kersten and Piersma 1987).

Many of the shorebird species that cross the Great Plains follow different routes during northward and southbound migrations. Elliptical clockwise migration patterns for north- and southbound migrants, in which individual birds follow inland pathways one direction and coastal the other, most likely evolved as an adaptation to the prevailing westerly and easterly winds (Bellrose and Graber 1963, McNeil and Cadieux 1972). Evidence from banding and morphological data confirms an elliptical migration pattern for Semipalmated Sandpipers (*C. pusilla*); many travel north through the central United States during the spring and return south along more eastern routes through James Bay and the Atlantic coast (Harrington and Morrison 1979, Gratto-Trevor and Dickson 1994). Other species that follow a similar elliptical route include White-rumped Sandpipers, Hudsonian Godwits (*Limosa haemastica*), Least Sandpipers, and Short-billed Dowitchers (*Limnodromus griseus*) (McNeil and Cadieux 1972). In the west, many species that migrate to the North Pacific region, especially long-distance migrants that breed at high latitudes and winter in the neotropics, cross the interior of North America during southward migrations (Gill et al. 1994). Some Western Sandpipers (*C. mauri*) travel north along the Pacific coast in spring and return south through the continental interior (Senner and Martinez 1982).

The elliptical features of this migration system and the long-distances covered by many species emphasize the interconnectedness of biopolitical regions in the maintenance of these species during all life history stages. The vulnerability of a species to population declines is intricately tied to the extent of human activities in the geographic regions and the habitats during both breeding and nonbreeding periods (Page and Gill 1994). Successful conservation of these long distance migrants requires coordination across regions. Thus, the Western Hemisphere Shorebird Reserve Network, established in 1985, was established to protect wetlands throughout the western hemisphere.

Population changes in many arctic and subarctic breeders are difficult to detect because of the paucity of information on distribution and abundances. One exception is the dramatic decline of the Eskimo Curlew (*Numenius borealis*), attributed to a combination of factors including hunting and alteration of breeding habitats (Page and Gill 1994). Mountain Plover (*Charadrius montanus*) declines are the most notable among the temperate upland nesting shorebirds (Knopf 1994). There is also evidence of significant declines in migrating populations of Black-bellied Plovers (*Pluvialis squatarola*), Whimbrels (*N. phaeopus*), Sanderlings (*Calidris alba*), Short-billed Dowitchers (Howe et al. 1989), Least Sandpipers, and Semipalmated Sandpipers (Morrison et al. 1994).

How Do Migration Patterns Differ between Great Plains and Coastal Sites?

In coastal systems, shorebirds concentrate in large numbers at sites of seasonally predictable and abundant food resources (Morrison 1984, p. 139, Morrison and Myers 1989, p. 85). Suitable foraging habitat occurs with the tidal cycle in intertidal zones of the Atlantic and Pacific coasts. Shorebird densities in these coastal areas are remarkable. Semipalmated Sandpipers roosting at the Bay of Fundy, Nova Scotia, and vicinity during fall migration reach densities of more than 100 birds/m^2, and peak abundances approached ca 800,000–1,000,000 birds (Mawhinney et al. 1993). Peak numbers of 1.5–2 million shorebirds have been estimated on tidal mudflats of the Copper River Delta, Prince William Sound, Alaska, during spring migration (Isleib 1979, Senner 1979), and peak numbers of spring migrating shorebirds in Delaware Bay exceed 216,000 (Clark et al. 1993).

Until recently, the scientific literature on shorebirds has predominately described migration from the coastal perspective, which emphasizes large numbers of birds in one site with relatively long stays, during which the birds "refuel" in preparation for flight to the next site or their breeding or wintering areas (Myers 1983, Senner and Howe 1984, Morrison and Myers 1989). As of 1984, of 58 major North American stopover or staging sites for shorebirds known, only 3 (5%) were in the Great Plains states, whereas 42 (72%) were in coastal habitats (Senner and Howe 1984).

Shorebirds traversing the Great Plains appear to be more dispersed and to occur in smaller numbers in local sites than birds migrating along the Pacific and Atlantic coasts (Skagen and Knopf 1993). There is ample evidence of small dispersed populations, including four published studies of migrants in interior wetlands in Oklahoma, Kansas, Texas, and Arkansas (Oring and Davis 1966, Schreiber 1970, Neill and Kuban 1986, Smith et al. 1991). These studies cite local peak numbers for individual species ranging from only 75 to 540 birds in spring and from 70 to 2,400 in the fall, and sums of peak counts of all species ranged from 280 to 3,000 (Table 10.1). Unpublished data are also available for additional wetlands that host relatively small numbers of birds in the Dakotas and Nebraska (Table 10.1).

An exception to this generalization is Cheyenne Bottoms Wildlife Management Area (WMA), central Kansas. Cheyenne Bottoms WMA has been designated as a hemispheric reserve by the Western Hemisphere Shorebird Reserve Network because of the large numbers of birds reported there (Senner and Howe 1984). Recent data reveal that there are substantial concentrations of shorebirds in other sites in the interior as well, notably in the Canadian Prairie provinces and the Prairie Potholes of the Dakotas (Table 10.1). During spring migration, shorebirds are abundant in the Canadian interior, with several sites exceeding peak numbers of 20,000 birds and a few exceeding 100,000 birds (Morrison et al. 1991; Table 10.1). At Minnewauken Flats, North Dakota, a conservative estimate of 80,000 birds, including 30,000 White-rumped Sandpipers, was made in late May 1993 (S.K. Skagen, unpublished data). More than 50,000 shorebirds, including nearly 15,000 White-rumped Sandpipers, were recorded at Dry Lake, Clark County, South Dakota, ín the spring of 1992. The predictability of habitat availability in the prairie sites, however, is low; suitable habitat conditions that allow for such concentrations do not occur each year.

Table 10.1 illustrates that some wetlands in the interior region host small numbers of birds whereas shorebird assemblages in other interior wetlands are

Table 10.1. Peak Numbers of the Most Abundant Shorebird Species (Species Code in Parentheses) and the Combined Peak Abundances of All Species of Shorebirds Surveyed at Several Sites throughout the Canadian and U.S. Interior and the Gulf of Mexico Coast during Spring and Fall Migrations[a]

Location	Peak Abundances in Spring		Peak Abundances in Fall		Source
	One Species	All Species[b]	One Species	All Species[b]	
Canadian prairie provinces					
Big Quill Lake, SASK	23,498 (SAND)	155,008			Morrison et al. 1991
Old Wives Lake, Chaplin Lake, SASK	51,654 (SAND)	124,165			Morrison et al. 1991
Pelican Lake, SASK	1,000 (MAGO)	75,000+			Morrison et al. 1991
Oak Hammock Marsh, MAN	7,000 (WRSA)	29,337			Morrison et al. 1991
Whitewater Lake, MAN	10,000 (WRSA)	23,068			Morrison et al. 1991
U.S. Interior—Great Plains					
Minnewauken Flats, Devil's Lake, ND	30,000 (WRSA)	80,798			S.K. Skagen, unpublished data
Cheyenne Bottoms WMA, KS	21,120 (BASA)	67,695	2,240 (LBDO)	10,170	G. Castro, personal communication
Cheyenne Bottoms WMA, KS	20,000 (BASA)	47,673	12,395 (LBDO)	37,273	S.K. Skagen, unpublished data
Dry Lake, Clark County, SD	14,898 (WRSA)	52,239			S.K. Skagen, unpublished data
Salt Plains NWR, OK	10,468 (SESA)	21,814	11,915 (LESA)	27,087	S.K. Skagen, unpublished data
Quivira NWR, KS	5,082 (WRSA)	15,633	1,975 (LBDO)	6,221	Skagen and Knopf 1993
Milwaukee Lake, SD	3,865 (WRSA)	15,178			S.K. Skagen, unpublished data
Lake Thompson, SD	1,536 (LBDO)	8,463			S.K. Skagen, unpublished data
Lake McConaughy, NE			5,000 (BASA)	8,685	S.K. Skagen, unpublished data
Arlington, TX, water treatment plant	540 (LESA)	2,734	2,400 (LESA)	3,057	Neill and Kuban 1986
Northwestern Arkansas	250 (SESA)	1,214	185 (KILL)	754	Smith et al. 1991
Norman, OK	215 (BBSA)	1,144	70 (LEYE)	382	Oring and Davis 1966
Northwestern ND	301 (SESA)	644	535 (WIPH)	1,452	M. DeLeon, personal communication
Clay County, NE			436 (PESA)	1,386	S.K. Skagen, unpublished data
Ellis County, KS	75 (KILL)	350	174 (KILL)	283	Schreiber 1970
Gulf Coast					
South central Louisiana, rice fields	3,348 (LEYE)	21,050	2,020 (WESA)	10,400	Rosenberg and Sillett 1991
Laguna Atascosa NWR, TX	13,433 (LESA)	65,464			S.K. Skagen, unpublished data
San Bernard NWR, TX	11,825 (LBDO)	60,960			R. Speer, personal communication
Harris/Waller counties, rice fields, TX	4,200 (LBDO)	17,692			S. K. Skagen, unpublished data

[a] All surveys were conducted between 1961 and 1993. See Appendix 1 for definitions of species codes.

[b] Peak abundances are calculated two ways. For all sites in the Canadian Prairie Provinces, Minnewauken Flats, ND, and Lake McConaughy, NE, peak abundance refers to the largest number of birds occurring at one time. For the remaining sites, peak abundance is the sum of individual species peaks recorded during several surveys.

often large. This table is not intended to be an exhaustive summary of shorebird occurrences nor does it portray the relative importance of individual sites.

The total number of birds that pass through a site during migration is not necessarily reflected by the peak number of birds at the site, but rather is a function of variables such as the duration of stay at the stopover and the total length of time that migration takes (Thompson 1993). The average stopover time for Semi-palmated Sandpipers staging in the Bay of Fundy in late summer is estimated at 15 days (Hicklin 1987), whereas stopover times during spring passage across the midcontinent average 4–7 days (Skagen and Knopf 1994b). Due to the higher turnover, the Plains, therefore, may host more birds than current figures suggest.

Various regions in the Plains appear to host different species, probably due to the variable migration strategies and distinct routes of these species. Long-distance migrants are prevalent in the central swath through the Plains states but are primarily absent from the intermountain regions (Fig. 10.2). Some species, such as Semipalmated Sandpipers, White-rumped Sandpipers, and Long-billed Dowitchers, are common in spring in several regions of the interior, ranging from the Gulf Coast of Texas to Saskatchewan (Skagen and Knopf 1993; S.K. Skagen, unpublished data). However, Dunlin (*Calidris alpina*) are prevalent along the Texas coast during the winter and spring (Withers and Chapman 1993; S.K. Skagen, unpublished data), underrepresented in the central Plains, and common again in the Prairie Pothole region of the Dakotas (S.K. Skagen, unpublished data). Sanderlings commonly occur along the Texas coast in winter and spring but only in small numbers in the central and northern Plains and are then highly concentrated again in Saskatchewan (Table 10.1).

Do different regions of the Plains provide stopover sites for different breeding populations? Banding and morphological evidence suggest that northbound Semipalmated Sandpipers crossing the central United States originate primarily from Alaskan and central Canadian arctic breeding grounds (Harrington and Morrison 1979). In autumn, the birds that stop over in the Plains primarily breed in Alaska, whereas the central Canadian breeding birds follow a more easterly route along James Bay and the Atlantic coast. Whether distinct regions within the Plains host Alaskan or central Canadian populations, or both, is not known.

The Dynamics of Change

The feature that distinguishes Great Plains shorebird migration patterns from coastal patterns is the tremendous variability in the availability of suitable habitat. In contrast to the coastal regions, the patterns of rainfall and hydrology in the Great Plains are highly dynamic and result in variability in both occurrence and condition of wetlands on every spatial and temporal scale (Fredrickson and Reid 1990, Skagen and Knopf 1994a, Laubhan and Fredrickson, this volume). Many shorebird species, with the exception of phalaropes and yellowlegs, generally use shallow water/wet mud areas that are relatively free from tall vegetation (Colwell and Oring 1988, Helmers 1992). In coastal intertidal regions, such habitat appears and disappears with the predictable tidal cycle. In the Plains, however, the water levels in both unmanaged and managed wetlands can change continually and rapidly (Skagen and Knopf 1994a); suitable shorebird habitat is constantly changing in its location and extent, as in a "shifting mosaic" (Bormann and Likens 1979, Baker 1989). On larger geographic scales, the effects of these dramatic water fluctuations may be modulated, because all wetlands combined have a higher probability of offering some habitat that is suitable for shorebirds. At any given point in time, the spatial pattern of wetland habitats that provide suitable shorebird resources is generated by a variety of processes, including weather, groundwater, hydrology, basin topography, and vegetation history.

Large permanent wetlands, such as Cheyenne Bottoms WMA, may provide the most predictable resources for interior migrants, but even they are less predictable than coastal intertidal areas. Habitat conditions were suitable for shorebirds at Cheyenne Bottoms WMA in only four of seven spring migrations, 1989–1995, and during those four seasons, numbers of shorebirds peaked at more than 20,000. During the other three spring seasons, however, shorebirds found little or no suitable habitat because conditions were either too dry or too wet; peak numbers did not exceed 1,000 birds in those years (G. Castro and H. Hands, personal communication; S.K. Skagen, unpublished data). Dry Lake Two, Clark County, South Dakota, had extensive suitable habitat in only one of three consecutive spring migrations, 1991–1993. Throughout the past several decades, Minnewaukan Flats, Devil's Lake, North Dakota, has undergone several periods of inundation with extremely high water, mixed with periods of drought and periods when habitat was suitable for shorebirds (R. Martin, personal communication). In late summer 1995, the water levels at Minnewauken Flats were ca 3.5 m above the levels in the spring of 1993 (when 80,000 shorebirds were recorded), totally inundating the expansive mudflat areas.

Vegetation encroachment is common in prairie wetlands and can severely reduce the area of habitat that is suitable for shorebirds. During the past few decades, cattail (*Typha* spp.) has steadily encroached on vast areas of unvegetated mudflats at Cheyenne Bottoms WMA, thereby eliminating considerable areas of suitable habitat for migrating shorebirds (H. Hands, personal communication). Although prescribed burning and mowing theoretically can turn wetlands back to earlier successional stages, thereby creating better habitat for shorebirds (Stone 1994), these techniques are extremely difficult if not impossible to carry out over large expanses of habitat (H. Hands, personal communication). The sustainability of Cheyenne Bottoms WMA as the premier midcontinental staging area for migrating shorebirds remains a challenge.

Great Plains wetlands characteristically are seasonably variable in the availability of wet mud/shallow water for shorebirds. Two hundred thirty-two wetlands in a 15-county region during three spring migration seasons were surveyed (about 76 wetlands were surveyed repeatedly) to describe the abundance and distribution of migrants in the Prairie Potholes, South Dakota. Habitat in many of the wetlands changed dramatically between seasons. For example, Dry Lake Two, Clark County, was wet, with little shallow water habitat in 1991 and 1993; however, in

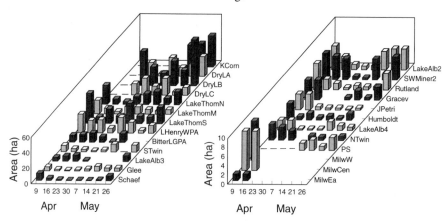

Figure 10.3. Estimated area (ha) of wet mud-shallow water habitat in 25 wetlands in the Prairie Pothole region of South Dakota throughout April and May 1992. Dashes mean that no surveys were conducted, and blanks represent that no suitable habitat was present during survey.

1992 wet mud/shallow water habitat (varying through time from 50 to 100 ha) was available to host a peak of nearly 40,000 migrants. At Milwaukee Lake, Lake County, 50–60 ha of suitable shorebird habitat was available throughout 1991, but high water flooded the flats in 1992 and 1993. Wetland habitats also changed markedly within seasons. We documented within season variability among a sample of 25 wetlands in the Prairie Potholes of South Dakota throughout the spring of 1992 (Fig. 10.3).

Besides the ecological dynamics of wetlands on the Great Plains, total numbers of wetlands in the region have declined to only a fraction of the wetlands that existed before human settlement. Estimates of wetland losses in the Great Plains states since European settlement range from 35 to 89% (Dahl 1990). In addition to wetland loss, the juxtaposition and quality of wetlands have been altered, wetland complexes have been fragmented and disrupted, and hydrology and resulting vegetation have been modified (Fredrickson and Laubhan 1994). Shallow ephemeral wetlands that provide habitat for many shorebirds are more prone to destruction than deep wetlands because they are less costly to drain (Laubhan and Fredrickson 1993). In some regions, only 50, 10, or as few as 1% of the presettlement wetlands remain. Reservoirs may ameliorate some losses of shorebird habitats (Taylor et al. 1992) as impoundments were being created while native wetlands were disappearing (Howe 1987).

Shorebird Responses to Dynamic Habitats

Wide-ranging species, such as migrating shorebirds, have evolved with the dynamic nature of wetland complexes across the entire Great Plains. Birds that

Table 10.2. Correlations between Area of Wet Mud/Shallow Water (<8 cm) Habitat and Numbers of Shorebirds Using Individual Wetlands in the Prairie Potholes Region of South Dakota[a]

Time Period	Number of Wetlands	r	P
Early May 1991	22	0.279	NS
Late May 1991	23	0.558	<.01
Early May 1992	53	0.447	<.001
Late May 1992	46	0.684	<.001
Early May 1993	27	0.551	<.001
Late May 1993	45	-0.020	NS

[a]Wetlands that had no wet mud/shallow water habitat were eliminated from the analyses.

exploit unpredictable resources must rely on flexible behaviors, such as opportunistic use and colonization of new sites, rather than fixed behaviors, such as fidelity to specific wetland sites. In the Plains, shorebirds exhibit the ability to colonize available habitat opportunistically, finding suitable habitat as it appears in a temporally dynamic and spatially complex landscape (Skagen and Knopf 1994a). On relatively small geographic scales and perhaps at larger scales, shorebirds are able to "track" available habitat. Tracking is supported by the statistically significant correlations between the amount of suitable habitat and numbers of shorebirds in seven of eight surveys at Quivira National Wildlife Refuge, Kansas (Skagen and Knopf 1994a), and in four of six surveys across 11 counties in the Prairie Potholes of South Dakota (Table 10.2).

There is little evidence of site fidelity between migration seasons in interior wetlands. Although recaptures of previously banded birds are certainly notable events, the probability of recaptures of Semipalmated Sandpipers, Western Sandpipers, and Least Sandpipers after the year of banding at Cheyenne Bottoms WMA, Kansas, was fairly low (1–1.7%) (Martinez 1979). Further, when habitat at Cheyenne Bottoms was not suitable for shorebirds, five birds banded previously at Cheyenne Bottoms were recaptured at Quivira National Wildlife Refuge, 30 km south (Skagen and Knopf 1994a). Greater site fidelity would be expected in habitats that are fairly constant or that are dynamic in a regular periodicity, such as intertidal areas, than in habitats that are highly dynamic, such as freshwater wetlands (Evans and Townshend 1986, Colwell and Oring 1988, Skagen and Knopf 1994a). Semipalmated Plovers (*Charadrius semipalmatus*), in fact, are highly faithful to a fall migration stopover site along the Atlantic coast (Smith and Houghton 1984), although this example may be extreme and not characteristic of all shorebirds (Evans and Townshend 1986).

Other examples of opportunistic use of ephemeral food resources include concentrations of Surfbirds (*Aphriza virgata*) and Black Turnstones (*Arenaria*

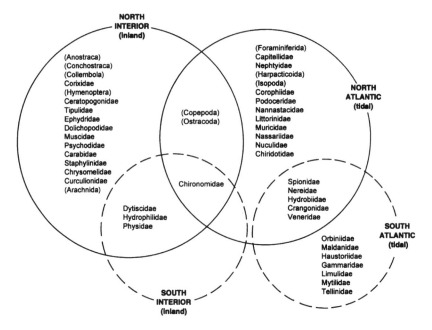

Figure 10.4. Families of invertebrates in diets of Semipalmated Sandpipers in four geographic regions. Circles delineate diet overlap between regions. Taxon names in parentheses are orders or subclasses. (From Skagen and Oman 1996.)

melanocephala) at massive but temporary concentrations of the roe of Pacific herring (*Clupea harengus pallasi*) in Prince William Sound, Alaska (Norton et al. 1990). Additionally, nonbreeding shorebirds in the paleoarctic appear to track the carrying capacities of coastal wetlands across a broad latitudinal range (Hockey et al. 1992).

Dietary flexibility of shorebirds also allows for the exploitation of the variable resources they encounter throughout the year. Such flexibility is advantageous to shorebirds that migrate across vast landscapes and inhabit a variety of wetland types, both inland and tidal. Across the western hemisphere, shorebirds (42 species) use a variety of invertebrates, including 12 phyla, 22 classes, 72 orders, 238 families, and 404 genera (Skagen and Oman 1996). The predominant phylum consumed by shorebirds in the interior regions is Arthropoda, especially the families Chironomidae, Tipulidae, Ephydridae, and Dolichopodida (Class Insecta, Order Diptera), Carabidae, and Chrysomelidae (Class Insecta, Order Coleoptera). Some shorebird species, such as the Semipalmated Sandpiper and Dunlin, shift their diets substantially as they move between different geographic regions, presumably to take advantage of the abundant taxa in each location (Fig. 10.4). Most shorebird species show little overlap in invertebrate taxa consumed between regions (Table 10.3), consistent with an opportunistic foraging strategy. However, diet similarities of species and guilds within regions and of locally coexisting species are relatively high (Table 10.3).

Table 10.3. Similarity Coefficients for Individual Species between Regions, for Pairs of Species within Regions, and for Pairs of Species and Foraging Guilds within Studies[a]

	C_j	C_s
Semipalmated Plover		
North Atlantic tidal–Central American tidal	0.09	0.16
North Atlantic tidal–North Interior inland	0.04	0.08
North Interior inland–Central American tidal	0	0
South Pacific tidal–Central American tidal	0.17	0.29
Semipalmated Sandpiper		
North Interior inland–South Interior inland	0.22	0.36
Interior inland–Atlantic tidal	0.06	0.12
North Atlantic tidal–South Atlantic tidal	0.18	0.30
Dunlin		
Pacific tidal–Pacific inland	0.10	0.18
Pacific tidal–Interior inland	0.05	0.10
North Pacific tidal–South Pacific tidal	0.31	0.47
Pacific inland–Interior inland	0.25	0.40
Within-region species pairs, all studies in region		
SEPL–SESA (North Interior, inland)	0.42	0.59
SESA–DUNL (North and South Interior, inland)	0.35	0.51
Within-study species pairs		
SEPL–SESA (North Interior, inland; $n = 2$)	0.64	0.75
SESA–DUNL (North/South Interior, inland; $n = 4$)	0.67	0.77
All species pairs within guilds ($n = 29$)	0.51	0.62
Guild pairs ($n = 28$)	0.55	0.67

[a] C_j, Jaccard; C_s, Sorenson (Magurran 1988). Species codes are SEPL (Semipalmated Plover), SESA (Semipalmated Sandpiper), DUNL (Dunlin). Modified from Skagen and Oman (1996).

In the prairie region, the greatest threat to shorebird migrations is the rapid loss of wetlands. A scenario of potential effects of habitat loss on shorebirds has been described by Sutherland and Goss-Custard (1991). They project that loss of habitat would lead to increased shorebird densities in other wetlands, with an increase in prey depletion rates and interference. Further, there would be redistribution of shorebirds to poorer sites and reduced intake rates of poorer competitors, perhaps resulting in decreased condition and increased mortality of poorer competitors. If competition during the nonbreeding season influences population regulation, new equilibrium population sizes may result (Sutherland and Goss-Custard 1991). Aggression (Hamilton 1959, Recher and Recher 1969, Harrington and Groves 1977, Burger et al. 1979, S.K. Skagen, unpublished data), size-related competition (Eldridge and Johnson 1988), and prey depletion (Schneider and Harrington 1981) have been documented at migration stopovers, suggesting that food resources may already be limiting. Loss of wetlands can have an effect on the abilities of shorebirds to find resources by increasing the search energy costs and decreasing success rates.

Lipid Profiles Give Insights into Migration Strategies

Profiles of body condition at interior wetlands throughout the migration season can yield insight into the strategies used by shorebirds crossing the Plains. Body fat is one index of body condition because migratory flight is thought to be fueled primarily by lipids, although the role of protein reserves is being examined (Davidson and Evans 1986, Piersma and Jukema 1990, Lindstrom and Piersma 1993). If migration strategies are similar to those described in coastal areas, shorebirds would stop in interior wetlands for fairly long periods of time and refuel to high threshold levels of fat needed for subsequent long flights. Lipid profiles at a stopover site that are consistent with this scenario would reveal a pattern of low initial body mass followed by steady refueling rates and high body mass at departure.

I developed profiles of body condition (as expressed by body fat) of two species of shorebirds, Semipalmated Sandpipers and White-rumped Sandpipers, at two wetland complexes in the interior. Calibration curves were developed to relate body fat to other measurements that are easily taken in the field. Predictive equations were generated from information from birds that were captured with mist nets at Quivira National Wildlife Refuge (38°10′ N, 98°40′ W), Stafford County, Kansas, in April and May 1992, and at Minnewaukan Flats (48°04′ N, 99°10′ W), the westernmost bay of Devil's Lake, Benson County, North Dakota. Measurements of each bird included body mass (0.1 g, Pesola), tarsus length (0.1 mm), wing length (flattened, 1 mm), total head length (0.1 mm), and culmen (0.1 mm). Total body electrical conductivity (TOBEC) was measured with an EM-SCAN SA-1 Small Animal Body Composition Analyzer following procedures described in Skagen et al. (1993). Birds were sacrificed using chloroform and stored frozen until laboratory analyses were conducted (described in Skagen et al. 1993). Morphological measurements and TOBEC readings were used to generate equations to predict fat mass of Semipalmated Sandpipers (20 collected in Kansas in 1992 and 29 collected in North Dakota in 1993) and White-rumped Sandpipers (21 from Kansas and 12 from North Dakota), following the protocol in Skagen et al. (1993). I assumed that because the input to the models crossed geographic areas and years, that I could apply these inclusive models to birds captured across the northern Plains in spring.

Predictive equations for body fat (FM) of Semipalmated Sandpipers are

$$FM = 15.622 + 0.794 \, BM - 0.142 \, W - 0.480 \, TH$$

and

$$FM = 8.676 + 0.839 \, BM - 0.108 \, W - 0.283 \, TH - 0.256 \, I_{LM}$$

Predictive equations used for White-rumped Sandpipers are

$$FM = 18.955 + 0.700 \, BM - 0.344 \, W$$

and

$$FM = 0.492 + 0.746 \, BM - 0.175 \, W - 0.100 \, I_{LM}$$

Table 10.4. Mean Body Fat (g) and Percentage Body Fat of Semipalmated Sandpipers and White-Rumped Sandpipers at Quivira National Wildlife Refuge, Stafford County, Kansas, in the Springs of 1990–1992 and in the Prairie Pothole Region (Clark, Kingsbury, and Lake Counties) of South Dakota in the Spring of 1992

| | Semipalmated Sandpipers | | | White-rumped Sandpipers | | |
Location	n	Mean Body Fat (sd; g)	% Fat (sd; %)	n	Mean Body Fat (sd; g)	% Fat (sd; %)
Kansas 1990	264	4.64 (2.63)	16.56 (7.48)	72	5.92 (2.89)	13.78 (5.79)
Kansas 1991	298	3.70 (1.82)	13.83 (5.82)	134	7.35 (4.58)	15.47 (8.16)
Kansas 1992	168	3.35 (1.98)	12.82 (6.75)	53	7.19 (4.09)	15.63 (7.35)
South Dakota 1992	198	4.75 (3.22)	15.99 (8.99)	64	8.64 (4.88)	17.96 (8.12)

where BM is total body mass, W is flattened wing length, TH is total head length, and I_{LM} is the TOBEC reading.

The Semipalmated Sandpipers used in generating these predictive equations averaged 6.17 g of fat (range, 1.24–12.24 g), which was equivalent to 20.7% body fat (range, 5.3–35.6%). Similarly, the White-rumped Sandpipers averaged 9.72 g of fat (range, 2.61–15.02 g), which was equivalent to 20.2% body fat (range, 6.9–29.2%). These equations predicted percentage fat with an average error of 2–3%. I applied these predictive equations to Semipalmated and White-rumped Sandpipers captured at Quivira National Wildlife Refuge, Stafford County, Kansas, in the springs of 1990, 1991, and 1992, and in the Prairie Pothole Region (Clark, Kingsbury, and Lake counties) of South Dakota, during the spring of 1992, using morphological measurements only when no TOBEC reading was available.

Average body fat of Semipalmated Sandpipers ranged between 3 and 5 g (12–17%), and average body fat of White-rumped Sandpipers fell between 7 and 9 g (13–18%) (Table 10.4). There was within-season variability in body condition (expressed as percentage body fat) of both species in all years at both sites (Figs. 10.5 and 10.6), although average body fat did not differ between seasons and sites. Estimated percentage fat ranged from 0 to 37% in Semipalmated Sandpipers and from 0 to 31.5% in White-rumped Sandpipers. Such variability has been observed in other studies. Mascher and Marcstrom (1976) reported lipid levels in migrating Dunlin ranging from 5 to 30%.

During each season/site, percentage of body fat of both species increased significantly (Figs. 10.5 and 10.6), indicating that the populations as a whole were able to put on body fat at the stopover sites, albeit slowly. Fattening rates of individual birds could not be computed from this data set, however, because of infrequent recaptures and because the turnover rate was fairly rapid within the

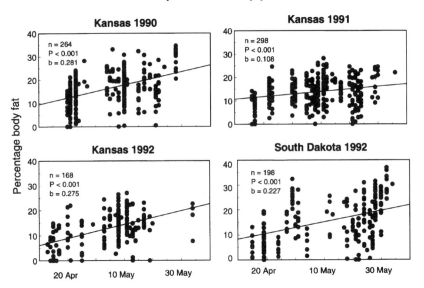

Figure 10.5. Body condition (expressed as percentage body fat) of Semipalmated Sandpipers captured during spring migration at Quivira National Wildlife Refuge, Stafford County, Kansas, in 1990, 1991, and 1992, and in the Prairie Pothole Region (Clark, Kingsbury, and Lake counties) of South Dakota, 1992.

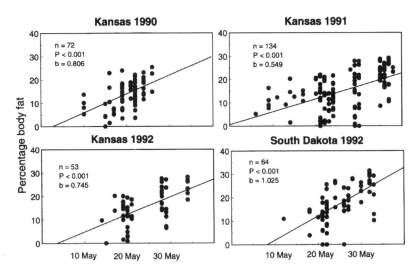

Figure 10.6. Body condition (expressed as percentage body fat) of White-rumped Sandpipers captured during spring migration at Quivira National Wildlife Refuge, Stafford County, Kansas, in 1990, 1991, and 1992, and in the Prairie Pothole Region (Clark, Kingsbury, and Lake counties) of South Dakota, 1992.

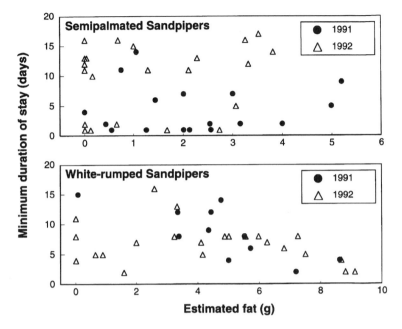

Figure 10.7. Relationship between estimated lipid reserves at capture and minimum duration of stay (MDS) of Semipalmated Sandpipers and White-rumped Sandpipers at Quivira National Wildlife Refuge, Kansas, in springs of 1991 (closed circles) and 1992 (open triangles). The correlation is statistically significant (F = 12.989, df = 1,9, P = .006) only for White-rumped Sandpipers in 1991. (Modified from Skagen and Knopf 1994b.)

flocks. Residency periods during spring migration stopovers at Quivira National Wildlife Refuge average 3.4-9.7 days for Semipalmated Sandpipers and 6.8-8.5 days for White-rumped Sandpipers and are highly variable (Skagen and Knopf 1994b). The Semipalmated Sandpiper populations sampled in this study had fattening rates, or daily gains in mass relative to lean body mass, of only 0.1-0.3%, and White-rumped Sandpiper populations gained 0.5-1% daily. These estimates are far below documented fattening rates of 1-3.6% for three species of shorebirds at coastal stopovers (Gudmundsson et al. 1991).

The estimated lipid reserves of both Semipalmated Sandpipers and White-rumped Sandpipers on arrival at the stopover site was highly variable, ranging from 0-20 % body fat (Skagen and Knopf 1994b). Body condition at arrival, however, bore no relationship to residency periods of Semipalmated Sandpipers in either 1991 or 1992 or of White-rumped Sandpipers in 1992 (Fig. 10.7). Only in 1991 did White-rumped Sandpipers that arrived with little body fat stay longer than birds that were fatter on arrival (Skagen and Knopf 1994b).

Skagen and Knopf (1994b) found no evidence that shorebirds remain at a migration stopover until they reach threshold lipid levels to allow for continuous

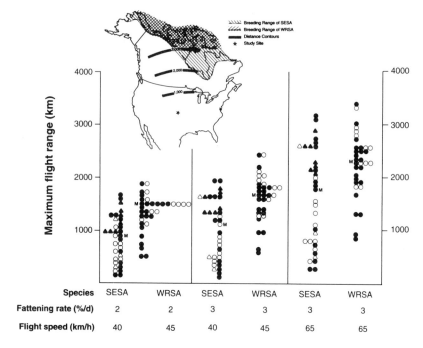

Figure 10.8. Estimated maximum flight ranges of Semipalmated Sandpipers (SESA) and White-rumped Sandpipers (WRSA) calculated according to Castro and Myers (1989), assuming still air conditions and three combinations of fattening rates (expressed as %/d relative to lean body mass) and flight speeds (km/h). Birds equipped with radio transmitters in 1991 are represented by open symbols and in 1992 by closed symbols. For Semipalmated Sandpipers, triangles represent females and circles represent males. Circles represent both sexes of White-rumped Sandpipers. Breeding ranges (Hayman et al. 1986) and distance contours from Quivira National Wildlife Refuge, Kansas, are designated. (From Skagen and Knopf 1994b.)

flights to breeding grounds, as assumed by Harrington et al. (1991). In general, estimated lipid levels at departure and subsequent flight ranges (Fig. 10.8) were highly variable. The variability in the fat levels in the profiles presented in this chapter are also consistent with this conclusion. Similarly, Page and Middleton (1972) found that many Semipalmated Sandpipers refueling in Ontario did not accumulate large fat reserves before departure.

It is not surprising that shorebirds in the midcontinent region depart stopovers with variable fat reserves. There are added costs to transporting more fat than is necessary to reach the next optimal place to stop, and a series of shorter flights are energetically cheaper than long flights (Piersma 1987). In addition, a bird's capacity to escape predation is adversely affected by high fuel loads (Hedenstrom 1992). However, a series of shorter flights would require more search time to look for several good feeding sites.

Flight Ranges on Departure from Migration Stopovers

Many shorebirds have been documented as flying long distances without stopping. Semipalmated Sandpipers fly nonstop for 3,200 km during transoceanic flights from Maine to South America (McNeil and Cadieux 1972, Dunn et al. 1988). Barter and Hou (1990) discuss three species of knots that are able to fly 4,500 to 5,500 km, and Jehl (1979) estimates flight ranges of nearly 6,500 km for Baird's Sandpipers. Estimated flight ranges of White-rumped Sandpipers traversing inland areas in spring are 1,100–1,600 km and over water in fall range from 3,200 to 4,200 km (McNeil and Cadieux 1972)

By contrast, estimated flight ranges of most individual sandpipers departing central Kansas were considerably shorter (Skagen and Knopf 1994b) (Fig. 10.8). The calculations of flight ranges were based on known parameters (fat mass at arrival and duration of stay) and on several assumptions (fattening rates of about 2–3%, calm air conditions, and flight speeds that represented the most economical cruising speed for birds of the appropriate body size). Estimates of percentage fat at departure ranged from 2 to 40% for Semipalmated Sandpipers and from 7 to 36% for White-rumped Sandpipers. Regardless of flight speed assumptions, most birds could not reach the breeding grounds in one jump, as previously proposed for White-rumped Sandpipers (Harrington et al. 1991). Rather, the flight ranges suggest that a myriad of stopover sites between Quivira NWR and the breeding grounds are certainly used.

Predictive Modeling

The status of shorebird migration resources in the interior Plains assessed at appropriate temporal and spatial scales would be invaluable in setting conservation goals. The effectiveness in using satellite imagery and remote sensing techniques to monitor the availability of shorebird habitat, however, has been mixed. In coastal areas, the classification of intertidal surface sediment type from LANDSAT TM imagery did allow for the prediction of shorebird distribution, explaining about 50% of the variation in shorebird numbers (Yates et al. 1993). This work was possible because images could be taken at low tide, thus exposing sediments. In interior regions, however, LANDSAT TM imagery was less useful in identifying priority breeding areas for shorebirds, primarily because of the irregular flooding (Gratto-Trevor 1993).

Remote sensing techniques are limited in their usefulness in identifying shorebird habitat because they cannot distinguish shallow water (1–8 cm deep) from deeper water, nor is the temporal resolution fine enough to document rapid changes in wetland condition. Predictive modeling of wetland systems to portray habitat availability during migration generally suffers from a paucity of information on local hydrology, soil type, wetland topography, and climate information and their relationships to wetland conditions. In the South Dakota data set described, wetland size was potentially useful as a predictor of the area of wet mud/shallow water habitat, accounting for about 50% of the variation, but only in years with intermediate water levels.

Several recent studies have attempted to model optimal fat loads and choices of stopover habitats by migrating birds (Gudmundsson et al. 1991, Lindstrom and Alerstam 1992, Weber et al. 1994). Hypothetically, such models may be useful in predicting the locations of optimal stopover sites. Optimal fat loads and choices of staging sites differ depending on whether the birds are adapted to minimize energy expenditure or time spent in migration (Gudmundsson et al.1991). If birds are minimizing time spent in migration, one would predict overloading of fat and bypassing of possible but low-quality staging sites.

These models may be more relevant to coastal areas than to the interior Plains. In some cases, unpredictability of resources clouds the model assumptions. The question, "Do birds skip stopover sites?" presupposes that birds have knowledge of where the next stopover site is, an unlikely assumption given the variability in habitat conditions in the Plains. Also, birds will bypass some potential sites in the Plains because there is more habitat in some years than can be used. Another question posed in these modeling exercises is "Should birds overload fuel?" Most birds apparently carry more fat than they need to reach the next stopover. Semi-palmated Sandpipers ranged up to 18% body fat, and White-rumped Sandpipers ranged up to 19% body fat at arrival in central Kansas. Overloading fat is an insurance against unpredictable food resources (Gudmundsson et al.1991). Again, to overload fuel presupposes that the birds know where the next stopover site is and how much fuel is required to reach it, a unlikely assumption in the Plains.

Predictive modeling can provide insights into the spacing and landscape configuration of wetlands in the Plains necessary to support migration needs of shorebirds as well as other groups of birds. Conclusions applicable to continental wetlands, however, must be derived from models with appropriate assumptions for the region. The spatial and temporal dynamics of wetland conditions, and assumptions of imperfect knowledge on part of the birds regarding the location and condition of stopover sites, must be incorporated into the models.

Management Implications

Unlike migrating songbirds that spread out along a wide migratory front, shorebirds are thought to concentrate along fixed pathways during migration and to rely on a few stopover areas where they can replenish fat reserves (Myers et al. 1987, Helmers 1993). The application of this paradigm to midcontinental migrations, however, is currently being challenged. In the Plains, there are no stopovers that can be relied on to such an extent, and even the most predictable sites are devoid of suitable habitat during some years and migration seasons. Many sites in the Plains cannot consistently provide the refueling resources necessary for long migratory jumps; certainly some sites can in some years, but not in all years. By necessity, then, conservation for shorebirds in the Plains must not only identify and protect the most predictable sites but must also account for unpredictability in patterns of resource availability.

Management of dispersed and dynamic wetland habitats is a challenge. Conservation of interior-migrating shorebirds requires available and abundant resources throughout all regions of the continental interior and the presence of alternative sites when traditional sites are lost. The minimum sizes of nature reserves can be defined in terms of minimum land areas that exhibit stable patch mosaics (Baker 1989). Similarly, the appropriate scale for managing continental stopover sites for shorebirds might be the number of wetlands that ensures a high probability of suitable shorebird habitat regardless of weather and water regimes during migration. Larger complexes of wetlands modulate local conditions in water fluctuations seasonally. Furthermore, such regional complexes should provide heterogeneous rather than homogeneous wetland depths to provide options under a broad array of climatic conditions. Because of the variability in body condition of birds throughout the Plains, these complexes need to be located along the migration route much as stepping stones, mimicking the conditions under which these birds evolved.

Ideally, management for shorebirds in the Great Plains will take place within the larger context of integrated wetland management, an emerging approach that incorporates the complexity in wetland systems (Laubhan and Fredrickson 1993), where resources will be provided for many species over a large region. The North American Waterfowl Management Plan (NAWMP), implemented in 1986 with the goal of protecting and enhancing habitat for waterfowl, has expanded its concern to many other taxonomic groups (Dickson and McKeating 1993, Streeter et al. 1993). Many of the wetland protection and enhancement efforts of the NAWMP and associated joint ventures have also benefited shorebirds and songbirds.

Habitat requirements of individual shorebird species and wetland management techniques to provide those requirements are presented by Helmers (1992,1993). Natural hydrologic regimes that include water level fluctuations within and among seasons and years are necessary to maintain wetland productivity (Fredrickson and Reid 1990). Management to stabilize water levels has undoubtedly resulted in a decrease of habitat quality. Staggered manipulations, or drawdowns, in an extensive complex of wetlands are a preferred management alternative because they provide shallow water throughout an entire migration period and help maintain healthy ecosystem functioning (Helmers 1993).

The flexible nature of food choice in shorebirds also has important implications for the management and restoration of wetland habitats. Because shorebirds use a broad range of invertebrate taxa, efforts to maintain vital food resources for shorebirds during all seasons should focus on understanding and maintaining hydrologic regimes and ecosystem processes that promote the growth and maintenance of invertebrate populations in general. Particular invertebrate taxa need not be targeted. Wetland management, restoration, and creation that enhances naturally occurring populations of invertebrates will undoubtedly be successful if the invertebrates are accessible to shorebirds.

Currently, we do not have information from remote sensing or predictive modeling that would allow us to evaluate if there are sufficient stopover resources for shorebirds each season in each year throughout the Plains. Possibly, as more refuge personnel have access to on-line services, a computer forum (supplemented with fax and telephones) could be established to gather immediate information on wetland conditions at strategic places across the Plains, thereby assessing and monitoring the availability of shorebird habitat. It is probable that suitable habitat is naturally available across the Plains during most years and that no increased intensive management need be conducted. This communication system would also detect "bottlenecks," or places in time with little or no suitable shorebird habitat, allowing more intensive and immediate management when needed.

Conclusions

Several species of transitory shorebirds migrate between arctic and subarctic breeding grounds to Central and South American wintering grounds, with stop-over sites in prairie wetlands to rest and replenish fat reserves. The existing paradigm of shorebird migration is built from a coastal model that emphasizes large numbers of birds in one site with relatively long stays during which the birds store large quantities of fat in preparation for one long jump to the next strategic site. The application of this paradigm to midcontinental migrations, however, is inappropriate. The feature that most distinguishes Great Plains shorebird migra-tion patterns from coastal patterns is the seasonal variability in the presence of suitable habitat. In response to this variability, shorebirds traversing the Great Plains find and use resources opportunistically and are able to "track" habitats as they appear in a spatially complex landscape. In the interior, shorebirds are more dispersed, occur in smaller numbers locally, and are spatially and temporally unpredictable in occurrence. The behavioral flexibility is complimented by dietary flexibility that favors the exploitation of a greater variety of resources.

Profiles of the body conditions of Semipalmated Sandpipers and White-rumped Sandpipers at interior wetlands yield insight into the migration strategies used by shorebirds crossing the Plains. During spring migration, there was tremendous within-season variability in body condition and no evidence that sandpipers depart stopovers at threshold lipid levels necessary for continuous flights to breeding grounds. Rather, the body conditions at departure and estimated flight ranges suggest that a myriad of stopover sites along a latitudinal gradient between the Gulf coast and the breeding grounds is probably used.

Conservation of interior-migrating shorebirds requires available and abundant resources throughout the central region of the continent and the presence of alternative sites when traditional sites are lost. The rapid loss of wetlands in the prairie region has detrimental effects on shorebird assemblages. Currently, we do not have information from remote sensing or predictive modeling that would allow us to evaluate if there are inadequate or marginal resources for shorebirds at some locations in the Plains. An extensive on-line communication system with state and national refuges and other entities may allow the assessment of habitat availability and the detection of energetic bottlenecks. Intensive and immediate management may be necessary only during these bottlenecks. Ideally, management for shorebirds in the Great Plains will take place within the larger context of integrated wetland management.

Appendix: Shorebird Species That Commonly Cross the Great Plains during Spring and Fall Migrations*

Alpha Code	Common Name	Scientific Name	Breed	Migration Distance
Family Charadriidae				
BBPL	Black-bellied Plover	*Pluvialis squatarola*		I
LEGP	Lesser Golden Plover	*P. dominica*		L
SNPL	Snowy Plover	*Charadrius alexandrinus*	B	S
SEPL	Semipalmated Plover	*C. semipalmatus*		I
PIPL	Piping Plover	*C. melodus*	B	S
KILL	Killdeer	*C. vociferus*	B	S
MOPL	Mountain Plover	*C. montanus*	B	S
Family Recurvirostridae				
BNST	Black-necked Stilt	*Himantopus himantopus*	B	S
AMAV	American Avocet	*Recurvirostra americana*	B	S
Family Scolopacidae				
GRYE	Greater Yellowlegs	*Tringa melanoleuca*		I
LEYE	Lesser Yellowlegs	*T. flavipes*		I
SOSA	Solitary Sandpiper	*T. solitaria*		I
WILL	Willet	*Catoptrophorus semipalmatus*	B	S
SPSA	Spotted Sandpiper	*Actitis macularia*	B	I
UPSA	Upland Sandpiper	*Bartramia longicauda*	B	I
WHIM	Whimbrel	*Numenius phaeopus*		I
LBCU	Long-billed Curlew	*N. americanus*	B	S
HUGO	Hudsonian Godwit	*Limosa haemastica*		L
MAGO	Marbled Godwit	*L. fedoa*	B	S
RUTU	Ruddy Turnstone	*Arenaria interpres*		I
REKN	Red Knot	*Calidris canutus*		I
SAND	Sanderling	*C. alba*		I
SESA	Semipalmated Sandpiper	*C. pusilla*		I
WESA	Western Sandpiper	*C. mauri*		I
LESA	Least Sandpiper	*C. minutilla*		I
WRSA	White-rumped Sandpiper	*C. fuscicollis*		L
BASA	Baird's Sandpiper	*C. bairdii*		L
PESA	Pectoral Sandpiper	*C. melanotos*		L
DUNL	Dunlin	*C. alpina*		I
STSA	Stilt Sandpiper	*C. himantopus*		L
BBSA	Buff-breasted Sandpiper	*Tryngites subruficollis*		L
SBDO	Short-billed Dowitcher	*Limnodromus griseus*		I
LBDO	Long-billed Dowitcher	*L. scolopaceus*		I
COSN	Common Snipe	*Gallinago gallinago*	B	S

*Breeding status is denoted by B (breed in the United States), and migration distances are denoted as S (short), I (intermediate), and L (long), categorized according to Figure 10.1.

Alpha Code	Common Name	Scientific Name	Breed	Migration Distance
AMWO	American Woodcock	*Scolopax minor*	B	S
WIPH	Wilson's Phalarope	*Phalaropus tricolor*	B	I
RNPH	Red-necked Phalarope	*P. lobatus*		I

Acknowledgments. I warmly thank Kelli Stone, Jeffrey Rupert, Steven Dinsmore, Robin Corcoran, Kathy Castelein, and Cynthia P. Melcher for assistance in field surveys, the assessment of body condition in shorebirds, and data management. Funding was provided by the Prairie Potholes Joint Venture of the North American Waterfowl Management Plan, Region 6, U. S. Fish and Wildlife Service, and the National Biological Service.

References

Baker, W.L. 1989. Landscape ecology and nature reserve design in the Boundary Waters Canoe Area, Minnesota. Ecology 70:23-35.

Barter, M., and W.T. Hou. 1990. Can waders fly non-stop from Australia to China? The Stilt 17:36-39.

Bellrose, F.C., and R.C. Graber. 1963. A radar study of the flight directions of nocturnal migrants. Proc. Int. Ornithol. Congr. 13:362-389.

Bormann, F.H., and G.E. Likens. 1979. Catastrophic disturbance and the steady state in northern hardwood forests. Am. Sci. 67:660-669.

Burger, J., D.C. Hahn, and J. Chase. 1979. Aggressive interactions in mixed-species flocks of migrating shorebirds. Anim. Behav. 27:459-469.

Castro, G.C., and J.P. Myers. 1989. Flight range estimates for shorebirds. Auk 106:474-476.

Clark, K.E., L.J. Niles, and J. Burger. 1993. Abundance and distribution of migrant shorebirds in Delaware Bay. Condor 95:694-705.

Colwell, M.A., and L.W. Oring. 1988. Habitat use by breeding and migrating shorebirds in southcentral Saskatchewan. Wilson Bull. 100:554-566.

Dahl, T.E. 1990. Wetland losses in the United States, 1780's to 1980's. U.S. Fish and Wildl. Serv., Washington, DC.

Davidson, N.C., and P.R. Evans. 1986. Prebreeding accumulation of fat and muscle protein by arctic-breeding shorebirds. Proc. Int. Ornithol. Congr. 19:342-352.

Dickson, H.L., and G. McKeating 1993. Wetland management for shorebirds and other species—experiences on the Canadian prairies. Trans. N. Am. Wildl. and Nat. Resour. Conf. 58:370-378.

Drent, R.H., and S. Daan. 1980. The prudent parent: energetic adjustments in avian breeding. Ardea 68:225-252.

Dunn, P.O., T.A. May, M.A. McCollough, and M.A. Howe. 1988. Length of stay and fat content of migrant Semipalmated Sandpipers in eastern Maine. Condor 90:824-835.

Eldridge, J.L., and D. H. Johnson. 1988. Size differences in migrant sandpiper flocks: ghosts in ephemeral guilds. Oecologia 77:433-444.

Evans, P.R., and D.J. Townshend. 1986. Site faithfulness of waders away from the breeding ground: how individual migration patterns are established. Proc. Int. Ornithol. Congr. 19:594-603.

Fredrickson, L.H., and M.K. Laubhan. 1994. Intensive wetland management: a key to biodiversity. Trans. N. Am. Wildl. Nat. Resour. Conf. 59:555-565.

Fredrickson, L.H., and F.A. Reid. 1990. Impacts of hydrologic alteration on management of freshwater wetlands. Pp. 71–90 *in* J.M. Sweeney, ed. Management of dynamic ecosystems. North Cent. Sect., The Wildl. Soc., West Lafayette, IN.

Gill, R.E., Jr., R.W. Butler, P.S. Tomkovich, T. Mundkur, and C.M. Handell. 1994. Conservation of North Pacific Shorebirds. Trans. N. Am. Wildl. Nat. Resour. Conf. 59:63–78.

Gratto-Trevor, C.L. 1993. Can Landsat TM imagery be used to identify priority shorebird habitat in the Mackenzie Delta lowlands? Abstract, American Ornithologists' Union symposium. Fairbanks, AK.

Gratto-Trevor, C.L., and H.L. Dickson. 1994. Confirmation of elliptical migration in a population of Semipalmated Sandpipers. Wilson Bull. 106:78–90.

Gudmundsson, G.A., A. Lindstrom, and T. Alerstam. 1991. Optimal fat loads and long-distance flights by migrating Knots *Calidris canutus,* Sanderlings *C. alba* and Turnstones *Arenaria interpres.* Ibis 133:140–152.

Hamilton, W.J., III. 1959. Aggressive behavior in migrant Pectoral Sandpipers. Condor 61:161–179.

Harrington, B.A., and S. Groves. 1977. Aggression in foraging migrant Semipalmated Sandpipers. Wilson Bull. 89:336–338.

Harrington, B.A., F.J. Leeuwenberg, S.L. Resende, R. McNeil, B.T. Thomas, J.S. Grear, and E.F. Martinez. 1991. Migration and mass change of White-rumped Sandpipers in North and South America. Wilson Bull. 103:621–636.

Harrington, B.A., and R.I.G. Morrison. 1979. Semipalmated Sandpiper migration in North America. Stud. Avian Biol. 2:83–100.

Hayman, P., J. Marchant, and T. Prater. 1986. Shorebirds: an identification guide to the waders of the world. Houghton Mifflin, Boston.

Hedenstrom, A. 1992. Flight performance in relation to fuel load in birds. J. Theor. Biol. 158:535–537.

Helmers, D.L. 1992. Shorebird management manual. Western Hemisphere Shorebird Reserve Network, Manomet, MA.

Helmers, D.L. 1993. Enhancing the management of wetlands for migrant shorebirds. Trans. N. Am. Wildl. Nat. Resour. Conf. 58:335–344.

Hicklin, P.W. 1987. The migration of shorebirds in the Bay of Fundy. Wilson Bull. 99:540–570.

Hockey, P.A.R., R.A. Navarro, B. Kalejta, and C.R. Velasquez. 1992. The riddle of the sands: why are shorebird densities so high in southern estuaries? Am. Nat. 140:961–979.

Howe, M.A. 1987. Wetlands and waterbird conservation. Am. Birds 41:204–209

Howe, M.A., P.H. Geissler, and B.A. Harrington. 1989. Population trends of North American shorebirds based on the International Shorebird Survey. Biol. Conserv. 49:185–199.

Isleib, M.E. 1979. Migratory shorebird populations on the Copper River Delta in eastern Prince William Sound, Alaska. Stud. Avian Biol. 2:125–129.

Jehl, J.R., Jr. 1979. The autumnal migration of Baird's Sanpdiper. Stud. Avian Biol. 2:55–68.

Kersten, M., and T. Piersma. 1987. High levels of energy expenditure in shorebirds; metabolic adaptations to an energetically expensive way of life. Ardea 75:175–187.

Knopf, F.L. 1994. Avian assemblages on altered grasslands. Stud. Avian Biol. 15:247–257.

Laubhan, M.K., and L.H. Fredrickson. 1993. Integrated wetland management: concepts and opportunities. Trans. N. Am. Wildl. Nat. Resour. Conf. 58:323–334.

Lindstrom, A., and T. Alerstam. 1992. Optimal fat loads in migrating birds: a test of the time-minimization hypothesis. Am. Nat. 140:477–491.

Lindstrom, A., and T. Piersma. 1993. Mass changes in migrating birds: the evidence for fat and protein storage re-examined. Ibis 135:70–78.

Magurran, A.E. 1988. Ecological diversity and its measurement. Princeton University Press, Princeton, NJ.

Martinez, E.F. 1979. Shorebird banding at the Cheyenne Bottoms Waterfowl Management Area. Wader Study Group Bull. 25:40–41.

Mascher, J.W., and V. Marcstrom. 1976. Measures, weights, and lipid levels in migrating Dunlins *Calidris alpina* L. at the Ottenby Bird Observatory, South Sweden. Ornis Scand. 7:49–59.

Mawhinney, K., P.W. Hicklin, and J.S. Boates. 1993. A re-evaluation of the numbers of migrant Semipalmated Sandpipers, *Calidris pusilla,* in the Bay of Fundy during fall migration. Can. Field-Nat. 107:19–23.

McNeil, R., and F. Cadieux. 1972. Fat content and flight-range capabilities of some adult spring and fall migrant North American shorebirds in relation to migration routes on the Atlantic coast. Naturaliste Can. 99:589–605.

Morrison, R.I.G. 1984. Migration systems of some new world shorebirds. Pp. 125–202 *in* J. Burger and B.L. Olla, eds. Shorebirds: migration and foraging behavior. Plenum Press, New York.

Morrison, R.I.G., R.W. Butler, H.L. Dickson, A. Bourget, P.W. Hicklin, and J.P. Goossen. 1991. Potential Western Hemisphere Shorebird Reserve Network sites for migrant shorebirds in Canada. Can. Wildl. Serv. Tech. Rep. Series 144. Canadian Wildlife Service, Ottawa.

Morrison, R.I.G., C. Downes, and B. Collins. 1994. Population trends of shorebirds on fall migration in eastern Canada 1974–1991. Wilson Bull. 106:431–447.

Morrison, R.I.G., and J.P. Myers. 1989. Shorebird flyways in the New World. Pp. 85–96 *in* H. Boyd and J.-Y. Pirot, eds. Flyways and reserve networks for water birds. Int. Waterfowl and Wetlands Res. Bur. Spec. Publ. 9, Gloucester, England.

Myers, J.P. 1983. Conservation of migrating shorebirds: staging areas, geographic bottlenecks, and regional movements. Am. Birds 37:23–25.

Myers, J.P., R.I.G. Morrison, P.Z. Antas, B.A. Harrington, T.E. Lovejoy, M. Sallaberry, S.E. Senner, and A. Tarak. 1987. Conservation strategy for migratory species. Am. Sci. 75:18–26.

National Geographic Society. 1981. National Geographic atlas of the world. Fifth Ed. National Geographic Soc., Washington DC.

Neill, R.L., and J.F. Kuban. 1986. Shorebird migration at Arlington, Texas: 1977–1986. Bull. Texas Ornithol. Soc. 19:13–20.

Norton, D.W., S.E. Senner, R.E. Gill, Jr., P.D. Martin, J.M. Wright, and A.K. Fukuyama. 1990. Shorebirds and herring roe in Prince William Sound, Alaska. Am. Birds 44:367–371.

Oring, L.W., and W.M. Davis. 1966. Shorebird migration at Norman, Oklahoma: 1961–63. Wilson Bull. 78:166–174.

Page, G.W., and R.E. Gill, Jr. 1994. Shorebirds in western North America: late 1800s to late 1900s. Stud. Avian Biol. 15:147–160.

Page, G.W., and A.L.A. Middleton. 1972. Fat deposition during autumn migration in the Semipalmated Sandpiper. Bird Banding 43:85–96.

Piersma, T. 1987. Hink, stap of sprong? Reisbeperkingen van arctische steltlopers door voedselzoeken, vetopbuow en vliegsnelheid. [Hop, skip or jump? Constraints on migration of arctic waders by feeding fattening, and flight speed.] Limosa 60:185–194.

Piersma, T., and J. Jukema. 1990. Budgeting the flight of a long-distance migrant: changes in nutrient reserve levels of Bar-tailed Godwits at successive string staging sites. Ardea 78:315–337.

Recher, H.F., and J.A. Recher. 1969. Some aspects of the ecology of migrant shorebirds. II. Aggression. Wilson Bull. 81:140–154.

Rosenberg, K.V., and T.S. Sillett. 1991. Shorebird use of agricultural fields and mini-refuges in Louisiana's rice country. Final report to the Louisiana Nature Conservancy, Baton Rouge.

Schneider, D.C., and B.A. Harrington. 1981. Timing of shorebird migration in relation to prey depletion. Auk 98:801–811.

Schreiber, R.K. 1970. Shorebird migration in Ellis County, Kansas: 1968. Trans. Kansas Acad. Sci. 73:11–19.

Senner, S.E. 1979. An evaluation of the Copper River Delta as critical habitat for migrating shorebirds. Stud. Avian Biol. 2:131–145.

Senner, S.E., and M.A. Howe. 1984. Conservation of nearctic shorebirds. Pp. 379–421 *in* J. Burger and B.L. Olla, eds. Shorebirds: breeding behavior and populations. Plenum Press, New York.

Senner, S.E., and E.F. Martinez. 1982. A review of Western Sandpiper migration in interior North America. Southwest. Nat. 27:149–159.

Skagen, S.K., and F.L. Knopf. 1993. Toward conservation of midcontinental shorebird migrations. Conserv. Biol. 7:533–541.

Skagen, S.K., and F.L. Knopf. 1994a. Migrating shorebirds and habitat dynamics at a prairie wetland complex. Wilson Bull. 106:91–105.

Skagen, S.K., and F.L. Knopf. 1994b. Residency patterns of migrating sandpipers at a midcontinental stopover. Condor 96:949–958.

Skagen, S.K., F.L. Knopf, and B.S. Cade. 1993. Estimation of lipids and lean mass of migrating sandpipers. Condor 95:944–956.

Skagen, S.K., and H.D. Oman. 1996. Dietary flexibility of shorebirds in the western hemisphere. Can. Field-Nat. 110(3): in press.

Smith, K.G., J.C. Neal, and M.A. Mlodinow. 1991. Shorebird migration at artificial fish ponds in the prairie-forest ecotone of northwestern Arkansas. Southwest. Nat. 36:107–113.

Smith, P.W., and N.T. Houghton. 1984. Fidelity of Semipalmated Plovers to a migration stopover area. J. Field Ornithol. 4:247–248.

Stone, K.L. 1994. Shorebird habitat use and response to burned marshes during spring migration in south-central Kansas. M.S. thesis. Oklahoma State University, Stillwater.

Streeter, R.G., M.W. Tome, and D.K Weaver. 1993. North American Waterfowl Management Plan: shorebird benefits? Trans. N. Am. Wildl. Nat. Resour. Conf. 58:363–369.

Sutherland, W.J., and J.D. Goss-Custard. 1991. Predicting the consequence of habitat loss on shorebird populations. Proc. Int. Ornithol. Congr. 20:2199–2207.

Taylor, D.M., C.H. Trost, and B. Jamison. 1992. Abundance and chronology of migrant shorebirds in Idaho. Western Birds 23:49–78.

Thompson, J.J. 1993. Modelling the local abundance of shorebirds staging on migration. Theor. Pop. Biol. 44:299–315.

Weber, T.P., A.I. Houston, and B.J. Ens. 1994. Optimal departure fat loads and stopover site use in avian migration: an analytical model. Proc. R. Soc. London 258:29–34.

Withers, K., and B.R. Chapman. 1993. Seasonal abundance and habitat use of shorebirds on an Oso Bay mudflat, Corpus Christi, Texas. J. Field Ornithol. 64:382–392.

Yates, M.G., J.D. Goss-Custard, S. McGrorty, K.H. Lakhani, S.E.A. Le V. Dit Durell, R.T. Clarke, W.E. Rispin, I. Moy, T. Yates, R.A. Plant, and A.J. Frost. 1993. Sediment characteristics, invertebrate densities and shorebird densities on the inner banks of the wash. J. Anim. Ecol. 30:599–614.

3. Conclusion

11. Conservation of Grassland Vertebrates

Fritz L. Knopf and Fred B. Samson

The Great Plains and Conservation History

The Great Plains grasslands of North America have historically been referred to as the western hemisphere counterpart of the Serengeti Plains of Africa, with herds of roaming ungulates including bison (*Bison bison*), elk (*Cervus elaphus*), deer (*Odocoileus* spp.), and pronghorn (*Antilocapra americana*) and an associated large carnivore assemblage including grizzly bear (*Ursus arctos*), gray wolf (*Canis lupus*), and coyote (*Canis latrans*). Native peoples lived in harmony within this landscape, growing vegetables on the central and eastern Plains and nomadically hunting the bison herds of the western Plains. Estimates of bison numbers have been as high as 60 million. Although we will never know for certain, surely they numbered in the tens of millions (Shaw 1995). The number of carnivores also is uncertain, but Native Americans noted that wolves alone killed one-third of all bison calves each year (De Smet 1905).

The lure of open spaces and the western frontier drew adventurous Easterners to the Plains. The first zoologist to cross the continent was J.K. Townsend who, in accompaniment of the noted botanist T. Nuttall, rode with the Wyeth expedition in 1832 (Townsend 1839). Many naturalists accompanied other exploration and survey expeditions onto the Plains. These included T. Say (Long Expedition, 1819–1820), S. W. Woodhouse (Creek Boundary Expedition, 1849–1850), E. Coues (Hayden Expedition, 1855–1857), and J.A. Allen (North Pacific Railroad Expedition, 1873) (Allen 1874, Coues 1874, James 1966, Tomer and Brodhead 1992). Others came not so much to describe the biota but to experience the thrill of the buffalo hunt, often joining Pawnee, Cheyenne, and other tribes on their hunts. Most significant of these was George Bird Grinnell.

George Bird Grinnell was the key individual in the founding of natural resource conservation in North America (Reiger 1975). From his early concerns over the demise of the bison herds (Grinnell 1873), Grinnell went on to develop the foundations of wildlife management, working closely with his good friend and political spearhead, President Theodore Roosevelt. Products of their activities included founding the U.S. Forest Service and U.S. National Park Service, the Boone and Crockett Club, the predecessor to The National Audubon Society, and state wildlife agencies. The essence of the Grinnell/Roosevelt team was their common love for the Great Plains and the American frontier. Modern-day conservation of natural resources grew from their collective experiences on the central and northern Great Plains and their shared vision that America's most endangered natural resources were those of the Plains. That vision was realized within their lifetimes with the demise of the bison and again a half century later during the nation's greatest conservation disaster, the dust bowl.

The tragedy of the dust bowl brought soil conservation into the forefront of Great Plains programs. Despite repeated pleas (Weaver 1954, Risser 1988) for attention, however, conservation programs have generally ignored the native terrestrial biota. Today, grasslands of the Great Plains remain the nation's most threatened ecosystem (Samson and Knopf 1994).

Ecological Drivers on the Great Plains

The science of ecology was born with the recognition of the orderly process of succession within biotic associations (Clements 1916). Beginning with primary invaders colonizing abiotic substrates, ecological succession fosters progressively more complex species assemblages and energy cycling within the biotic food chain. Succession ultimately approaches a state generally known as a "climax" biota that was once believed to represent long-term stability and enhanced diversity in ecosystems (Brookhaven National Laboratory 1969). More recently, the role of major disturbances on the landscape has been recognized as critical to maintaining the health of ecosystems (Pickett and White 1985, Turner 1987). In grasslands specifically, recent evidence also indicates that biotic diversity begets stability in periods of disturbance (Tilman and Downing 1994). Probably the most "disturbed" North American ecosystem historically was the Great Plains.

The forces of ecological disturbance on the Great Plains have been drought, fire, and grazing. These forces played major roles in directing evolution of the grassland biota. In that sense, they are more "drivers" of the ecosystem than "disturbances" per se (Evans et al. 1989), and it is the prevention of drought (via irrigation), the suppression of fire, and the removal of grazers that represent the true ecological disturbances of prairies. The interaction of these ecological drivers at varying intensities and scales are fundamental to maintaining landscape heterogeneity and biotic diversity within native grasslands (Collins and Barber 1985, Coppock and Detling 1986, Collins 1987, 1992, Howe 1994). Historically, drought has been a relatively universal ecological driver across the Great Plains, with grazing being the secondary driver on the westerly shortgrass prairie and fire the secondary driver on the easterly tallgrass prairie.

Drought

Periodic drought and fire are primarily responsible for the development of the grasslands (Weaver 1954, Anderson 1990), although each individually is mostly inadequate to maintain a grass landscape (Anderson 1982). In North America, the prevailing westerly winds rise to cross the Rocky Mountains, dropping moisture in passing and creating a rainshadow on the western Plains. Precipitation in Montana, Wyoming, Colorado, New Mexico, and the Oklahoma/Texas panhandles comes primarily from vernal thunderstorms; native grasses generally do not green until late May. The northward movement of Gulf of Mexico moisture results in increasing precipitation eastward. Thus, average annual precipitation increases from west to east across the Great Plains (Parton et al. 1981, Risser et al. 1981). In addition, relative humidity increases and wind speed, solar radiation, and potential evapotranspiration decrease from the western to the eastern Plains (Risser 1990).

The average rates of precipitation, however, are only partial drivers of evolutionary processes on the Great Plains. Rather, the inherent unpredictability of precipitation among years also has driven evolutionary processes within the biota (Mock 1991). Many changes in vegetation attributed to grazing may, in fact, be driven as much by drought (Branson and Miller 1981, Branson 1985). Infrequent, severe drought can cause massive local extinctions of annual forbs and grasses that have invaded stands of perennial species, and recolonization of those sites is quite slow (Tilman and El Haddi 1992).

Fires

Historic fires were ignited primarily by lightening in summer thunderstorms (Higgins 1984) and by native peoples (Higgins 1986) to protect villages from wildfire or to attract herbivores such as bison and pronghorns that respond positively to greening grasses after a burn (Coppock and Detling 1986, Higgins 1986, Shaw and Carter 1990). Early expeditions in the tallgrass and mixed-grass regions ignited the prairie during westward movements to ensure that nutritious forage would be available for their horses on the return journey (Tomer and Brodhead 1992, p. 205).

The role of fire as an ecological driver has been well researched on these prairies (Bragg and Steuter 1996, Steinauer and Collins 1996). Fire invigorates stands of grasses by recycling nutrients and destroying invading woody species, thus resulting in increased production. Destruction of the litter layer by summer fires especially opens the stand to seedling establishment, which favors cool-season grasses that enhance plant species richness and landscape heterogeneity. Both the role of historical fire on the shortgrass prairie (Wenger 1943) and its value as a management tool are less well understood.

Grazing

Grazing is the third major ecological driver on the Great Plains. Whereas dominant tallgrass prairie species such as big (*Andropogon gerardi*) and little (*Schizachyrium scoparium*) bluestems and indiangrass (*Sorghastrum nutans*) decrease under regimes of prolonged grazing, dominant shortgrass species such as blue (*Bouteloua gracilis*) and side-oats (*B. curtipendula*) gramas and buffalograss (*Buchloe dactyloides*) increase (Weaver 1954). In the tallgrass prairie, the behavior of grazing animals favors among-site heterogeneity of vegetation, especially where grasses are also subjected to periodic fire (Glenn et al. 1992). In the shortgrass prairie, which evolved with a major herbivory driver, heavy grazing promotes homogeneity of the landscape and inadequate grazing pressure results in enhanced heterogeneity (Larson 1941, Milchunas et al. 1988) (Fig. 11.1). In the latter situation, dominant grasses are stimulated by grazing and grow rapidly afterward, thus maintaining a competitive edge over invading grasses and forbs. Basal cover of grasses increases and cover of forbs decreases after grazing.

Heterogeneity within shortgrass landscapes historically was favored by the nomadic nature of the large herds of bison creating differential grazing pressures locally. The major promoter of natural heterogeneity in the shortgrass landscape,

Figure 11.1. Intensive grazing pressure of herbivores on the shortgrass prairie favors grazing-adapted grasses and promotes more homogeneous vegetative landscapes (Campbell County, Wyoming) (Photograph by F.L. Knopf.)

however, remains the activities of prairie dogs (*Cynomys* spp.). Although drastically reduced in numbers (Summers and Linder 1978), prairie dogs still create a landscape patchwork of intensively grazed islands and disturbed soil surface (Whicker and Detling 1988). The grazing impacts and surface disturbances of prairie dogs were enhanced by the behavioral attractiveness of "towns" to bison and pronghorn, which preferentially forage and loaf on such sites (Coppock et al. 1983, Coppock and Detling 1986, Krueger 1986).

The Prairie Landscape in 1996

The arrival of European descendents on the North American grasslands drastically altered the face of the landscape as well as ecological relationships within the biota. The overwhelming influence has been to modulate the inherent range of natural variation in the ecological drivers of the prairies. Water management in the shortgrass and mixed-grass regions has locally removed the threat of periodic drought, resulting in increased cultivation and a westward extension of cereal grain agriculture. Fire suppression in the tallgrass and mixed-grass prairie has led to loss of species richness and, in the case of species like the blowout pentstamen (*Penstemon haydenii*) in the Nebraska Sandhills, the potential extinction of species.

Cultivation and residential and industrial development have obliterated potential habitats for many vertebrate species locally. Total losses of native prairie range from 20% of shortgrass prairie in Wyoming to greater than 99% of tallgrass prairies in Illinois and Iowa (Table 11.1). Overall, estimates of conversion of native prairie to either cropland or pastureland (seeded with non-native, tame grasses) in the United States range from 29% of shortgrass, 41% of mixed-grass, and more than 99% of tallgrass landscapes (U.S.Department of Agriculture 1987). Pastureland provides surrogate prairie habitat for some vertebrate species of the eastern Plains (Herkert 1993, 1994).

The loss of native grasslands as potential vertebrate habitats is even more devastating as remnant grasslands become more and more fragmented and isolated. The effects of fragmentation are threefold. First, many species of vertebrates require large, intact parcels of grassland for survival and reproduction (Samson 1980, Herkert 1994). As remnants decrease in size, these area-sensitive species are progressively extirpated locally. Second, as remnants become more isolated, the probability of colonization/recolonization of a patch decreases with distance from another patch (Kaufman and Kaufman, this volume). Third, populations in isolated patches suffer from genetic inbreeding and accelerated rates of genetic drift (Benedict et al. 1996).

Fragmentation is not specifically a cultivation issue. Throughout the Great Plains, tree plantings have resulted in a patchwork pattern of forest and grass, creating a pastoral landscape. Windbreaks are interspersed across the former grass landscape to the point that 3% of the Great Plains is now forested (Friedman et al., this volume). Trees are currently being planted at the rate of 20.7×10^6 per year

Table 11.1. Estimated Current Area, Historic Area, and Percentage Decline of Tallgrass, Mixed-Grass, and Shortgrass Prairies[a]

	Historic (ha)	Current (ha)	Decline (ha)
Tallgrass			
Manitoba	600,000	300	99.9
Illinois	8,900,000	930	99.9
Indiana	2,800,000	404	99.9
Iowa	12,500,000	12,140	99.9
Kansas	6,900,000	1,200,000	82.6
Minnesota	7,300,000	30,350	99.6
Missouri	5,700,000	30,350	99.5
Nebraska	6,100,000	123,000	98.0
North Dakota	1,200,000	1,200	99.9
Oklahoma	5,200,000	NA[b]	NA
South Dakota	3,000,000	449,000	85.0
Texas	7,200,000	720,000	90.0
Wisconsin	971,000	4,000	99.9
Mixed grass			
Alberta	8,700,000	3,400,000	61.0
Manitoba	600,000	300	99.9
Saskatchewan	13,400,000	2,500,000	81.3
Nebraska	7,700,000	1,900,000	77.1
North Dakota	13,900,000	3,900,000	71.9
Oklahoma	2,500,000	NA	NA
South Dakota	1,600,000	NA	NA
Texas	14,100,000	9,800,000	30.0
Shortgrass			
Colorado	NA	NA	NA
Oklahoma	1,300,000	NA	NA
Saskatchewan	5,900,000	840,000	85.8
South Dakota	179,000	NA	NA
Texas	7,800,000	1,600,000	80.0
Wyoming	3,000,000	2,400,000	20.0

[a]From Samson and Knopf 1994.
[b]NA, not available.

(Griffith 1976), many being exotics and mostly subsidized by state forest agencies (Olson and Knopf 1986). Fire suppression has led to brush and tree encroachment along the periphery of the Plains (Bird 1961, Pulich 1976).

The historic prairie landscape included patchy-to-linear stands of deciduous trees along streambanks. These associations were more common in the east and became infrequent moving west, especially on the southern Plains. Woody riparian associations also were common along the foot of the Rocky Mountains, where streams flowed seasonally into the Plains, some as headwaters emptying into larger systems as the Arkansas, Canadian, and Platte rivers and others drying or

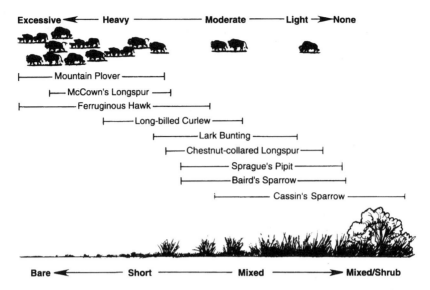

Figure 11.2. Distributions of endemic birds of prairie uplands on a shortgrass/mixed-grass and historical grazing pressure continua across the western landscapes of the Great Plains. (From Knopf 1996.)

seeping into the soils. These riparian forests on the Great Plains have become more extensive with fire suppression on the east (Abrams 1986, Rothenberger 1989) and ecological succession after water management on the west (Wedel 1986, Knopf and Scott 1990), resulting in ribbons of deciduous vegetation slicing the grassland sea. Similarly, clearing of the eastern deciduous forest for hayfields has created a patchwork of trees and grasslands through the historic prairie peninsula on to New England (Askins 1993). These trends will continue as population pressures homogenize landscapes across the nation.

The estimated tens of millions of bison on the western Plains were replaced by an estimated 45 million cows and an equal number of domestic sheep by 1890 (Fedkiw 1989). Replacement of bison with cattle has not, however, had a major impact on western grasslands (Hartnett et al., this volume), but management of cattle with fences has created endless homogeneous landscapes by removing the differential intensities of grazing among sites that historically created the mosaic of habitats needed to support many species (Knopf 1996a). Fences are used to regulate grazing pressure more precisely, primarily to reduce soil erosion while simultaneously maximizing long-term vegetative productivity of a site under the general range management paradigm of "take half and leave half" of the vegetation. Grassland birds (Wiens 1973), and especially the primary endemic species (Knopf 1996b), evolved within a gradient of differentially grazed landscapes (Fig. 11.2). The uniformity of grazing management on the Great Plains probably has a more negative effect on endemic avian assemblages than the actual presence of livestock or the consequences of grazing (Knopf 1996a).

Figure 11.3. Seasonal wetlands, some present less than 1 year in 10, provide critical foraging areas for long-distance migrants such as the Long-billed Dowitcher (*Limnodromus scolopaceus*) that cross the Great Plains to reach their breeding habitats within the arctic circle (Larimer County, Colorado) (Photograph by F.L. Knopf.)

Prairie streams had a strong riffle/pool structure that resembled more a series of seasonally connected small ponds or lakes during periods of low flow (Brown and Matthews 1996). Size of pools increased and length of riffles generally decreased moving down the drainage; all except the Missouri River periodically may have become intermittent in periods of drought. Today, water diversion and ground-water pumping have accentuated the intermittency of these streams on most of the Great Plains (Samson et al. in press).

A less noticeable, but equally pervasive, threat to native fishes has been the rampant accidental and deliberate introduction of alien (North American species native to biogeographic provinces other than the Great Plains) and exotic (species from other continents) fishes into native streams. Ross (1991) reported that more than three of every four introductions of exotic fishes resulted in declines in populations of indigenous species. Introductions of fishes indigenous to other biomes, however, may represent a larger threat to Plains fishes than introductions of game fishes (Fausch and Bestgen, this volume). Thus, both the loss of habitat and fish introductions seem critical to protecting the simplistic indigenous fish assemblages of the western Plains.

Across the northern Great Plains, historic natural wetlands have been destroyed at an alarming rate. Estimates of wetland loss range from 86% in tallgrass prairie states (Illinois and Iowa) to 40% in Montana (Dahl 1990). These losses are of most conservation concern in the Rainwater Basin and Sandhills wetlands of Nebraska and the Prairie Pothole region that extends from the Dakotas into Manitoba and Saskatchewan as prairie-parkland wetlands. Drainage of wetlands and conversion of the landscape to row cropping continues to destroy these major breeding grounds for waterfowl populations (Betheke and Nudds 1995). Smaller complexes to the south in Kansas and Oklahoma (e.g., Cheyenne Bottoms, Kansas [Zimmerman 1990]) are less extensive but especially critical to support transcontinental migrations of both waterfowl and other wetland species that breed in the Arctic but winter in South America (Skagen, this volume) (Fig. 11.3).

Focusing Conservation Action

Societies can protect portions of an ecosystem in nonuse preserves, but in the long term, partnerships addressing issues on lands managed for timber, grazing, and other commodities will play a major role in the conservation of the native biota (Raven 1990). A framework for conservation is developing for the Great Plains (Risser 1996). This framework is built on the principles that (1) past and current research is providing more sophisticated (especially geospatial) information to define and address conservation issues, (2) prairie conservation must be based on addressing multiple uses, (3) both short- and long-term economic values of prairies must be secured, and (4) conservation actions must extend across property lines. Prototypes for conservation action are currently seen in successful partnerships such as the North American Waterfowl Plan's Joint Ventures (Kresl et al. 1996), the Western Governor's Association's Great Plains Partnership (Clark 1996), and Canada's Prairie Conservation Action Plan (Dyson 1996). Each of these programs emphasizes proactive conservation through the building of working partnerships between the private sector and federal, state, and county agencies to define and address conservation issues.

Given the evolving foundations for progressive conservation on the Great Plains, a major hurdle remains to integrate the variety of social, economic, and biological issues into working conservation plans (Clark 1996, Johnson and Bouzaher 1996). However, scientists have had difficulty setting the biological priorities for the conservation of diversity (Roberts 1988), and the profession lacks any unified approach to evaluating conservation strategies (Erwin 1992). Much of the confusion can be attributed to the need to assess conservation actions across multiple biological, temporal, and spatial scales (Knopf and Smith 1992).

The embryonic science of ecosystem management (Samson and Knopf 1996) offers much hope for proactive conservation of biological diversity in the future. To date, however, hard science has played only a minor role in the conservation of biological diversity (Weston 1992), and the cost and time constraints necessary to understanding all aspects of ecosystem function are prohibitive. Samson and

Knopf (1993) suggest focusing ecosystem management with a mechanistic approach that (1) understands the differences between alpha and beta diversity, (2) defines the biotic integrity of an ecosystem, (3) protects ecological processes maintaining that integrity, and (4) assures that those processes are sustainable through time.

Diversity and Integrity

The extensive grasslands of North America have a long evolutionary history since they arose with increasing aridity more than 30 million years ago in the Miocene (Howe 1994, Brown and McDonald 1995). In more recent times, the Great Plains have experienced a series of glacial advances since the Pleistocene that have fostered periodic biogeographic barriers to dispersal of vertebrates. The recurring barrier favored accelerated adaptive radiation of forest species to the east and west and, through instability, resulted in the evolution of comparatively few species on the grasslands themselves (Mengel 1970, Wells 1970, Axelrod 1985).

Modern settlement and development of the Great Plains have been accompanied by a trend of ecological generalist species invading the Great Plains from contiguous biomes, with examples ranging from the obvious house mouse (*Mus musculus*) to more subtle range extensions such as those of the least shrew (*Cryptotis parva*) and Blue Jay (*Cyanocitta cristata*) (Knopf and Scott 1990, Benedict et al. 1996, Rabeni 1996). Thus, modern vertebrate assemblages contain more species now than historically. Whereas one would suspect enhanced faunas in heavily developed locales such as the current avifauna of suburban Tucson, Arizona, which is 95% synthetic (Emlen 1974), it is surprising that nearly 90% of the current breeding avifauna of rural northeastern Colorado has colonized the area since the turn of the century (Knopf 1986).

Conservation of biological diversity of the Great Plains has generally been hampered by a tendency to confuse species richness (alpha diversity) and species diversity (beta and gamma diversity) topically. The importance of Plains riparian areas has been promoted because of the habitats they provide for more species than surrounding grasslands (e.g., Tubbs 1980). Emphasizing the total number of species locally is short sighted, especially in riparian vegetation (Knopf and Samson 1994), and ultimately becomes counterproductive when viewed at larger scales (Murphy 1989, Martin 1992). The problem in using the total number of species lies in the observation that a large number of ecological generalists can overwhelm the unique elements of the vertebrate biota. As in northeastern Colorado, the collective vertebrate biota of the Great Plains comprises many species that are either ecological generalists (occurring across many biotic provinces of North America) or are peripheral to their main geographic distribution in contiguous biomes.

Evolutionarily, a small set of species and abiotic processes structures ecosystems across spatial and temporal scales. In his classic synthesis, Holling (1992) compared vertebrate assemblages of Canadian shortgrass prairies and boreal forests and suggested that only a few processes over a limited range of scales

Table 11.2. Numbers of Vertebrate Species Recorded on the Great Plains, Number That Are Endemic to the Great Plains, and Number with Strongly Western Great Plains Affinities

	No. Species	No. (%) Endemic	No. (%) Western	References
Fish	250	34 (14)	0 (0)	Cross et al. 1986, Rabeni 1996, Bestgen, personal communication
Amphibians	34	2 (6)	1 (50)	Corn and Peterson 1996, Corn, personal communication
Reptiles	90	8 (9)	1 (17)	Corn and Peterson 1996, Corn, personal communication
Birds	604	12 (2)	12 (100)	Mengel 1970
Mammals	138	16 (12)	11 (69)	Benedict et al. 1996

uniquely characterize ecosystems and the morphology of their animals. Ecosystems are controlled and governed by a few species and abiotic processes that ultimately structure the landscape. Holling referred to this as the *Extended Keystone Hypothesis.* Certain species reflect ecological processes within an ecosystem, and those are the species historically indigenous to that system. These species evolved within the system and are referred to as the ecological endemics.

The degree of endemism (gamma diversity) among the vertebrate classes on the Great Plains ranges from 2% of bird species to 14% of fish species (Table 11.2). Of the endemic vertebrates, the aquatic classes (fish, amphibians) and reptiles (including three aquatic species also) are primarily species of the eastern Great Plains. Alternatively, the terrestrial classes (birds and mammals) are primarily western derived. Vertebrate endemism in the Great Plains follows the west-east gradient of increasing moisture, the primary ecological driver of the region. Thus, reducing drought effects is conservationally correct on the eastern Plains and ecologically short sighted on the western Plains.

Sustaining Ecological Processes

The first sign of degradation of an ecosystem appears at the population level of sensitive species (Odum 1992). The endemic vertebrates of the Great Plains can be argued to be the most sensitive to changes in the ecological drivers of the biotic province (Knopf and Samson 1996). Thus, they become indicators of ecosystem health. Rather than monitor and research the ecology of 1,116 species of fish, amphibians, reptiles, birds, and mammals, ecosystem health can be tracked by programs that monitor only 72 species (Table 11.3). Prolonged declines of some species would trigger research to define causes of those declines—a highly cost-effective approach relative to funding research on all 1,116 species and their supporting floral and invertebrate biota. The long-term monitoring of the 72 vertebrate species would not be conducted across the entire Great Plains. As noted,

Table 11.3. Endemic Vertebrate Species of the Great Plains

Fish[a]
- Acipenseridae
  - Pallid sturgeon *Scaphirhynchus albus*
- Cyprinidae
  - Western silvery minnow *Hybognathus argyritis*
  - Sturgeon chub *Macrhybopsis gelida*
  - Sicklefin chub *M. meeki*
  - Redspot chub *Nocomis asper*
  - Red River shiner *Notropis bairdi*
  - Wedgespot shiner *N. greenei*
  - Blacknose shiner *N. heterolepis*
  - Kiamichi shiner *N. ortenburgeri*
  - Ozark shiner *N. ozarcanus*
  - Peppered shiner *N. perpallidus*
  - Duskystipe shiner *Luxilus pilsbryi*
  - Bleeding shiner *L. zonatus*
  - Slim minnow *Pimephales tenellus*
- Ictaluridae
  - Ozark madtom *Noturus albater*
  - Checkered madtom *N. flavater*
  - Ouachita madtom *N. lachneri*
  - Neosho madtom *N. placidus*
  - Caddo madtom *N. taylori*
- Amblyopsidae
  - Ozark cavefish *Amblyopsis rosae*
- Cyprinodontidae
  - Plains topminnow *Fundulus sciadicus*
- Centrarchidae
  - Ozark bass *Ambloplites constellatus*
- Percidae
  - Arkansas darter *Etheostoma cragini*
  - Arkansas saddled darter *E. euzonum*
  - Yoke darter *E. juliae*
  - Yellowcheek darter *E. moorei*
  - Niangua darter *E. nianguae*
  - Paleback darter *E. pallididorsum*
  - Stippled darter *E. punctulatum*
  - Orangebelly darter *E. radiosum*
  - Missouri saddled darter *E. tetrazonum*
  - Bluestripe darter *Percina cymatotaenia*
  - Longnose darter *P. nasuta*
  - Leopard darter *P. pantherina*

Amphibians
- Pelobatidae
  - Plains spadefoot *Spea bombifrons*
- Ranidae
  - Plains leopard frog *Rana blairi*

Reptiles
- Emydidae
  - Texas map turtle *Graptemys versa*
  - Ornate box turtle *Terrapene ornata*
- Iguanidae
  - Dunes sagebrush lizard *Sceloporus arenicolus*
- Scincidae
  - Prairie skink *Eumeces septentrionalis*

Reptiles (*continued*)
- Colubridae
  - Brazos water snake *Nerodia harteri*
  - Concho water snake *N. paucimaculata*
  - Plains garter snake *Thamnophis radix*
  - Lined snake *Tropidoclonion lineatum*

Birds
- Accipitridae
  - Ferruginous Hawk *Buteo regalis*
- Charadriidae
  - Mountain Plover *Charadrius montanus*
- Scolopacidae
  - Long-billed Curlew *Numenius americanus*
  - Marbled Godwit *Limosa fedoa*
  - Wilson's Phalarope *Phalaropus tricolor*
- Laridae
  - Franklin's Gull *Larus pipixcan*
- Motacillidae
  - Sprague's Pipit *Anthus spragueii*
- Emberizidae
  - Cassin's Sparrow *Aimophila cassinii*
  - Baird's Sparrow *Ammodramus bairdii*
  - Lark Bunting *Calamospiza melanocorys*
  - McCown's Longspur *Calcarius mccownii*
  - Chestnut-collared Longspur *C. ornatus*

Mammals
- Canidae
  - Swift fox *Vulpes velox*
- Mustelidae
  - Black-footed ferret *Mustela nigripes*
  - Spotted skunk *Spilogale putorius*
- Antilocapridae
  - Pronghorn *Antilocapra americana*
- Bovidae
  - Bison *Bison bison*[b]
- Sciuridae
  - Black-tailed prairie dog *Cynomys ludovicianus*
  - Franklin's ground squirrel *Spermophilus franklinii*
  - Richardson's ground squirrel *S. richardsonii*
  - Thirteen-lined ground squirrel *S. tridecemlineatus*
- Geomyidae
  - Plains pocket gopher *Geomys bursarius*
- Heteromyidae
  - Olive-backed pocket mouse *Perognathus fasciatus*
  - Plains pocket mouse *P. flavescens*
  - Hispid pocket mouse *Chaetodipus hispidus*
- Muridae
  - Northern grasshopper mouse *Onychomys leucogaster*
  - Plains harvest mouse *Reithrodontomys montanus*
  - Prairie vole *Microtus orchrogaster*
- Leporidae
  - White-tailed jackrabbit *Lepus townsendii*

[a]Systematics of fish follow Robins et al. (1991).
[b]Not identified as grasslands endemic by Benedict et al. (1996).

the aquatic and semiaquatic species tend to be on the eastern Plains and the terrestrial species on the western Plains. Early detection of dysfunction in an ecological driver at an area would permit early corrective action and ensure ecosystem sustainability through time.

Monitoring unique vertebrate elements of the Great Plains as indicators of ecosystem health would operate on the assumption that major losses within lower biotic forms as invertebrates would produce detectable changes in vertebrate species. Certainly, declines in insectivorous vertebrates regionally would stimulate stepdown inquiries into the status of invertebrate indicator assemblages (Arenz and Joern 1996) that, in turn, may reflect changes in vegetative components of an ecosystem. Plant/invertebrate-grazing communities are the first to adjust to ecosystem disturbance and likely possess some resiliency to compensate impacts and, thereby, buffer consequences to vertebrates.

The use of narrow endemic species as biological indicators of the health of the Great Plains would help avert environmental "train wrecks" (Stone 1993) and simultaneously be compatible with other environmental legislation such as the Endangered Species Act (ESA). Endemic species, with their limited distributions and narrower ecological tolerances, dominate the national list of endangered species. Simply stated, endemic species are less resilient. Using research on endemic species to secure the future of all species is a proactive approach to keeping species from being listed under the ESA. Once listed, recovery of an endangered species is comparatively costly and often confrontational with human development of landscapes. Conservation actions using endemic vertebrate species as indicators of ecosystem health will ultimately minimize species loss, political confrontation, and economic compromise on the Great Plains. Implementing such an approach ensures that the North American Great Plains will continue to be the frontier of conservation theory and practice.

References

Abrams, M.D. 1986. Historical development of gallery forests in northeast Kansas. Vegetatio 65:29–37.

Allen, J.A. 1874. Notes on the natural history portions of Dakota and Montana territories. Proc. Boston Soc. Nat. Hist.17:33–86.

Anderson, R.C. 1982. An evolutionary model summarizing the roles of fire, climate, and grazing animals in the origin and maintenance of grasslands: an end paper. Pp. 297–307 in J.R. Estes, R.J. Tyrl, and J.N. Brunken, eds. Grasses and grasslands. Univ. Oklahoma Press, Norman.

Anderson, S.L. 1990. The historic role of fire in the North American grassland. Pp. 8–18 in S.L. Collins and L.L. Wallace, eds. Fire in North American tallgrass prairies. Univ. Oklahoma Press, Norman.

Arenz, C.L., and A. Joern. 1996. Prairie legacies—invertebrates. Pp. 91–110 in F.B. Samson and F.L. Knopf, eds. Prairie conservation: preserving North America's most endangered ecosystem. Island Press. Covelo, CA.

Askins, R.A. 1993. Population trends in grassland, shrubland and forest birds in eastern North America. Curr. Ornithol. 11:1–34.

Axelrod, D.I. 1985. Rise of the grassland biome, central North America. Biol. Rev. 51:163–201.

Benedict, R.A., P.W. Freeman, and H.H. Genoways. 1996. Prairie legacies—mammals. Pp. 149–167 *in* F.B. Samson and F.L. Knopf, eds. Prairie conservation: preserving North America's most endangered ecosystem. Island Press, Covelo, CA.

Betheke, R.W., and T.D. Nudds. 1995. Effects of climate change and land use on duck abundance in Canadian prairie-parklands. Ecol. Appl. 5:588–600.

Bird, R.D. 1961. Ecology of the aspen parkland of western Canada in relation to land use. Can. Dept. Agric. Res. Branch Publ. 1066, Ottawa.

Bragg, T.B., and A.A. Steuter. 1996. Prairie ecology—the mixed prairie. Pp. 53–66 *in* F.B. Samson and F.L. Knopf, eds. Prairie conservation: preserving North America's most endangered ecosystem. Island Press, Covelo, CA.

Branson, F.A. 1985. Vegetation changes on western rangelands. Soc. Range Manage. Monogr. 2. Denver, CO.

Branson, F.A., and R.F. Miller. 1981. Effects of increased precipitation and grazing management on northeastern Montana rangelands. J. Range Manage. 34:3–10.

Brookhaven National Laboratory. 1969. Diversity and stability in ecological systems. Brookhaven Symp. Biol. 22. Upton, NY.

Brown, A.V., and W.J. Matthews. 1996. Stream ecosystems of the central United States. Pp. 89–116 *in* C. Cushing, K.W. Cummins, and G.W. Minshall, eds. River and stream ecosystems. Elsevier Press, Amsterdam.

Brown, J.H., and W. McDonald. 1995. Livestock grazing and conservation on Southwestern rangelands. Conserv. Biol. 9:1644–1647.

Clark, J.S. 1996. Great Plains partnership. Pp. 169–174 *in* F.B. Samson and F.L. Knopf, eds. Prairie conservation: preserving North America's most endangered ecosystem. Island Press, Covelo, CA.

Clements, F.E. 1916. Plant succession: an analysis of the development of vegetation. Carnegie Inst. Publ. 242. Washington, DC.

Collins, S.L. 1987. Interaction of disturbances in tallgrass prairie: a field experiment. Ecology 68:1243–1250.

Collins, S.L. 1992. Fire frequency and community heterogeneity in tallgrass prairie vegetation. Ecology 73:2001–2006.

Collins, S.L., and S.C. Barber. 1985. Effects of disturbance on diversity in mixed-grass prairie. Vegetatio 64:87–94.

Coppock, D.L., and J.K. Detling. 1986. Alteration of bison/prairie dog grazing interaction by prescribed burning. J. Wildl. Manage. 50:452–455.

Coppock, D.L., J.K. Detling, J.E. Ellis, and M.I. Dyer. 1983. Plant–herbivore interactions in a North American mixed-grass prairie. I. Effects of black-tailed prairie dogs on interseasonal aboveground plant biomass and nutrient dynamics and plant species diversity. Oecologia 56:1–9.

Corn, P.S., and C.R. Peterson. 1996. Prairie legacies—amphibians and reptiles. Pp. 125–135 *in* F.B. Samson and F.L. Knopf, eds. Prairie conservation: preserving North America's most endangered ecosystem. Island Press, Covelo, CA.

Coues, E. 1874. Birds of the Northwest: a hand-book of ornithology of the region drained by the Missouri River and its tributaries. U. S. Dept. Inter., Geol. Surv. Misc. Publ. 3 [1974 reprint].

Cross, F.B., R.L. Mayden, and J.D. Stewart. 1986. Fishes in the western Mississippi Basin (Missouri, Arkansas and Red rivers). Pages 363–412 *in* C.H. Hocutt and E.O Wiley, eds. The zoogeography of North American freshwater fishes. John Wiley & Sons, New York.

Dahl, T.E. 1990. Wetlands losses in the United States, 1780's to 1980's. U.S. Dept. Inter. Fish and Wildl. Serv., Washington, DC.

De Smet, P.-J. 1905. Life, letters, and travels of Father Pierre-Jean De Smet. H.M. Chittenden and A.T. Richardson, eds. Arno Press, New York [1969 reprint].

Dyson, I.W. 1996. Canada's Prairie Conservation Action Plan. Pp. 175–186 *in* F.B. Samson and F.L. Knopf, eds. Prairie conservation: preserving North America's most endangered ecosystem. Island Press. Covelo, CA.

Emlen, J.T. 1974. An urban bird community in Tucson, Arizona: derivation, structure, and regulation. Condor 76:184-197.

Erwin, T.L. 1992. An evolutionary basis for conservation strategies. Science 253:750-752.

Evans, E.W., J.M. Briggs, E.J. Finck, D.J. Gibson, S.W. James, D.W. Kaufman, and T.R. Seastedt. 1989. Is fire a disturbance in grasslands? Proc. N. Am. Prairie Conf. 11:159-161.

Fedkiw, J. 1989. The evolving use and management of the nation's forests, grasslands, croplands, and related resources. U.S. Dept. Agric., For. Serv. Gen. Tech. Rep. RM-175.

Glenn, S.M., S.L. Collins, and D.J. Gibson. 1992. Disturbances in tallgrass prairie: local and regional effects on community heterogeneity. Landscape Ecol. 7:243-251.

Griffith, P.W. 1976. Introduction to the problems. Shelterbelts on the Great Plains. Proc. Gr. Plains Agr. Counc. 78:3-7.

Grinnell, G.B. 1873. Buffalo hunt with the Pawnees. For. Stream 1:305-306.

Herkert, J.R. 1993. Habitat establishment, enhancement and management for forest and grassland birds in Illinois. Ill. Div. Nat. Heritage Tech. Publ. 1. Springfield, IL.

Herkert, J.R. 1994. The effects of habitat fragmentation on midwestern grassland bird communities. Ecol. Appl. 4:461-471.

Higgins, K.F. 1984. Lightning fires in North Dakota grasslands and in pine-savanna lands of South Dakota and Montana. J. Range Manage.37:100-103.

Higgins, K.F. 1986. Interpretation and compendium of historical fire accounts in the northern Great Plains. U. S. Fish and Wildl. Serv. Res. Publ. 161. Washington, DC.

Holling, C.S. 1992. Cross-scale morphology, geometry, and dynamics of ecosystems. Ecol. Monogr. 62:447-502.

Howe, H.F. 1994. Managing species diversity in tallgrass prairie: assumptions and implications. Conserv. Biol. 8:691-704.

James, E. 1966. Account of an expedition from Pittsburgh to the Rocky Mountains. March of Am. Facsimile Ser. 65. Univ. Microfilms, Ann Arbor, MI.

Johnson, S., and A. Bouzaher, eds. 1996. Conservation of Great Plains ecosystems: current science, future options. Kluwer Acad. Press, Dordrecht, The Netherlands.

Knopf, F.L. 1986. Changing landscapes and the cosmopolitism of the eastern Colorado avifauna. Wildl. Soc. Bull. 14:132-142.

Knopf, F.L. 1996a. Perspectives on grazing nongame bird habitats. Pp. 51-58 in P.R. Krausman, ed. Rangeland wildlife. Soc. Range Manage., Denver, CO.

Knopf, F.L. 1996b. Prairie legacies—birds. Pp. 135-148 in F.B. Samson and F.L. Knopf, eds. Prairie conservation: preserving North America's most endangered ecosystem. Island Press, Covelo, CA.

Knopf, F.L., and F.B. Samson. 1994. Scale perspectives on avian diversity in western riparian ecosystems. Conserv. Biol. 8:669-676.

Knopf, F.L., and F.B. Samson. 1996. Conserving the biotic integrity of the Great Plains. Pp. 121-133 in S. Johnson and A. Bouzaher, eds. Conservation of Great Plains ecosystems: current science, future options. Kluwer Acad. Press, Dordrecht, The Netherlands.

Knopf, F.L., and M.L. Scott. 1990. Altered flows and created landscapes in the Platte River Headwaters, 1840-1990. Pp. 47-70 in J.M. Sweeney, ed. Management of dynamic ecosystems. North Central Section, The Wildl. Soc., West Lafayette, IN.

Knopf, F.L., and M.H. Smith, chairs. 1992. Biological diversity in wildlife management. Trans. N. Am. Wildl. Nat. Resour. Conf. 57:241-342.

Kresl, S.J., J.T. Leach, C.A. Lively, and R.E. Reynolds. 1996. Working partnerships for conserving the nation's prairie pothole ecosystem: the U.S. Prairie Pothole Joint Venture. Pp. 203-210 in F.B. Samson and F.L. Knopf, eds. Prairie conservation: preserving North America's most endangered ecosystem. Island Press. Covelo, CA.

Krueger, K. 1986. Feeding relationships among bison, pronghorn, and prairie dogs: an experimental analysis. Ecology 67:760-770.

Larson, F. 1941. The role of the bison in maintaining the short grass plains. Ecology 21: 113-121.

Martin, T.E. 1992. Landscape considerations for viable populations and biological diversity. Trans. N. Am. Wildl. Nat. Resour. Conf. 57:283–291.

Mengel, R.M. 1970. The North American central Plains as an isolating agent in bird speciation. Pp. 280–340 *in* W. Dort and J.K. Jones, eds. Pleistocene and recent environments of the central Great Plains. Univ. Kansas Press, Lawrence.

Milchunas, D.G., O.E. Sala, and W.K. Lauenroth. 1988. A generalized model of the effects of grazing by large herbivores on grassland community structure. Am. Nat. 132:87–106.

Mock, C.J. 1991. Drought and precipitation fluctuations in the Great Plains during the late nineteenth century. Great Plains Res. 1:26–57.

Murphy, D.D. 1989. Conservation and confusion: wrong species, wrong scale, wrong conclusions. Conserv. Biol. 3:82–84

Odum, E.P. 1992. Great ideas in ecology for the 1990s. BioScience 42:542–545.

Olson, T.E., and F.L. Knopf. 1986. Agency subsidization of a rapidly spreading exotic. Wildl. Soc. Bull. 14:492–493.

Parton, W.J., W.K. Lauenroth, and F.M. Smith. 1981. Water loss from a short-grass steppe. Agric. Meteorol. 24:97–109.

Pickett, S.T.A., and P.S. White, eds. 1985. The ecology of natural disturbance and patch dynamics. Academic Press, New York.

Pulich, W.M. 1976. The Golden-cheeked Warbler. Texas Parks and Wildl. Dept., Austin.

Rabeni, C.F. 1996. Prairie legacies—fish and aquatic resources. Pp. 111–124 *in* F.B. Samson and F.L. Knopf, eds. Prairie conservation: preserving North America's most endangered ecosystem. Island Press, Covelo, CA.

Raven, P.H. 1990. The politics of preserving biodiversity. BioScience 40:769–774.

Reiger, J.F. 1975. American sportsmen and the origins of conservation. Winchester Press, Hampton, NJ.

Risser, P.G. 1988. Diversity in and among grasslands. Pages 176–180 *in* E.O. Wilson, ed. Biodiversity. Nat. Acad. Press, Washington, DC.

Risser, P.G. 1990. Landscape processes and the vegetation of the North American grassland. Pp. 133–146 *in* S.L. Collins and L.L. Wallace, eds. Fire in North American tallgrass prairies. Univ. Oklahoma Press, Norman.

Risser, P.G. 1996. A new framework for prairie conservation. Pp. 261–274 *in* F.B. Samson and F.L. Knopf, eds. Prairie conservation: preserving North America's most endangered ecosystem. Island Press, Covelo, CA.

Risser, P.G., E.C. Birney, H.D. Blocker, S.W. May, W.J. Parton, and J.A. Wiens. 1981. The true prairie ecosystem. Hutchinson Ross. Stroudsburg, PA.

Roberts, L. 1988. Hard choices ahead on biodiversity. Science 241:1759–1761.

Robins, E.R., R.M. Bailey, C.E. Bond, J.R. Brooker, E.A. Lachner, R.N. Lea, and W.B. Scott. 1991. Common and scientific names of fishes from the United States and Canada. Am. Fisheries Soc. Spec. Publ. 20. Bethesda, MD.

Ross, S.T. 1991. Mechanisms structuring stream fish assemblages: are there lessons from introduced species? Environ. Biol. Fishes 30:359–368.

Rothenberger, S.J. 1989. Extent of woody vegetation on the prairie in eastern Kansas, 1855–1857. Proc. N. Am. Prairie Conf. 11:15–18.

Samson, F.B. 1980. Island biogeography and the conservation of prairie birds. Proc. N. Am. Prairie Conf. 7:293–305.

Samson, F.B., and F.L. Knopf. 1993. Managing biological diversity. Wildl. Soc. Bull. 21:509–514.

Samson, F.B., and F.L. Knopf. 1994. Prairie conservation in North America. BioScience 44:418–421.

Samson, F.B., and F.L. Knopf, eds. 1996. Readings in ecosystem management. Springer-Verlag, New York.

Samson, F.B., F.L. Knopf, and W. Ostlie. In press. Grasslands. *In* Status and trends of the nation's natural resources. U.S. Natl. Biol. Serv., Washington, DC.

Shaw, J.H. 1995. How many bison originally populated western rangelands? Rangelands 17:148–150.

Shaw, J.H., and T.S. Carter. 1990. Bison movements in relation to fire and seasonality. Wildl. Soc. Bull. 18:426–434.

Steinauer, E.M., and Scott L. Collins. In press. Prairie ecology—the tallgrass prairie. Pp. 39–52 in F.B. Samson and F.L. Knopf, eds. Prairie conservation: preserving North America's most endangered ecosystem. Island Press, Covelo, CA.

Stone, R. 1993. Babbitt shakes up science at Interior. Science 261:976–978.

Summers, C.A., and R.L. Linder. 1978. Food habits of the black-tailed prairie dog in western South Dakota. J. Range Manage. 31:134–136.

Tilman, D., and J.A. Downing. 1994. Biodiversity and stability in grasslands. Nature 367:363–365.

Tilman, D., and A. El Haddi. 1992. Drought and biodiversity in grasslands. Oecologia 89:257–264.

Tomer, J.S., and J.J. Brodhead, eds. 1992. A naturalist in Indian Territory. The journals of S.W. Woodhouse, 1849–1850. Univ. Oklahoma Press, Norman.

Townsend, J.K. 1839. Narrative of a journey across the Rocky Mountains to the Columbia River. H. Perkins, Philadelphia (republished in 1978 as Across the Rockies to the Columbia. Univ. Nebraska Press, Lincoln).

Tubbs, A.A. 1980. Riparian bird communities of the Great Plains. Pp. 413–433 in R.M. DeGraff and N.B. Tilghman, tech. coords. Management of western forests and grass-lands for nongame birds. U.S. Forest Service Gen. Tech. Rep. INT-86, Ogden, UT.

Turner, M. G., 1987. Landscape heterogeneity and disturbance. Ecological Studies Vol. 64. Springer-Verlag, New York.

U.S. Department of Agriculture. 1987. Basic statistics 1982 national resources inventory. U.S. Dept. Agric., Soil Cons. Serv. Bull. 790. Washington, DC.

Weaver, J.E. 1954. North American prairie. Johnsen Publ. Co., Lincoln, NE.

Wedel, W.R. 1986. Central Plains prehistory: holocene environments and culture change in the Republican River basin. Univ. Nebraska Press, Lincoln.

Wells, P.V. 1970. Historical factors controlling vegetation patterns and floristic distributions in the central Plains region of North America. Pp. 211–221 in W. Dort and J.K. Jones, eds. Pleistocene and recent environments of the central Great Plains. Univ. Kansas Press, Lawrence.

Wenger, L.E. 1943. Buffalo grass. Kansas Agric. Exp. Sta. Bull 321. Manhattan, KS.

Weston, D. 1992. The biodiversity crisis: a challenge for biology. Oikos 63:29–38.

Whicker, A.D., and J.K. Detling. 1988. Ecological consequences of prairie dog distur-bances. BioScience 38:778–785.

Wiens, J.A. 1973. Pattern and process in grassland bird communities. Ecol. Monogr. 43: 237–270.

Zimmerman, J.L. 1990. Cheyenne Bottoms: wetlands in jeopardy. Univ. Kansas Press, Lawrence.

Index

Entries occurring in figures are followed by an f; those occurring in tables, by a t.

Ecological Studies

Ecological Studies

Ecological Studies